Physician
to the
Gene Pool

Genetic Lessons and Other Stories

James V. Neel, M.D., Ph.D.

John Wiley & Sons
New York   Chichester   Brisbane   Toronto   Singapore

To Priscilla with love

Her patience, understanding support, and participation in these intellectual peregrinations has been unfailing for 50 years.

This text is printed on acid-free paper.

Copyright © 1994 by John Wiley & Sons, Inc.

All rights reserved. Published simultaneously in Canada.

Reproduction or translation of any part of this work beyond that permitted by section 107 or 108 of the 1976 United States Copyright Act without the permission of the copyright owner is unlawful. Requests for permission or further information should be addressed to the Permissions Department, John Wiley & Sons, Inc., 605 Third Avenue, New York, NY 10158-0012.

Library of Congress Cataloging-in-Publication Data:

Neel, James V. (James Van Gundia), 1915-
    Physician to the gene pool : genetic lessons and other stories /
James V. Neel.
        p.   cm.
      ISBN 0-471-30844-7
      1. Neel, James V. (James Van Gundia), 1915- . 2. Geneticists-
-United States—Biography. 3. Medical genetics—United States-
-Biography. I. Title.
RB155.N44 1994
575.1'092—dc20
      [B]   93-36614
                                                     CIP

Printed in the United States of America

10 9 8 7 6 5 4 3 2 1

Contents

Preface / *vii*

1. **Beginnings** / *1*
2. **Medical Interlude** / *21*
3. **Thalassemia and Sickle-Cell Anemia** / *37*
4. **Organizing the Atomic Bomb Studies** / *57*
5. **A First Appraisal of the Bombs' Genetic Effects** / *75*
6. **When Cousins Marry** / *95*
7. **Amerindian Studies: Testing the Water** / *117*
8. **On to the Yanomama** / *133*
9. **Health and Disease in Recently Contacted Amerindians** / *149*
10. **Tribal Demography** / *173*
11. **Genetic Differences Between Villages: Stepping Stones to Evolution** / *191*
12. **Spontaneous Mutation** / *211*
13. **New Approaches to the Effects of the Atomic Bombs** / *229*
14. **What Is a Gene?** / *249*
15. **Defining the Gene Pool** / *265*
16. **Just Too Many People** / *281*
17. **Some Longer-Range Problems for the Gene Pool** / *301*
18. **Assessment of Genetic Risk** / *317*
19. **Genetic Medicine for Populations** / *341*
20. **Genetic Medicine for Individuals** / *361*
21. **Physician to the Gene Pool** / *389*

Notes / *399*

Index / *433*

Preface

Of all the stunning scientific developments during the last 50 years, those in the field of genetics must surely be ranked among the top few. And among all the organisms whose genetics are now better understood as a consequence of these developments, humans stand out. This is in part because of the unity of genetics: genetic breakthroughs in any other species almost always have transfer value to humans. To have been a practicing geneticist during this period and to have participated in some small way in these advances has been an exciting intellectual privilege. Society, through the largesse of the research grant system, has provided geneticists as a group with a freedom to pursue their scientific inclinations that, in past times, was accorded to only a very esoteric few. I am of the school that believes that along with this support and freedom there comes an implicit social contract. That contract requires the geneticist to be ever sensitive to the societal implications of his new knowledge. The contract involves a two-way exchange: the scientist informs society as to what genetics has to offer, and society in turn must decide along which of the many possible avenues of progress it will proceed.

The avalanche of discovery in recent years has been such that even geneticists in the thick of it—let alone the public—are unable to keep abreast. They try. But whereas when I came on the scene before World War II it was expected that you would annually attend the meeting of the American Association for the Advancement of Science and maybe one or two other professional society meetings, nowadays there is an interesting workshop, congress, or symposium almost every week, and the parade of guest lecturers through a research-oriented university is dizzying. In this over-heated scientific atmosphere, it is usually the glitziest developments that are brought to the public attention. The increasing competition for research dollars encourages universities and individual scientists to push the more spectacular developments forward. I have participated in some of this glitz, and know how easy it is to over-simplify—and even over-dramatize—a development, the better to make a point. A balanced outlook is not easy to maintain, either for scientists or those engaged in funding scientific research.

The implications of all this new genetic knowledge for human affairs are considerable. Inevitably, there are differences of opinion among geneticists. In the final six chapters, I use some rather strong words concerning my priorities in the application of genetic principles to human affairs. Indeed, laying out these priorities is the primary objective of this book. The viewpoint I espouse is at present somewhat heterodox. I do not, for instance, accord the prospects of human gene therapy the high priority it now receives in many quarters. As I have developed that philosophical position, it has been impressed upon me how much it has been shaped by my conditioning in various genetic trenches. Some of these experiences are rather foreign to contemporary geneticists working in a well-appointed clinic or laboratory. It seemed only appropriate that I lay this background out; gradually what was meant to be a minimal account of my life in genetics grew into the first 13 chapters. Between the first 13 and the last six chapters, there is a "hinge" of two chapters which, for those readers short on genetic background, briefly summarizes the complexity that we now know the human genetic material to have. Some understanding of this complexity is necessary if arguments as to how we use genetic knowledge are to have any rigor. The reader may wish to think of this book therefore as a series of inter-related essays, some of them several chapters in length, all of them providing material to be drawn upon in the final essay on the situation in which humankind now finds itself.

I have attempted to hold referencing to a minimum, drawing on review and monographic summaries where possible. Direct quotes have been referenced, as have, on occasion, articles by myself which substantially advance a thought touched on in the text. In addition, I sometimes reference very recent developments which have not yet found their way into the review literature.

Many persons have given freely of their time to critique this manuscript. I particularly acknowledge the helpful comments of Brian Athey,

Preface

Caleb Finch, Ronald Freedman, Martha Haviland, Sun-Chien and Betty Hsiao, Bettyann and Daniel Kevles, Myron Levine, Priscilla Neel, William Provine, and Richard Tashian. I also owe a major debt to "my" editor at John Wiley & Sons, William Curtis.

The author (right) with W. P. Spencer in 1954, examining some of Spencer's *Drosophila* strains (photo courtesy of the College of Wooster).

1

Beginnings

The happening that was to determine my intellectual orientation for the next 62 years occurred toward the end of my first semester at the College of Wooster. I was an avid collector of insects during my boyhood and early teens, so it seemed natural to sign up for General Biology when I first enrolled at Wooster at age 16. The course was winding down when we reached the chapter on heredity in Baitsell's *Textbook of General Biology*. I can still recall my sense of excitement when I glimpsed for the first time the orderly genetic processes that undergird the tremendous diversity of nature. From that moment, I knew how I wanted to spend the rest of my life.

My enrollment at Wooster was much more an act of necessity than of discriminating planning. My father had died of pneumonia six years earlier (in 1925). He had been a metallurgist and executive at a steel mill—by all measures an unusually fine person. The memories of him that stand out are of our fishing trips and of him down on his knees, teaching me to box after I came home much the worse for an alley fight in Detroit (a not uncommon event in those days). Until his death, mine had been a secure and uneventful childhood, nested within what appears to have been an unusually harmonious marriage. With his death, we left our home in Detroit, and mother, with three children, of whom I was the eldest, decided to settle in Wooster, Ohio, midway between McKeesport, Pennsylvania, where many members of his family lived, and Sycamore, Ohio, her family's home. Both of my parents had been raised on farms, and I spent most of my summers through high school doing chores on my maternal grandfather's farm.

Had the economic pattern of the 1920s continued, Dad would have left his family very adequately provided for, but he could not foresee the

Great Depression, ushered in by the stock market collapse of 1929. In 1931, when I was ready for college, it was a piece of unusual good fortune that Wooster, one of the better small colleges in Ohio, was only five blocks away. For financial reasons, there was never any real question of going elsewhere. Throughout college I lived at home and earned my tuition, principally through a variety of odd jobs in the Department of Biology at the College (which was itself under great financial pressures) funded by the depression-inspired Works Progress Administration (WPA). There were also significant at-home responsibilities to a mother, loving, supportive, but always demanding a best effort, who saw in her children a means toward fulfillment denied her by father's early death.

The Spencer Influence

Planned or not, in retrospect, it is not clear how much more appropriate any other college than Wooster could have been to my nascent intellectual needs. This was because of a single individual, Warren P. Spencer. A graduate of Wooster himself, Spencer was to spend his entire academic career there, aside from graduate work at Ohio State, sabbaticals at the California Institute of Technology and the University of Texas, and participation during World War II in the genetic research program of the Rochester, NY, branch of the Army's Manhattan Engineering District (as the atomic bomb project was called). Spencer had become interested in *Drosophila* when he began his graduate work at Ohio State, in 1923. At a time when others were taking advantage of the short life cycle and chromosomal simplicity of *Drosophila* to explore the mechanisms of inheritance, Spencer turned to the study of the genetic attributes of wild populations of fruit flies of many species. As much as any geneticist of his time, he demonstrated, through inbreeding, the tremendous store of hidden, genetically determined variation in physical traits in free-living fruit fly populations; this, of course, decades before biochemical techniques led to the realization that the amount of genetic variation "out there" strained our concepts of how natural selection might work. He also clearly recognized the tremendous swings in fly population sizes in the annual seasonal cycle, and how the little accidents determining which *Drosophila* populations survived the population bottlenecks these swings created might have a profound influence on the genetic make-up of the next and subsequent generations of the population. Years later, students of human genetics would recognize just how important these bottlenecks might be in human evolution.

2

Spencer, who was 33 when I first met him, did all this while on a teaching schedule that any of us in academia today would consider crushing. He probably put in a full 40-hour week at teaching, preparation of lectures, student counseling, faculty activities, etc. Science was for evenings and weekends. When he settled down for an evening of work, it was seldom for less than 6 hours. He frequently worked until 2 or 3 A.M., then taught a 7:30 A.M. class. Yet, even while crowding all this in, he never gave a student with an honest question the brush off. His growing reputation in the next two decades led to invitations to join institutions where there was much more research time but he preferred to remain at Wooster. The fates were to be unkind to this remarkable human being: in 1957, while at the University of Texas preparing for a trip to various Pacific islands to collect *Drosophila,* he suffered a major cerebral hemorrhage, from which he never recovered sufficiently to resume his professional obligations. He died in 1969.

Because the specialty courses in biology at Wooster were taught in alternate years, I had to wait until my junior year for the formal course in genetics taught by Spencer, contenting myself in the meantime with the usual undergraduate courses in biology and chemistry. The next year I undertook a supervised research project. Spencer had noticed that some stocks of *Drosophila funebris,* one of the several species of fruit flies he kept in his laboratory, showed extra bristles and disturbances in wing venation; but even when he inbred the lines in which this occurred, in order to create so-called "pure" lines, i.e., homozygous with respect to the genes responsible for the traits, the variability in the expression of the traits continued. He suggested I try altering in various ways the flies' developmental environment, to see how this would influence the expression of the traits. Temperature and the availability of food were two simple interventions which I could try.

When a person reviews his professional life, giving himself a little time and quiet to recall more than the obvious, the landmarks may not be those that would emerge in the usual brief, formal biography. My first genetic epiphany was the very discovery of genetics. My second occurred that evening in Spencer's laboratory when I had my first experimental results. Temperature and culture conditions did influence—and strongly—how the genes responsible for these traits expressed themselves. And as I looked at my tally sheets, there crept over me, like a spreading blush, the realization that for a brief moment, I had an insight into the workings of nature not shared by anyone else—my own private bit of "truth." It wasn't much, but I had created a new fact, to be chinked in among the growing body of data on how to influence gene expression. I was enthralled—hopelessly hooked—by the challenge of understanding the genetic basis of variation and how that genetic basis (the genotype) interacted with the multitudinous influences we subsume under the term environment to provide the traits of the mature organism (its phenotype.)[1]

I graduated from the College of Wooster in 1935. That summer I worked for the Cleveland Museum of Natural History as a trailside naturalist, maintaining (and living in) a little natural history museum located in the heart of a wonderfully unspoiled beech–maple forest. My boss, George Williamson, was one of the early ecologists, especially interested in the role of beetle larvae in disposing of dead trees, and I learned a lot about beetles that summer. I suspect most of us, looking back on our early years, can recall one or several brushes with disaster, from which, almost unbelievably, we emerged unscathed. One day I was working about 30 feet up on a dead tree, stripping bark, looking for beetle larvae, when the old climbing belt (which came with the job) gave way, and I arched backward toward the ground. Miraculously, my fall was broken at the fifteen foot level by the crown of a small tree, and, when I was finally able to pick myself up from where I sprawled on the ground, I was essentially uninjured except for a deep gash in my left leg, caused by my right climbing spike as I flailed about in mid-air. It was about then I began to nurture a life-long respect, reinforced from time to time by other experiences, for the narrowness of the margin that so frequently separates our unfolding lives from disaster.

On to Graduate School

Early in my senior year of college, Spencer and I began to talk of postgraduate education in genetics. Since I would be dependent on my own financial resources, the graduate school I chose obviously had to be one that offered a teaching assistantship or other employment on which I could survive. Spencer suggested I accompany him to the meeting of the Genetics Section of the American Association for the Advancement of Science, in Pittsburgh, over the Christmas holidays of 1934. There I had my first exposure to some of the names associated with the scientific papers I was beginning to read, and several opportunities to discuss graduate school. That spring—still exploring graduate school possibilities—I gave my own first scientific paper, at the annual session of the Ohio Academy of Science in Columbus, Ohio, meeting in the process Lawrence Snyder, author of the popular text on genetics we had used at Wooster and also a possible mentor for graduate studies at Ohio State. In the end, I chose the University of Rochester for my graduate studies.

4

The reasons for that choice were two-fold: the offer of a relatively generous teaching assistantship, by the standards of the times, and the presence there of Curt Stern. During the 20 years from 1915 to 1935, the principal vehicle for the advancement of genetic knowledge had undoubtedly been the fruit fly. It was small, easily manipulated, rapidly reproducing, and had only four pairs of chromosomes, which made assigning genes to chromosomal locations relatively easy. Stern had begun to work with *Drosophila* in 1924, when he came to this country from his native Germany on a fellowship permitting a year of study with T. H. Morgan and his associates at Columbia University (Morgan received a Nobel Prize in Medicine in 1933 for the seminal contributions to genetic thought resulting from the studies of *Drosophila* performed by himself and such students/colleagues as Calvin Bridges, A. H. Sturtevant, and H. J. Muller.) Stern rapidly emerged as a principal contributor to the *Drosophila* literature. Back in this country on a second fellowship in 1932–33, he reluctantly decided that the events in Nazi Germany precluded any hope of an academic career there for one of Jewish extraction, and decided to stay on in the United States, finding a position at Rochester in 1933. I was to be his first Ph.D. student. The fates were also to be cruel to Stern. At the age of 67 he developed Parkinsonism, and for the next 12 years a superb intellect slowly deteriorated. In my later years, the examples of the early and/or tragic deaths of my father, Spencer, and Stern, plus the perspective engendered by fieldwork in some rather difficult places, were to be very helpful, after a hard day in the academic trenches, in sustaining the aequanimitas extolled by William Osler in an essay whose reading should not be, as it is, so largely restricted to physicians.[2]

Stern's encyclopedic knowledge of genetics—he seemed never to forget anything he had ever heard or read—and his analytic mind, which stripped a problem of unnecessary encumbrance and held up the essentials for consideration, were awe-inspiring. Fortunately, this formidable intellect was matched by a rare degree of gentleness and tact. A single example will suffice. During my first year of graduate work, I made a very stupid scientific comment. Stern straightened me out but then—I must have looked very crestfallen—added: "Jim, it's alright. In Germany we have a proverb: Great men make great mistakes!"

Rochester has a small Department of Biology—nine people, split between an old guard and a new guard, the latter brought in a few years earlier when the University of Rochester came into the funding that enabled it to expand from a provincial little college to a substantial university. B. H. Willier, a distinguished embryologist, was its chairman. Because of my interest in how genes find expression, I was to have as much training in embryology as in genetics. There was a minimum of formal instruction for graduate students, most of the teaching coming through seminars in which students participated rather than from didactic lectures by our professors. Stern, out of the German tradition, believed that when you began graduate work, you had the necessary basics behind you. He could not possibly know how naive I was! There was only one other genetic-resource person in the Department, Don Charles, a young mouse geneticist. Years later, when as the chairman of a department of human genetics in a large university setting I fretted over the adequacy of our instructional program, it helped my perspective to think back to those earlier, simpler days.

In the summer of 1937, I had the obligatory student experience at the Marine Biological Laboratory in Woods Hole, Massachusetts. In addition to participation in the world famous course in the embryology of marine animals, I continued a collaborative relationship begun earlier in the year with Ernst Hadorn, a Swiss spending a Rockefeller Fellowship at Rochester. We were involved in the transplantation from one *Drosophila* larva to another of the very small gland that controls pupation. This was one of the first experimental studies on the action of insect hormones; we wanted to know when in larval life the gland became active and how quantitative the effects of the hormone were. We also wondered whether the hormone, later termed ecdysone, was species-specific. An unexpected opportunity to test the latter question came when we heard that a dead whale had washed up on the north shore of Cape Cod. When we judged that it should be suitably "ripe," we collected large quantities of decaying whale meat swarming with blow-fly maggots. For convenience, back in the laboratory we kept our "samples" in half-pint milk bottles with cotton plugs in the apertures. As the maggots approached pupation, that transitional stage between maggot and adult, we transplanted glands using microtechniques from the more advanced blowfly maggots into chronologically less mature (and much smaller) *Drosophila* larvae. These latter pupated on a greatly accelerated schedule: the hormone was not species-specific.[3]

This was the first (and last) time we worked with blowfly larvae, and we were unaware of one very signal fact in their life cycle. As the time for pupation approaches, they have a powerful urge to leave the decaying meat in which they have lived to pupate at a considerable distance from it. One maggot pushing on a cotton plug can't get far, but 50 or 100 maggots pushing in concert is something else. The morning following our small discovery described above, we came to the laboratory to find platoons of maggots marching up and down the corridors of the Marine Biological Laboratory in search of a quiet place to pupate. Not until we had captured every last one of the escapees were we again on good terms with our colleagues.

I was waiting tables that summer in return for board and room. Halfway through the summer, I heard that there was need for someone to wash the bottles used to cultivate *Drosophila* in the laboratory assigned to the California Institute of Technology group. I applied for the job and was accepted. So much for waiting tables, which I despised. The summer's visitors in the laboratory included T. H. Morgan (now largely back to his first scientific love, embryology), A. H. Sturtevant, and a younger pair of geneticists, B. Ephrussi and G. Beadle; the latter shared a Nobel Prize with E. Tatum in 1958 for the demonstration that genes work through the control of enzymes. At that time, Ephrussi and Beadle were also engaged in transplantation experiments involving *Drosophila*, transferring bits of tissue (the imaginal discs) from one fly to another to see how autonomous they were in development. It was a heady experience for a young student to listen, as he washed the bottles, to the uninhibited repartee of that group.

Back in Rochester that fall, I was soon well into my thesis problem. Finished in 1939, it consisted of two parts. The pattern of hairs—large and small—on the body of a fruit fly is normally very reproducible; this pattern is significantly disturbed by certain mutant genes. Now, the manner in which "pattern" is controlled genetically, be it the coil of a snail's shell, the complexity of a tiger swallowtail's wing, or the symmetry of the human face, is a basic biological question. The first part of my thesis was an effort to determine if the new patterns of abnormal bristles resulting from a variety of mutations were related to the preexisting norm. They were—the mutations did not obliterate the basic pattern but added a variation to it. The second part of the thesis was more demanding. Again I studied temperature effects on bristle expression, but now in a much more sophisticated fashion. At higher temperatures, flies homozygous for a variety of mutations causing extra bristles show fewer abnormal bristles. But at higher temperatures the flies are smaller. I could also show that at any given temperature, smaller flies had fewer bristles. It was thus possible to dissect the "temperature effect" on bristle number into a portion due to temperature per se and a portion due to the positive relationship between body size and bristle number, and give mathematical expression to each component. In those days of early efforts to understand the control of gene expression, this revealed that "temperature effects" were more complicated than we had previously suspected.[4]

I celebrated the completion of graduate work with attendance at the Seventh International Congress of Genetics at Edinburgh in August of

1939, presenting a paper on my thesis work. The cheapest passage over I could find was on a little Norwegian freighter. As the imminence of war became clear, the Congress began to deteriorate. My return passage, on a British passenger ship, was canceled, and I found emergency transportation on an American freighter, the City of Flint, which had fitted one of its holds up as a dormitory of sorts. Our second day out, we answered an SOS from a British passenger ship, the Athenia, which had been torpedoed just 10 hours after the declaration of World War II, the first of the 1994 ships lost over the next 5 years, 8 months, in what became known as the Battle of the Atlantic. We took on 200 survivors aboard a ship not designed to carry passengers. Living conditions were a bit strained for the next ten days, until we reached the nearest port, Halifax, Nova Scotia. We then proceeded to our original destination, Boston. Even to someone as immersed in his science as I was, it was clear that major developments were shortly to restructure our lives.

First Thoughts About Human Genetics

Although engrossed in my *Drosophila* thesis project, I began, in my last year of graduate work, to wonder whether I wanted "always" to work

Life boat crammed with survivors of the torpedoed Athenia approaching the City of Flint, two days following the outbreak of World War II (1939).

with fruit flies. Part of the uncertainty was intellectual, the feeling (shared by others) that those genetic issues on which the fruit fly was best fitted to shed light were becoming exhausted, and it was time to look around. My commitment to genetics didn't waver, but what kind of genetics? The other source of uncertainty was more emotional—a vague sense that I would like to apply my background in experimental genetics to the study of human genetics, a field then badly stigmatized in this country by the many incredibly sloppy and biased studies of the 1920s and 1930s that had provided the "scientific" justification for much of the American eugenics movement.

With this thought in mind, I approached Stern with the suggestion that we devote an upcoming, semester-long genetic seminar to the subject of human genetics. Stern quickly acceded—I suspect he was incubating the same thought—and gathered together the papers he considered worth reviewing, which we covered over a four-month period. First we reviewed what was known of the human chromosomes (at that time thought to consist of 24 pairs, rather than the 23 pairs we know now). Then we considered the special statistical techniques necessary to analyze human genetic data. These were important because in studying the genetic basis for various human diseases, one usually becomes aware of a relevant family only after the occurrence of at least one case of a disease in the family. This fact, termed the ascertainment bias, results in a distortion of genetic ratios and is especially important in the study of recessively inherited disease.The English school of biometrical geneticists (L. T. Hogben, R. A. Fisher, L. S. Penrose, J. B. S. Haldane, and their students) had been especially active in developing these statistical methods. This was only the first of the special problems that arise in the study of human genetics that do not exist for the experimentalist, who can arrange crosses (read "marriages") at his convenience. A second difficulty we considered was the special problem involved in determining from human data whether the genes for two traits were linked, i.e., on the same chromosome, a difficulty again due to the fact that the matings most critical in reaching a conclusion could not be arranged. When we moved to the specifics of human genetics, Stern had been able to locate for our consideration no more than two dozen papers dealing with specific inherited diseases that attempted to treat the inheritance of human traits with a rigor approaching that in experimental genetics. Finally, we discussed the potential usefulness of identical twins, each with the same set of genes, in the study of human inheritance. Since they arise from the early splitting of a single fertilized egg, differences between members of an identical twin pair should shed light on the potential lability in genotype expression.[5]

Although we stayed away from considering any of the then current, uncritical perversions in the name of eugenics of the very inadequate data base for human genetics, we were well aware of what had transpired at home and abroad. At home, an enthusiastic but poorly informed eugenics movement, largely driven by those we would today call WASPS (White, Anglo-Saxon, Protestant) had, even considering the times, oversimplified genetic issues to a point where most professional geneticists had distanced themselves from the movement. Abroad, Hitler's policies in the name of eugenics (the full scope of which did not become generally known until after World War II) had created great disquietude. For a group of students imbued with the rigorous experimental approach, the seminar was a sobering experience. Without question there was much that could be learned concerning human genetics, but how could one cut through both the methodological and sociological impediments to rigorous studies? There was another consequence of that seminar—I was becoming firmly convinced that *if* I was serious in my interest in human genetics, I must somehow obtain medical training, both for the valuable background it would provide and because without it I would be greatly handicapped in personal involvement in certain types of research.

That fourth year of graduate study I did something very uncharacteristic of students writing a thesis: I enrolled for medical human anatomy. I already had done medical school biochemistry and physiology, simply for their background value. I'm not quite sure why I did anatomy, other than as an affirmation of my interest in the human aspects of genetics. I was not yet thinking that seriously of medical school. I had no money. In retrospect, it was a most fortunate decision: when I did formally enter medical school, the completion of this array of course work meant I could be admitted with second year status.

Two Years at Dartmouth

In 1939, with the end of my thesis in sight, it was time to plan my next step. I would have preferred a postdoctoral fellowship, and applied for one of the prestigious National Research Council Fellowships. Several months before the awards were to be announced, Stern learned of a position for a beginning geneticist at Dartmouth College and after we had discussed it, recommended me. Every graduate student considers a postdoctoral fellowship the next step in advancing a proper career. Dartmouth—which I soon visited for an interview—was obviously a most attractive setting for one coming off the thesis push but, equally obviously, scarcely a powerhouse intellectual environment for an ambitious young man. But in 1939, decent positions with an academic future, i.e., in the tenure track, were not common. Dartmouth made me an offer. Stern and Willier both thought it wise to accept, and in the fall of 1939, straight off the return from England on the City of Flint, I arrived at Dartmouth to teach courses in Introductory Biology and Genetics. As I settled in, another of Stern's German aphorisms—meant to be comforting, as I had early expressed disappointment in scratching my fellowship application—was ricocheting through my neuronal circuits: "Remember, Jim, it's not the cage that determines how sweetly the canary sings!"

The Dartmouth interlude was notable for two events. First, the menage of six young bachelors of which I found myself a part contained two former members of the Dartmouth ski team, whose urgings and then rigorous instruction resulted in my introduction, as soon as there was a decent snow base, to an exhilarating activity. The second event was more professional. Within one of my long-inbred lines of normal *Drosophila melanogaster*, I suddenly began to detect variant flies in relatively greatly increased numbers, variants which for the most part transmitted their characteristics to their offspring, so that they could be attributed to recent mutations, i.e., newly arisen, transmissible changes in the genetic material. Shortly, I had an epidemic of mutations on my hands, but confined to this one line. There had been several previous reports of such events, but considerable skepticism persisted concerning the validity and magnitude of these happenings. Mutation was admittedly an unpredictable event, but, like the ionizations resulting from the decay of a radioactive isotope, was not thought to cluster in time-space. There was a tendency among geneticists to suspect poor fly husbandry as the cause of these events, i.e., contamination of one line by other lines in the laboratory. Fortunately for me, this could not explain my findings: I had brought a relatively few strains from Rochester to Dartmouth, and the mutants I was detecting were clearly different from anything already in the laboratory.

The first challenge was to document the phenomenon; the second was to understand it. The first was relatively easy. On very conservative grounds—I felt I had to be supercritical—I estimated the mutation rate I was observing to be some five times greater than normal. In retrospect I believe this was *very* conservative. This estimate did not encompass the relatively large number of abnormally appearing flies that were sterile, so that the genetic basis of their abnormality could not be checked out. The second challenge was never to my mind met satisfactorily. This was to attempt to determine if the effect could be related to any specific one of *Drosophila*'s four chromosomes. This required developing lines of *Drosophila* containing specifically (and only) each of the four different chromosomes from this particular strain, the remaining chromosomes of each line being from a standard strain. Now, mutations, even with the increase exhibited by this line, remained a rare event. One has to examine a very considerable number of flies to distinguish between lines with and without increases in mutation rates of the magnitude in question. In the end, despite a flat-out effort, I couldn't get the numbers I needed for a conclusive test of various hypotheses; the evidence was suggestive but by no means compelling that the effect was associated with the presence of the third chromosome of the unusual strain.[6]

Until now, my research interests had centered on how the expression of genes can be modified. The study of mutation was a new theme, one which in time would become my principal scientific activity. The phenomenon of mutation is critical to all biology, providing as it does the raw material on which the evolution of species is ultimately based. The possibility that it might occur in bursts, which if frequent enough could be responsible for most of this raw material, was not then in the mainstream of genetic thought. It is now. Subsequent investigators have established at least three bases for bursts of mutation. The first is the occurrence of transposons, genetic elements that can move from one chromosomal site to another, sometimes producing mutations at their new sites of insertion. Postulated in the 1940s by Barbara McClintock, to explain a series of puzzling phenomena in corn (a postulate for whose later validation she received the Nobel Prize in Medicine in 1983), the activation of latent transposons (sometimes referred to as jumping genes) could explain the type of phenomenon I observed in *Drosophila*. A second possible explanation was based on the sudden appearance in that strain of one or more mutator genes, i.e., genes of fixed position that can significantly alter the mutation rates of many other genes. Finally, a third cause for a phenomenon of the type I was observing is hybrid dysgenesis, so called because of the appearance of increased numbers of mutations (as well as increased sterility and evidence of increased chromosomal exchanges) in the offspring of crosses between selected lines of *Drosophila*. Clearly, I was not dealing with this third phenomenon, but it could have been either of the other two.

An NRC Fellowship

In the winter of 1941, I again applied for an NRC fellowship, choosing Columbia University as the place I would like to go to and Professors

Theodosius Dobzhansky and L. C. Dunn as my preceptors. Both were very well established geneticists, Dobzhansky working with *Drosophila,* Dunn with mice. Both were also, it developed, wonderful human beings for whom I came to have the warmest possible feelings. Another reason for the choice of Columbia was that I wanted a taste of big-city life. As regards the latter, one year was enough. When, some years later, I was offered a position at the University of Michigan, in Ann Arbor, one of the dominant factors in my decision to accept was the juxtaposition of a good large university with a small town; I find I am not drawn to big-city life.

The fellowship application was successful. I went to Columbia following a summer at the Carnegie Institution Laboratory for the Study of Experimental Evolution in Cold Spring Harbor, Long Island, New York. I had been there the previous summer, and found it a very pleasant setting for work and play. This summer it seemed all work, as I attempted to capitalize scientifically on the unexpected opportunities growing out of my epidemic of mutations. Miloslav Demerec, who in 1943 became director of the Lab, was my sponsor both summers. Because of his own interests in mutation, he was most generous with time and the necessary resources. Each morning, as he made his rounds, he poked his head into my laboratory to inquire in his rich Slavic accent "how everything is?"

My epidemic of mutations was attracting a little attention. This summer I was given larger quarters, in Jones Laboratory. H. J. Muller, who later received a Nobel Prize for his demonstration of the mutagenic effects of X-rays, was across the hall. To my left was Bentley Glass' laboratory; obliquely that of Ed Lewis. All of us were working with *Drosophila*; it was a convivial summer, of the type for which Cold Spring Harbor was famous. I recall several conversations with Barbara McClintock, who was then deeply immersed in her classical studies of jumping and mutable genes in corn, but I don't recall that we made the connection (which I now believe may be appropriate) between her findings and my own.

From September of 1941 until the following June, I worked at Columbia on understanding the mutational epidemic in my fruit flies, as well as several other problems with *Drosophila*. World War II was on its way; there was the sense of a "last fling" at science. On December 4, 1941, just three days before the attack on Pearl Harbor, I wrote the University of Rochester, requesting admission as a second-year medical student. I had taken the long-contemplated step, but now with the additional motivation that our involvement in what looked to be a long war seemed inevitable, and a physician would be much more useful than a geneticist.

On December 23, 1941, just 19 days later, Dean George Whipple of the University of Rochester Medical School wrote to tell me I was admitted, albeit my unusual background raised some problems as to just where I would fit in. Anyone who has struggled to gain admission to medical school recently will recognize from the speed of action and the relaxed attitude toward my patchy background, that medical schools—at least this one—went about their business somewhat differently in those days.

There remained only the question of how to finance all this. I was reluctant to incur debts because I was strongly committed to the relatively unremunerative academic life. Casting about, I recalled having heard vaguely of the Cramer Fellowships that Dartmouth sometimes made available to its graduates, for various types of advanced study. A Dartmouth graduate I was not, but I did have ties to the school which I had been careful not to sever. Technically I was on a leave of absence, and, when I made the decision to finish my medical education, had asked for an extension of that leave, with the thought of possibly pursuing my career in medical genetics through the facilities of the two-year medical school and the Hitchcock Clinic at Dartmouth. After minimal negotiations, Dartmouth granted me a Cramer Fellowship of $1,000, to advance my studies in genetics—a most liberal interpretation of what I was up to. On March 18, I received another piece of good news—I was awarded a tuition scholarship from the University of Rochester. I was on my way, even if just barely.

A Look at the Eugenics Record Office

With the decision regarding medical school in hand, my thoughts increasingly converged on the question of how best to proceed to make significant studies in human genetics. During my NRC fellowship, I had begun to take an interest in the Eugenics Record Office at the Carnegie Laboratory. The Carnegie Laboratory, whose hospitality I have already described, had been founded in 1904, when C. B. Davenport, at age 38 already a well-established figure in American biology, obtained funding from the Carnegie Institution of Washington, one of the first major philanthropic foundations in the United States, to establish a Station for the Experimental Study of Evolution at Cold Spring Harbor.

Although the program was at first indeed directed toward sound experimental studies, Davenport being one of the early proponents of Mendelian genetics in the United States, Davenport's interests and great energy soon focused on problems in human genetics. The laboratory was by the standards of the day well funded, but, in 1910, Davenport obtained additional funding for a Eugenics Record Office from Mrs. E. H. Harriman. The Office was to be located on 75 acres of land which she purchased, immediately adjacent to the Carnegie labs. The program there, at first involved in state-of-the-art studies of such relatively clearly defined and definable human traits as color blindness, eye color, and Huntington chorea, rapidly became increasingly pre-occupied with such vague and poorly understood traits as nomadism, alcoholism, insanity, and feeblemindedness. Most of these studies were performed by volunteers of class I social standing who were given a summer's training, then turned loose. There has been no better critique of the quality of these studies than the one written in 1913 by the English eugenicist Heron:[7]

> We believe that those who dispassionately consider the papers discussed in this criticism (predominantly Eugenics Record Office Investigations) must conclude with the present writer that the material has been collected in an unsatisfactory manner, that the data have been tabled in a most slipshod fashion, and that the Mendelian conclusions drawn have no justification whatever. The authors have in our opinion done a disservice to knowledge, struck a blow at careful Mendelian research, and committed a serious offense against the infant science of Eugenics. Every piece of unthorough work "dominates" in research, for it begets its likes; others find it equally easy to reach similar conclusions by loose methods applied to inadequate data. They await the same chorus of praise from an uninstructed press, and from those whose passion it is to tickle the taste of the moment. (p. 61)

In the 1920s Davenport and his associates, particularly H. H. Laughlin, became increasingly involved in eugenic schemes to limit the reproduction of persons with traits such as were mentioned in the previous paragraph, as well as to control the immigration into the United States of ethnic groups judged to be genetically inferior on the basis of continuing "studies" which by today's standards were grossly biased. Apparently completely oblivious of and/or immune to legitimate criticism such as that quoted above, Davenport and Laughlin progressively adopted positions with less and less scientific validity. Laughlin became an important advisor to the House of Representatives Committee on Immigration and Naturalization in the 1920s, and his prejudices concerning the intrinsic merits of various ethnic groups were translated into immigration quotas. The activities of the Office, ceded to the Carnegie

Institution in 1918, in time became so embarrassing to the Institution that when Davenport retired as director of the Office in 1934 he was not replaced, and its funding was gradually phased out. Although I recall meeting Davenport the summer of 1940, by 1942, when I began to take an active interest in this paper-jammed building, it was devoid of staff and Davenport (who died in 1944) was in total retirement. (Those interested in more extensive treatments of the place of the Office in the eugenics movement in the U.S. and abroad should consult Haller,[8] Ludmerer,[9] Kevles,[10] and Allen.[11]) But regardless of the disfavor into which it had fallen, the Office was still the principal repository of data on human heredity in the country, and, poised as I was in 1942 to gamble my professional future on studies of human genetics, it seemed to me imperative that I obtain some firsthand knowledge of just what the Office's files contained. With its staff gone, this was no easy task.

In addition to supervising the studies by his fieldworkers, Davenport had devised a form entitled "Record of Family Traits," which, widely distributed upon request to individuals and to groups caught up in the eugenics movement, were upon completion returned to the office, where they occupied drawer after drawer. All this material was more-or-less indexed on some three quarters of a million file cards. Earlier I mentioned my discovery of genetics in my first year of college. Somewhere in the desultory reading that followed over the next several years, I learned of the Eugenics Record Office, because I still have the very eugenically oriented packet which, upon expressing my interest in genetics, the Office sent me in 1933. It included, in addition to some very eugenic propaganda, the 14-page record of family traits mentioned above, which each respondent was requested to complete in whatever way seemed appropriate to him/her. A more uncritical approach is difficult to imagine—and this at a time of first rate work in experimental plant and animal genetics.

After several days in the summer of 1942 of not very well organized casting about the Eugenics Record Office, trying to discover a masterplan, by chance I saw a file labeled "red hair." This was well known to be familial, but had the clear advantage, over most of what had occupied the Office's attention, of an—almost—objective endpoint. It seemed, from the manner in which redheaded children were born to parents neither of whom had hair of that color, that it could be a recessive trait. A critical test of this hypothesis should be provided by marriages in which both parents were redheaded: only redheaded children should result. From the files I extracted data on the outcome of 26 marriages between two redheads. The reports were from a variety of sources, lay and professional, some almost anecdotal. When the results were added,

16

there were 101 redheaded offspring and 13 with hair of some other color. This was too many nonredheads for the recessive hypothesis but no other hypothesis fitted better. There was of course no hope of examining the apparently exceptional, nonredheaded children. With some trepidation—because the data were certainly substandard—I published a little note in 1943 (my first in human genetics) suggesting red hair color in general behaved like a recessive trait.[12] It was a good day when in 1952 T. E. Reed published the results of a really proper study, which confirmed this suggestion.[13]

Many of the senior geneticists with whom I had discussed a move into human genetics, geneticists for the most part studying *Drosophila,* had almost perceptibly recoiled, a reflex conditioned by the generally poor quality of "research" on human heredity in the U.S., as mentioned above, and the opprobrium which the eugenics movement in the U.S. and Germany had rightfully earned. Today, most geneticists are unaware that euthanasia for the "hereditarily defective" was legalized in Germany in 1939, ultimately resulting in some 70,000 deaths, exclusive of Jewish deaths in the holocaust; at the time, these facts were closer to the genetic conscience. There was a further reason for the reaction of my colleagues. Humans, as viewed through the eyes of the *Drosophila* geneticists, were not a favorable object for genetic study. Their generation time was too long, they had too few offspring, genetically interesting matings could not be arranged, and they had so many chromosomes compared to *Drosophila*'s four pairs that establishing gene-linkage groups (i.e., what genes were on the same chromosome) was bound to be difficult. Almost alone among the *Drosophila* geneticists with whom I spoke, Stern and Dobzhansky were staunchly supportive of my decision.

Only as I browsed through the files of the Eugenics Record Office, however, did the magnitude of what had to be done before human genetics could be a respected discipline become brutally apparent. In the seminar on human genetics with Stern, we had considered only the best. Here, by contrast, to be as charitable as possible, was a cross section of what had been done. The concerns expressed by so many of my friends and advisors over the parlous intellectual state of human genetics in the United States had been fully deserved. It was a real gamble to believe I could bring the rigor of *Drosophila* genetics into this arena. At that time, there was only a small—very small—handful of persons in this country engaging in careful studies on medical genetics, most notably William Allan, Madge Macklin, and Lawrence Snyder. I wasn't going to have a lot of company in this gamble. Today, profound and exciting new insights into the genetics of our species are announced almost weekly in the popular press, and human genetics has become Big Science with the announcement of a Genome Project to determine the precise molecular nature of all the human chromosomes. Human geneticists of the current generation cannot imagine how I agonized over the decision to turn my back on the hard genetics of *Drosophila* and enter the soft and tainted field of human genetics.

There was a companion issue, not entirely unrelated. One of the joys of *Drosophila* genetics, as noted above, was the ease with which the flies could be manipulated and the speed with which one got results; the life cycle at room temperatures was only 10–12 days, and the progeny of a single pair numbered in the hundreds. How well would a certain natural impatience of mine to get on with the job, confirmed by this grounding in *Drosophila* genetics, adjust to working with an organism with as long a life cycle as man, whose matings one certainly didn't manipulate? Some things I clearly wasn't going to change, but thoughts about how best to turn the structure of medicine to advantage for genetic studies were already on my mind.

Mentors

Almost every scientist who writes an autobiographical note feels absolutely compelled to mention how critical a role one's mentors and associates have played. More important than force of intellect and mastery of the field is their attitude toward science and the way they approach and think about a scientific problem. Fluidity of thought—the absence of the authoritative Geheimrat approach—and the readiness to question current dogma (even the dogma one has helped establish) are to me the most essential attributes of a preceptor.

One of the coveted honors of our times for a U.S. scientist is election to the National Academy of Sciences. Stern, Willier, Dunn, Dobzhansky, Demerec, Muller, Glass, and Lewis all achieved this recognition and much more. Spencer in another, more academically favorable setting would surely have gained that recognition. Hadorn, back in Switzerland, achieved similar prominence in European zoological science. Only in retrospect has it become clear to me how very fortunate I have been in having them as mentors and early associates. It was heady company for a young scientist, but it was also demanding company, which did not suffer fools gladly. In such company one learned to think before one spoke. One paradox of the times I especially recall. My second summer at Cold Spring Harbor, in 1941, coincided with Salvador Luria's first summer there, and the smallness of that setting brought us together frequently. Luria, a physician who had fled Europe the preceding fall, was in the early stages of developing those reductionist genetic strategies employing bacteriophage that brought him, in 1969, with Delbruck, a Nobel Prize. The irony that I was simultaneously fumbling with thoughts of bringing *Drosophila*-type rigor into genetic studies of that most intractable of all organisms, man, is not lost on me.

The rather austere house, a former interns residence, located about a block from the University Hospital, where human genetics at the University of Michigan got its start (1946).

2

Medical Interlude

On July 6, 1942, aged 27, I went back to school, resigning my National Research Council fellowship two months before its termination. It was not easy adjusting to medical school after the taste of the research life afforded by the preceding year. Medical education, even at a school as enlightened as Rochester, requires absorbing an enormous amount of factual information. So does public expectation—in the emergencies of surgery, obstetrics, pediatrics, and internal medicine, the patient does not much appreciate a physician who drops out of the action to look up the correct procedure. What's wanted is a properly programmed individual, able to run through the appropriate medical checklist mentally, make a decision, and tick off a properly weighted list of options.

So I buckled down, and, in due time, there was light at the end of the tunnel. Given the accelerated schedule of medical education obtaining during the war, medical school graduation was scheduled for September 1944. Accordingly, in the spring of 1944 I faced a decision concerning postgraduate medical training. Here again the decision was not too difficult—by that time Bill Valentine and I were deep in the research outlet to be described shortly, an effort to turn the clinical observations on Cooley's anemia made by M. M. Wintrobe, W. Dameshek, and M. B. Strauss into precise Mendelian genetics. Accordingly, when Rochester offered me a position as Intern in Internal Medicine, it seemed entirely appropriate to stay on, and then for an additional nine months as well, as Assistant Resident. (During the war period, medical "years" became nine months.)

Marriage, War Years Style

Priscilla and I were introduced by mutual friends in October of 1942. Fresh from a degree from the Smith School of Architecture in Cambridge, Massachusetts, she had very recently arrived in Rochester, to be employed as a mechanical designer (radar) at Stromberg-Carlson Co. Mother Nature smiled, and by the next spring we were planning a June wedding. At that time all medical students not exempted for physical reasons had been given military commissions and placed in the inactive reserve. Then, someplace in the Pentagon labyrinth, someone made the decision that this was inappropriate, that we should be decommissioned and placed on active duty as privates, assigned to duty in medical schools. Shortly after that solid piece of information was received, a rumor spread through our student ranks, that if you went on active status as an unmarried private and subsequently married, you could not draw the regularly authorized dependent allotment. That allotment—$50 a month—represented a substantial sum for financially strapped medical students like ourselves. Not to receive it was like losing a dowry. The rumor reached such a level that Associate Dean George Berry called a meeting of all students, at which point, taking cognizance of what we had all heard, he suggested that any of us planning to be married in the near future had best do it right now. Priscilla and I were married a week later, on May 6, 1943, the couple who had introduced us serving as best man and matron of honor. I was in the midst of a student rotation in obstetrics. A compassionate friend took my calls on our wedding night, but the next night, I never got home. I'm afraid Priscilla feels that set the tone for our married life.

The month following our marriage, along with those of my classmates who held reserve Army commissions, I reported briefly to Fort Dix for the transformation to buck private. This at least solved the problem of how to finance the rest of medical school. Then it was back to school, with the added minor annoyance of trying to conform to military requirements at the same time. As might be imagined, the military establishment encountered certain difficulties in imposing the mentality of privates on medical students. In order to generate some military structure, "cadet officers" were appointed. The good Lord in his time and wisdom will let me know how I came to be company commander. The Army thought we should drill, and drill we did, myself barking out the orders in my best drill sergeant's voice. One rainy day we were drilling in a small gymnasium and, at the critical moment, I could not summon the command "to the rear, march." My quick-witted classmates, never ones to miss an opportunity, marched on into a brick well where they piled up, moaning and groaning, feigning all manner of injury, while the regular Army sergeant major standing beside me, who was assigned to make military men of us, aged perceptibly as he wondered how he would ever explain this carnage to headquarters.

House Officer

The life of a house officer, as postgraduate medical trainees in a hospital are called, is always busy, but during the war, with so many of the hospital staff away in the service, it was especially so. We worked very hard, the load lightened by those situations where you felt your intercession had really made a difference in the patient's long-range outlook. A medical internship at Strong Memorial Hospital included rotations through psychiatry and infectious disease services, and it is with those services I associate my most traumatic moments.

One day I admitted to the infectious disease service a young woman with a diagnosis of florid, secondary stage syphilis, a stage when the disease is highly contagious. It was my responsibility to draw a blood sample before treatment was initiated. I put on gloves. During the war years, needles and syringes were in short supply and were used and reused and reused yet again. At the sink, where I proposed to wash the syringe and needle out, the syringe slipped, and suddenly that very infectious needle was well embedded in my left forefinger. You can imagine with what attention over the next several weeks I watched for the appearance of the chancre that would signal an infection. Nothing happened. There were serious professional occupational hazards long before the current AIDS epidemic.

Another day we admitted to Psychiatry a manic young man, all 200+ highly muscular pounds of him very belligerent. He was isolated with special precautions, to be attended to by the two male orderlies assigned to the floor. A student nurse, breaking the rules, answered a request, and suddenly she was screaming in his bear hug. I was the only male (other than patients) on the floor at the time, and rushed my 135 pounds into a diversionary attempt before she was injured. The ruse worked; he released her, and fortunately for me, just as we were joined in battle, the two orderlies appeared.

The Decision to Go to Michigan

It was clear to all of us physicians granted temporary deferrals from active service during the war effort that our call-up would come sooner or later. This did not keep us from trying to plan ahead for life after the service. Now came another time for decision. Although I had made the commitment to prepare for a career in human genetics by acquiring a medical education, I assure you there had been anxieties involved in abandoning a good start in *Drosophila* genetics to embark upon a career in the then discredited field of human genetics. Now came some further anxieties about a further career switch. I liked medicine. Maybe I should stay with it, using my genetic background as a kind of subspecialization. The financial and professional uncertainties of attempting to create a new field, especially now that our firstborn had arrived, weighted heavily. Again, some pacing of the floor. I decided to continue with the gamble.

Now I had to try to find a postwar setting in which to further that decision. A return to Dartmouth remained a possibility, although, because of the duration of my absence, I had long since been forced to sever the formal ties. The other principal thread from the past that might carry over into the future was the Eugenics Research Office. Demerec, now Director of the Carnegie Institution's activity at Cold Spring Harbor, was interested in reactivating the Office, and had asked me to submit for consideration by the Institution a program in human genetics that carried no obligations to the past, which request I had met in 1944. Then, in 1945, two unexpected possibilities surfaced, one a position at the Fels Research Institute in Yellow Springs, Ohio, the other at the University of Michigan in Ann Arbor.

The opening at Michigan had a number of intriguing aspects. In 1940, Lee R. Dice, an eminent zoologist–ecologist at the University, had been funded by the Rackham Foundation of the University to bring into being a research and counseling program in human genetics. Dice never had any pretensions to being a professional geneticist, but he recognized the need for a source of information for persons with genetic problems. (Largely for his role in the development of such services, he was elected the third president of the American Society of Human Geneticists, formed in 1948). A central aspect of the program was a Heredity Clinic, at which patients with genetic diseases would be studied and counseled. Because of the patient contacts, it was desirable that the clinic be an out-patient component of the University Hospital. This status was approved by the Executive Committee of the Medical School in December of 1940. Space was found in an old, white frame house near the hospital. Dice quickly recruited Charles Cotterman, who had obtained his Ph.D. with Lawrence Snyder at Ohio State, and C. Nash Herndon, a physician protégé of William Allan at Bowman Gray Medical School. Although the Medical School had not initiated the idea, it was supportive. It was a great stroke of luck that a young member of the Department of Ophthalmology, Harold Falls, became intensely interested in hereditary eye diseases; he handled the hospital contacts. Given the times, this was an impressive beginning. Unfortunately, it was not to last: Nash was soon recalled to higher responsibilities at Bowman Gray, Charley Cotterman was soon in the military service, and Harold Falls found himself overburdened by the excessive demands created by the absence of so many ophthalmologists doing military service during the war.

Dice contacted me in the fall of 1945 as a possible replacement for Herndon. My view of the future of human genetics demanded a close association with a medical center of excellence, which the University of Michigan Hospital–Medical School certainly was. One of the troublesome aspects of the possible association with a reoriented Eugenics Record Office was its remoteness from medical facilities (and patients). Furthermore, my earlier experience at the Office had convinced me that although there were in its files a few pedigrees of genuine genetic value, most of the material was worthless; there was little in the way of a legacy from the past to be realized there. The position at Michigan came closest to meeting the favorable circumstances I had envisioned from the first as necessary to a sound program in human genetics, especially since I would have a tie-in appointment with the Department of Internal Medicine. The salary, to be sure, was minimal and the position precarious, depending on grant funds whose continuity was not assured. I visited Ann Arbor to inspect the situation. With respect to physical facilities, the building that housed the activity had formerly been an interns' quarters but had finally been declared unsuitable for that purpose. In those days, any building unfit for interns was not much of a building. Once again, I swallowed hard, remembering Stern's aphorism about the sweetness of a canary's song not being determined by the cage to which it was confined. On the plus side, Ann Arbor was an ideal setting in which to live, raise a family, and work. I agreed to come to Michigan as soon as my military commitments were met. I've been there ever since.

Military Service

The war ended while I was an Assistant Resident in Medicine at Strong Memorial Hospital. Priscilla and I remember V-J night very clearly. We were living in an apartment over Charley's Red and White grocery store, a location that resulted in an amazing display of small animal life on our walls come evening. On V-J night, we were awakened by two drunks under our window, arguing as to whether our building really was on fire. (For you Rochesterians, we faced on busy Genesee Street, just across the Genesee River from the Medical Center.) Others must have noticed the flames; as we trundled our firstborn, daughter Frances, aged two months, down the stairs, we met the firemen on their way up with the hoses. The fire had begun in the apartment next to ours, which, when the door was broken down, revealed its occupant, a young woman, to have been overcome by smoke. She was taken to "my" hospital; the next day I stopped by to assess the damages and talk with her mother, a nurse who had been working the night shift when the fire broke out. The admission note, written by an intern friend riding the wagon that night, began: "Patient picked up in hallway of dirty tenement building. . . . " We had no place to go but up. (Yes, the young woman recovered nicely.)

Despite the war's end, the military had a continuing need for physicians, and it was expected and only proper that those of us deferred during the war years would take the place of the physicians being demobilized in large numbers. The most likely fate for a first-lieutenant-to-be like myself was assignment to a Station Hospital in one of the occupied areas—not a medically exciting prospect, since challenging medical problems simply passed through on their way to a General Hospital, leaving behind the soldiers with respiratory infections, venereal disease, gastrointestinal upsets, and the like. But with the end of the war came a local revelation. We all knew that a major medical research activity had come into being at the University of Rochester Medical Center during the war, involving many of the faculty but also a number of individuals recruited specifically for the activity. Stern was involved, joined by Spencer; it had been an occasional pleasure during my medical training to slip away from clinical matters for genetic discussions. Neither of them could talk about what he was doing, and I never probed, although judging from the people involved in the total program, I had my suspicions. At war's end, it was revealed that this activity had been a principal focus for research on the somatic and genetic effects of radiation sponsored by the Manhattan Engineering District, as the atomic bomb project was officially termed. With this revelation, a highly improbable military assignment—almost a fantasy—began to intrude into my thoughts concerning military service. The newspapers and popular journals abounded with speculation concerning the genetic implications of the atomic bomb explosions over Hiroshima and Nagasaki. Surely, somewhere in the United States, someone was thinking about follow-up studies. The commanding officer of the Rochester Manhattan District unit was Col. Stafford Warren, former chairman of the Department of Radiology at Rochester; his adjutant was an internist, Capt. Joe

Howland, a Rochester product with whom I was casually acquainted. One day I spoke to Joe: if Col. Warren knew of any long-range planning within the military regarding follow-up studies of Hiroshima and Nagasaki, might he throw in the hopper the name of Lieutenant-to-be Neel, as someone interested in the genetic aspects.

Some three months after that conversation concerning an assignment in Japan, I finished my scheduled hospital training, and in due time my military orders arrived. There was nothing exotic about them; I was ordered to temporary duty, first at Waltham General Hospital and then at Murphy General Hospital, both near Boston. This was a stroke of luck, for we had agreed that while I was in the service, Priscilla would return with our firstborn to her mother's, near Boston. Presumably I would get my regular assignment as soon as this temporary duty was completed; I had no real expectation of anything out of the ordinary. I never did receive that "regular" assignment; on October 29, 1946, orders reached me at Murphy General, directing me to report at once to the Medical Department at the Pentagon. In the military scheme of things, there is nobody much more superfluous, unloved, and unwanted than a first lieutenant on temporary assignment in a large military hospital. I could not help but observe how my status changed when orders came to report directly to the Pentagon. At such times it is wise not to be too explicit. Murmuring only something about a "special assignment, can't talk," I left the post for the railroad station in the colonel's car.

Incredibly, my talk with Joe Howland had resulted in an assignment to the Manhattan Engineering District, with orders to serve as support personnel to two civilian radiobiologists scheduled to survey Hiroshima and Nagasaki and advise the National Academy of Sciences concerning a program of studies on the delayed effects of the atomic bombs. Exactly 24 days later I was on my way to Japan. In the interval, I had divided my time in Washington between the Pentagon and the National Academy of Sciences. At the Pentagon, I had feverishly read up on radiation effects, including as many of the still top-secret documents about the effects of the atomic bombs on the people and the buildings of Hiroshima and Nagasaki as I could get to (I'd been given a top security clearance.) At the Academy, which was in the process of assuming responsibility for follow-up studies in Hiroshima and Nagasaki, I had a similarly intense short course, as I tried to understand the bomb's structure.

My experiences in Japan will be treated in Chapters 4, 5, and 13. Here let me note only that after a 16-month immersion, in Japan and Washington, in organizing follow-up studies on the effects of the bomb, I was free to assume the position waiting at the University of Michigan and resume what I have termed my "medical interlude."

Subsequent Medical Training and Activities at the University of Michigan

In March 1948, my immediate commitments to the genetics program that I had developed in Japan finished, I proceeded straight from Japan to the University of Michigan, rendezvousing with an eight-months pregnant Priscilla coming from Boston. She held the stork at bay long enough for us to settle into a small apartment. Then, now with a son as well as a daughter, it was time to initiate the great experiment in human genetics. However, although my primary role at Michigan was to bring up a strong program in human genetics, it had been agreed that I would involve myself in such academic medical activities as were necessary to my rather catholic view of what was essential to being a well-rounded human geneticist. To that end, along with my other responsibilities, during the next three years I rotated intermittently through a series of the specialty clinics maintained by the Department of Medicine, clinics, for instance, devoted to such areas as hematology, endocrinology, or diseases of the chest. In due time, my background was such that I felt comfortable functioning in the role of attending physician on the general medical wards of the University of Michigan Medical Center.

My "practice" of medicine has been limited to the clinics and wards of a teaching hospital specializing in tertiary care. An attending physician on the wards of a university hospital can easily acquire a very different outlook on the practice of medicine from that of the solo physician in a smallish town. Historically, the University of Michigan Medical Center has functioned as a court of last appeal for the citizens of Michigan afflicted with difficult medical problems. In consequence, a relatively high proportion of the patients whom we saw on the Medical Service were the diagnostically difficult or the therapeutically resistant. Patients with uncomplicated heart attacks or pneumonia are usually treated in the local community and seldom reach a university hospital such as ours. Of necessity, we taught our students and house officers (interns and residents) on the assumption that if we developed in the student the thought processes to manage the difficult and unexpected, he or she would do well with the ordinary. This may be true from the technical standpoint, but I am not sure it properly prepares the student for the emotional and psychological demands of day-to-day practice.

Genetic Counselor

Concurrently with the medical activities described in the previous section, I had the responsibilities for the Heredity Clinic mentioned earlier.

It was one of the first in the country. (We have never settled whether ours antedated a similar clinic at the University of Minnesota. If so, we were *the* first in the U.S.) Technically it was one of the many specialty clinics of the University of Michigan Medical Center; our patients were either referred to us from within the Medical Center or came to us directly, upon the suggestion of their physician. The Clinic's responsibilities were varied. Sometimes the patient was an individual aware of possibly hereditary disease in his/her family, concerned about the chances of developing the disease. Other times the "patient" was a set of parents, to whom had been born a child with a real or fancied genetic problem; they were concerned about the likelihood of a recurrence. Occasionally the question was the risks of a consanguineous marriage. In each instance, we provided the most accurate information currently available. A different type of problem arose when an agency arranging adoptions would consult us as to the ethnicity of a child who had become its ward. The usual question in this situation involved a very young child allegedly of Negro–Caucasian ancestry; our responsibility was to predict its probable appearance as it matured, as a guide to the agency in arranging placement.

Some three or four percent of all children are found to have a serious congenital defect at birth, or during infancy and childhood exhibit mental retardation, severe defects of vision or hearing, or other handicaps not readily diagnosed at birth. There is also a considerable spectrum of inherited diseases with onset later in life. Given something like three children per family, then roughly one family in ten is touched by such a tragedy. Many families accept the event stoically, but for others it is a happening whose possible recurrence is a source of deep concern. Dice had been prescient in discerning the role of a genetic counseling service in a medical center.

Anxious to avoid the taint of eugenics, we were as nondirective as one can be in counseling situations, contenting ourselves with explaining the genetic facts and the probabilities as clearly and simply as possible. But it was a little difficult to stay uninvolved when, as so often happened, the patient asked: "Doctor, what would you do if you were me?" Quite aside from providing simple odds, counseling had two functions. It contributed to the patient's understanding of what was usually an obscure and troubling happening. It also frequently provided reassurance, since couples often came to see us with quite unjustified fears as to the probability of recurrence of a congenital defect that had already afflicted one of their children.

At that time, a real shortcoming of genetic counseling in the case of recessively inherited disease was that it took the birth of an affected child to bring the family to counseling. For recessively inherited disease, the expectation in the usual situation, when both parents are normal-appearing carriers of the gene, is that one in four of the children will be affected. Because of the small size of American families, usually there is only one affected child in a sibship. In only about 20% of all sibships in which both parents are carriers are there two or more affected children. Thus, if the first counseling is only after the birth of an affected child, most of the children ever to be affected will already have been born before the parents seek advice. A similar problem exists where parents are carriers of a dominantly inherited disease that does not find expression until late in life.

The "ideal" counseling situation would involve the ability to detect the carriers of the genes responsible for severe defects *before* reproduction, with the further ability to diagnose the children who will be defective *prenatally*, with abortion possible if so requested by the parents. Most parents would then "try again," the odds the next time being for a normal child. We will discuss later the amazing progress in carrier detection and prenatal diagnosis. In the meantime, I note how little patience I have with "right-to-lifers," who would, given these technical developments, still force upon enlightened parents the care of a child afflicted with severe genetic disease when it would have been possible, following an abortion, for the parents to produce a normal child in the next pregnancy. I wonder if the antiabortionists truly understand the adverse impact of a severely defective child on the nuclear family. How the prohibition of *early* abortion can be twisted into the will of a loving and caring God eludes me.

As genetics has developed, so have the functions of a genetics service. It is now less a matter of dispensing probabilities and more a matter of active involvement in the diagnosis–treatment process. The precise role of the genetics service varies from medical center to medical center. The recognition of new genetic syndromes has so multiplied in recent years that the medical geneticist may be the staff person most competent to reach a correct diagnosis on a handicapped child, and the correct diagnosis is the key both to prognosis and counseling. The genetics service may also be responsible for the cytogenetic studies which, if conducted prenatally on cells obtained through amniocentesis, can result in the prenatal diagnosis of Down syndrome and other diseases due to chromosomal abnormality (with elective abortion). Biochemical studies on cells obtained prenatally also often provide the basis for the diagnosis of certain inherited diseases. Further, the service may be responsible for the specialized biochemical studies of newborn infants, children, and adults that result in the diagnosis of certain genetic diseases and also can, in some instances, identify which members of the family carry particular recessive genes. More recently, these biochemical tests have been extended to the DNA level. In short, medical genetics is now as "high-tech" an area as many of the other rapidly evolving areas of medicine, and beginning to participate in the same issues of resource allocation we shall discuss next.

Retreat from Clinical Medicine

I enjoyed clinical medicine, both for the diagnostic challenge and the satisfaction in relieving suffering, and hope I was a competent and compassionate physician. But increasingly, as the genetic side of my career unfolded, I was impressed by the amount of physician effort which, often at great and ever increasing expense, was devoted simply to prolonging a human life whose mission was clearly accomplished, often under circumstances for which the patient had no great enthusiasm. Do not misunderstand: There is no more worthy and noble effort than to bring relief to a suffering patient. But the geneticist in me was gradually getting the upper hand. In 1958, for a symposium sponsored by the American College of Physicians,[1] I found myself writing:

> Man, after something like 5,000,000 precarious years on this planet, has now succeeded in grasping firmly the reins of his own destiny. With this development come awesome responsibilities, in which no group shares to a greater extent than does the medical profession, unless it be the nuclear physicist, whose dawning appreciation of what he has done is so forcibly presented in Snow's novel, "The New Men." Who of us has not with pride identified himself with Sir Luke Filde's wonderful portrait of a physician keeping lonely vigil by the bedside of a sick child? The physician's devotion to the individual patient is still unchallenged in a world of rapidly changing values. But, with signs that the human species as a whole may be confronted with medical problems no less serious than those of the child in Filde's portrait, who in this time of transition is keeping similar vigil over the species? We are all familiar with the complex apparatus of modern medicine, which briskly and efficiently swings into action when a patient with congenital heart disease enters hospital for surgery, or a patient with an acute renal

shutdown comes in for dialysis. But, in our concern for the individual, have we forgotten to set up the team which has as its concern the species as a whole?

The Harvey Society of New York City each year sponsors a series of some eight lectures. They honored me with an invitation for one such lecture during the 1960–61 series. I consider that lecture, entitled, "A Geneticist Looks at Modern Medicine," my formal farewell to the role of attending physician in The University of Michigan Hospitals. The decision had three components. First, I had since 1956 been Chairman of a newly created Department of Human Genetics at Michigan. To do well, the Department needed more of my time than it was getting. Second, the content of medicine was changing rapidly and to keep up required a major effort. I was becoming concerned that I soon might not be able to meet the needs of patients or medical students adequately. But there was a third reason, philosophical, with which I was having increasing difficulty. Let me quote from my closing remarks in that lecture:[2]

My thesis this evening has been that an increasing proportion of medical effort is being directed toward the terminal ramifications of human pathology at a time when the biological challenge to our species is certainly assuming new forms and may be at a maximum. Any evaluation of the true nature of this challenge draws heavily upon the concepts of population genetics. An attempt has been made to indicate some of the vast areas of ignorance regarding human population genetics which must be filled in if this challenge is to be properly understood—areas not only presenting a legitimate field of inquiry for medicine, but whose exploration often requires specialized medical knowledge.

But what if these research needs were met, what then? How would we utilize the knowledge? This brings us to one of the thorniest problems ever tackled by the human mind. Few would dispute the thesis that the more we know concerning the nature of man, the more rational our therapeutic approaches. But going beyond this, how can we utilize detailed knowledge of the genetic structure of populations? It would, to begin with, be obvious folly on my part to predict the utilization of knowledge and wisdom still to be acquired. I hope I have succeeded in indicating to you this evening just how much we geneticists have yet to learn. . . .

However, we are already far enough along to begin to come

The Lawrence D. Buhl Building, dedicated in 1963, the first of the modern facilities that the Department of Human Genetics eventually occupied.

to grips with some questions of principle, chief among which is: To what extent can and should man, into whose hands scientific advances have placed awesome responsibilities, now assume a role in the guidance of human evolution? The temptation to the premature and procrustean use of half-knowledge will be strong and indeed exists already. I suggest there may be a more immediate approach, supplemental rather than truly alternative. Instead of trying to tailor man to developments only dimly visualized, is not the more immediate and pressing task that of developing the culture best fitted to man's needs? The term culture is here used, in a very broad sense, as simply the milieu in which we function. There are, I submit, certain "runaway" aspects to the present rate of cultural change. Man's ingenuity is far outstripping his ability to foresee the consequences of his actions. We are now at the stage where we can and must begin to shape our cultural evolution consciously, thereby avoiding a clash between genetic man and cultural man. It can be taken as an article of faith in the scientific method that the more we learn of man's inborn attributes, the better we can develop an appropriate cultural and physical environment.

The past 30 years have only accentuated the trend of modern medicine with which I was concerned in those two presentations. On the one hand, persons over age 65, the heavy consumers of medical services in this country, increased from 20.1 million to 30.1 million in the interval 1970–1990, and are projected to increase to 47.3 million by 2020, an increase quite disproportionate to population growth. On the other hand, medical technology, for diagnosis and intensive care, has proliferated at an exponential rate. So have family and patient expectations concerning the type of general support services received if one is hospitalized. More and more, physicians have become the custodians of death delayed in older citizens rather than the champions of life unfolding. This is not a pejorative statement. On two occasions in the last 10 years I have been very grateful, in life-threatening situations, to be the recipient of the most advanced medical technologies. But it is a simple fact that, overall, the activities of medicine are less and less relevant to the gene pool of the next generation.

A tangible result of these developments is that expenditures for medical care are the fastest growing segment of the national economy, as expressed as a fraction of the gross national product. The total cost of medical care in the U.S. is difficult to estimate but for 1992 has been set at about 870 billion dollars, representing about 14% of the gross national product. This cost has risen some 10% annually over the last ten years. It is clear that unless somehow checked, with the continuing advances in medical technology, medical costs under the current system can only continue to climb. At the same time, substantial numbers of Americans are not receiving adequate care for standard medical problems, some 37 million persons being without the health insurance that renders medical care affordable. What is particularly worrisome is that a disproportionate number of those below 20 are without adequate health care. Medical costs are perceived by government and industry alike as "runaway," and their control has become a high priority.

The current turmoil over how to extend medical coverage to all while at the same time containing the costs of medical care to a reasonable fraction of the GNP will persist for some years. In the course of this debate, society must address the apportionment of services, medical and paramedical, between the oncoming, genetically significant young and our loved and respected but genetically spent old. This effort, if driven by thought rather than the voting power of senior citizens, will bring us back to the challenge, mentioned in that Harvey Lecture, of designing, from conception onward, a culture (medical and otherwise) more attuned to our unfolding insights into genetic variation and how its expression is modified than is the case at present. I examine this issue in some depth in Chapter 19.

Although following that Harvey Lecture of 1960 I would continue to be involved with genetic counseling, which comprehensively done, absolutely requires a sound medical background, and would consult on patients with certain genetic diseases, I would not again function as an attending physician in a hospital setting, even though I have the title of Professor of Internal Medicine. By this time my interests were clearly converging on the gene pool, rather than on the individual and his genes. The studies on the genetic effects of the atomic bombs and on consanguinity effects, to be described later, both "gene pool" studies, were well under way. It was time to be honest with myself, to follow my inclinations in the same way that, 20 years earlier, I had opted for human genetics.

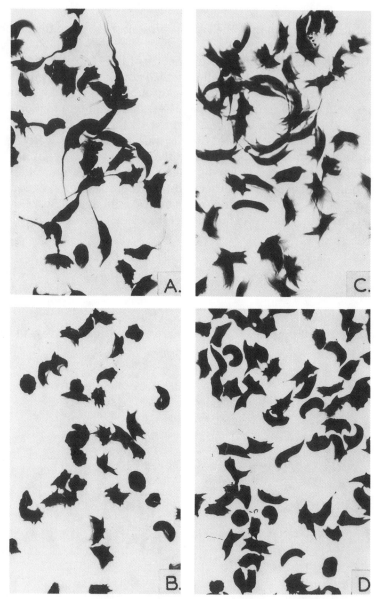

Stained, deoxygenated red blood cells from children with sickle-cell anemia (A and C) and from one of their parents with the sickle-cell trait (B and D). The sickling is more easily induced and is more severe as the photographs show in persons with sickle-cell anemia, resulting in more rapid red blood cell destruction than in carriers or unaffected individuals.

3

Thalassemia and
Sickle-Cell Anemia

The life of a medical student is a busy one and was more so during the war years when vacations disappeared and a general shortage of physicians created additional medical student responsibilities in those areas where they could function effectively. I certainly had no anticipation of time for research as a medical student, but—

Thalassemia

Early in my third year of medical school, I encountered for the first time a child with Cooley's anemia, a rare, chronic, severe anemia usually fatal during the first two decades of life. Recognized as a distinct entity in 1925 by a Detroit pediatrician, T. B. Cooley, it was known in the United States to occur predominantly in persons of Italian and Greek ancestry. For this reason it was also called Mediterranean anemia. The basic defect seemed to be an ability to make hemoglobin. Repairing that evening to Max Wintrobe's superb textbook of hematology (first edition),[1] I read, after his description of the hematological features of the disease, the following:

> We have encountered several cases of mild anemia which bear a strong resemblance to Mediterranean anemia as just described . . . . The similarity of this condition to Cooley's anemia is so great that it seems justified considering it as representing a benign form of the fatal anemia which Cooley described in infants and young children. Because of its benign character,

37

those possessing the "trait" live to adult life and perhaps pass the trait to their offspring. In two families we found that both parents of a child with typical Cooley's anemia showed the characteristics above described . . . . Similar observations have been made in Greece . . . . Under the title "target cell anemia" and "familial microcytic anemia" cases similar to those described above have been reported from Boston. . . . (p. 517)

This is one of those moments that glow in the recollection. It really didn't require a great deal of genetic insight to respond at once to these sentences with a hypothesis: there was a gene that in single dose (the heterozygous condition) produced a mild anemia but in double dose (the homozygous condition) resulted in the much more severe condition. So there I sat, my first major problem in medical genetics rapidly shaping up. There was an additional reason for pursuing this particular hypothesis: even at that stage in my biomedical education I realized that we could conveniently characterize blood cells better than any other cell of the body. This, in the vernacular, was a "clean" trait, about which we should be able to be precise and objective. It met the criteria I had set for my studies on the genetics of humans.

Testing such a hypothesis is in principal straightforward. One conducts detailed hematological studies on as many of the members of a family in which the disease occurs as possible. If the hypothesis was correct, there were three expectations: 1) both parents of a child with the severe disease should show the mild condition; 2) among the unaffected siblings of a child with the severe disease there should be children with the mild condition and also completely normal children, in the ratio of 2:1; and 3) one in four of all children born to carrier parents should show the severe condition. In testing for the third expectation we had to make allowance for the fact that we approached families for study only if *at least one* affected child had already been born. This fact introduced what we call an ascertainment bias, for which a statistical correction was necessary before one could accurately test for the 3:1 ratio of Mendelian inheritance.

To test the hypothesis, I had to have a base of operations, namely, access to a laboratory equipped for hematological studies. With some trepidation, I approached J. S. Lawrence, Professor of Medicine and the hematologist on the Rochester faculty. He was known to be a rather particular man and rightly so—the last thing one needs in a well-run hematology lab is a loose cannon mixing up the solutions, exhausting the special glassware, etc. To my great relief, Lawrence—who became very supportive of the work—agreed quite readily, and then came a surprise: he informed me that someone else had approached him for facilities to study the milder condition. It was Bill Valentine, then an intern. We met and made the obvious decision to combine forces. Again the fates had been kind: Bill is a person of uncommon ability, later to become a distinguished hematologist and academic physician, with election to the National Academy of Sciences—a rare honor for a medical scientist.

Our plan was simple: comb the hospital records for cases of Cooley's anemia, then seek out the families of these patients for hematological study. We sensed we would get better cooperation if we went to most of the families in their homes rather than asking them to come to hospital. So we put into two doctor's bags what we needed to draw blood samples and prepare blood films for large families and then, because neither of us had cars, became experts on the public transportation systems of Rochester.

We pursued this problem together for the next several years. I have already mentioned how short-staffed hospitals were during the war years, with corresponding additional responsibilities impacting on house officers and even medical students. Fortunately for us, even during the war years, the Medicine Service at Rochester maintained the remarkably humane institution of Night Float, a physician who went on call at 10 p.m. to admit emergencies and handle unexpected developments in patients already admitted. In general, the "day shift" seldom finished before 10 p.m. themselves, and were often there much later, straightening out the day's events.

A problem with our interest in Cooley's anemia was that it was only after-hours, so to speak, that we could carry out our blood studies (doing blood counts, staining and examining blood films, and testing red blood cell fragility). There was no such thing as released time for research. Even if a family of interest could be induced to come to the hospital on weekdays, we didn't begin our special studies of their blood samples until after the day's work was done. But at least, because of the Night Float, we had our nights largely free. If we used our weekend "off-call" to visit a family and obtain blood samples, there went the weekend, also often far into the night, because, if certain observations were to be valid, they had to be made on fresh samples. The tradition launched on my second night of marriage was well reinforced during those years.

In due time we had sufficient data for a significant test of the hypothesis. Everything fit. Meantime, the disease had not gone unnoticed. William Dameshek, an astute hematologist who was responsible for some of the findings from Boston cited earlier by Wintrobe, had published another paper in 1943 that, reviewing his own studies, said it clearly: "This suggests that the mild disorder is readily transmitted as a Mendelian dominant, but that the severe disorder requires the presence of homozygous genes."[2] In fact, digging back into a paper published in 1938 by the Greek hematologist Caminopetros (written in French in a journal seldom seen in the U.S.), we found a passage which was very close to the central idea. What we had done was to prove the concept beyond reasonable doubt.[3]

Now there arose a question of terminology. What was a convenient and proper designation for two hematological conditions related in this fashion? Earlier, Professors G. H. Whipple and W. L. Bradford of our own institution, in reporting on some cases of the severe anemia seen at Strong Memorial/Rochester Municipal Hospitals, had suggested the term "thalassemia" in an effort to avoid the medical scourge of diseases bearing their discoverer's name. Although their suggestion had not been enthusiastically embraced, it was a good point of departure: we designated the severe disease "thalassemia major" and the mild one "thalassemia minor," and that is how they are generally known today. (In the current terminology, the disease we studied would be termed β-thalassemia, associated as this type is with a defect in the β-polypeptide of hemoglobin.)

Before our studies were finished, we had identified 11 sibships into which one or more children with thalassemia major had been born to persons of Italian origin in the city of Rochester during the preceding 15 years. In Rochester, "Italian origin" implied provenance from the lower third of the Italian peninsula or Sicily. Since we could estimate the number of persons of Italian ancestry in Rochester and the numbers of children born to them during the time span from which our patients were drawn, and since we understood the precise genetic basis for the condition, we could by virtue of a very simple genetic principle estimate the frequency of the heterozygous carriers of the gene with thalassemia minor. When there are only two alternative forms of a particular gene in a population's gene pool, say A and a, the relative frequencies of AA, Aa, and aa individuals can be easily predicted. If p is the frequency of the A gene and q the frequency of a ($p + q = 1$), then these proportions are $p^2AA:2pqAa:q^2aa$. This relationship is known as the Hardy–Weinberg Law, after the British mathematician and German physician who independently worked it out early in this century. Our studies had provided an estimate of the frequency of the aa type (thalassemia major) or q^2. It was 1 in 2400 births. From this it was no trick at all to deduce that $2pq$, the frequency of Aa carriers, should be about 4% of the population.[4]

As a result of this simple calculation, we were immediately in hot water with the hematologists. This carrier state, while seldom resulting in overt disease, is often characterized by a mild anemia, resembling that produced by iron deficiency. The hematologists were sure they couldn't have missed so many people with thalassemia minor. It wasn't until several years later, when almost precisely this frequency of thalassemia minor began to be revealed by surveys in southern Italy by E. Silvestroni, that our figure was accepted. These carrier individuals had often been simply misdiagnosed as exhibiting iron deficiency anemias that somehow failed to respond to the usual treatment with tablets of ferrous sulfate!

An Essay on "Carrier States"

Earlier I discussed briefly the applications of genetic counseling. One of the standard problems in dealing with a family in the genetic counseling setting was to know who in the family carried the gene of concern and who was free of it. With thalassemia the situation was suddenly different. Now, by the presence of a mild condition, we could with near 100% accuracy not only identify those in a family who were the trait bearers but, in principal, predict which marriages were at risk for producing a child with the fatal disease. The implications of this for genetic counseling were obvious, so obvious that in 1946, while completing my residency at Rochester, it seemed worth surveying the genetic literature to determine the number of other inherited entities for which one could detect with some accuracy the asymptomatic carriers of disease predispositions. That review, published in 1947, contained 18 examples of varying validity. Some are still accepted but others have not stood the test of time.[5] Today, of course, such a list would be very different: most of the 200 or so recessively inherited biochemical disorders delineated since then have a clearly defined carrier state.

Whatever the value of that review for genetic science, for me it was a very good thing to have done: it directed my attention to my next major interest, sickle-cell anemia. I had seen only a single patient with this disease while at Rochester; the city did not have a large African-American population. The disease takes its name from the characteristic shape the red blood cells assume when deprived of oxygen (see figure on page 36); it is a chronic, severe anemia, often fatal almost as early as thalassemia major. Periodically there are "crises" of blood destruction and pain, which bring the patient to hospital. The prevailing concept at the time was that the sickling tendency was genetically determined, caused by a gene extremely variable in its manifestations, usually producing the harmless sickle-cell trait but sometimes the severe disease. Having studied thalassemia, I was prepared for the by now obvious alternative interpretation—that the sickle-cell trait was the heterozygote and the sickle-cell anemia the homozygote for the same gene. I had tested both parents of our single Rochester patient; both had the sickle-cell trait. So far, so good. It seemed worth including this disease in my list of conditions for which a carrier state could be identified, but first I had to review the medical literature. The results were unsettling. I could find reports of 31 family studies in which both parents of a child with sickle-cell anemia had been tested for the sickling tendency. In 7 families neither parent had the sickling tendency, in 9 families both parents had the tendency, and in 15 families one parent was reported to show the trait, the other not! This fitted with neither hypothesis. Fortunately, by then I had a reasonable skepticism of the reliability of many laboratory tests, including the test for sickling. Nine families in which both parents sickled, plus the one I studied, seemed too many if only a single gene was necessary. I suggested the homozygous–heterozygous hypothesis in my review on carrier states, but that was all I could do about the subject for a while. By then my time in Rochester had expired and the Army was ready to exercise the option that resulted in my assignment to Japan.

The Genetics of Sickle-Cell Anemia

The experience with that one case of sickle-cell anemia in Rochester got me intensely interested in the disease. Once back from Japan and settled into Ann Arbor in 1948, I was ready to study the genetics of sickle-cell anemia properly. There was a large African-American population in nearby Detroit and environs. The National Institutes of Health responded favorably to an application for modest funding, submitted while I was still in Japan. I had two objectives. I had the sense that too much of our thinking about the situation was colored by the hospital setting. My first objective was to determine whether, when the disease was presumably quiescent—with the patient at home, not in "crisis"—I could by standard hematological techniques demonstrate a clear distinction between those with only the trait and those with the anemia. Second, assuming this was indeed the case, would careful genetic studies such as those carried out on thalassemia reveal that the anemia and the trait were related as homozygote and heterozygote?

The rather strange results from testing both parents of children with sickle-cell anemia, mentioned earlier, had convinced me that the first step was to improve the reliability of the test for the sickling phenomenon. This done, I proceeded to contact every family with sickle-cell anemia ever seen at the University Hospital in Ann Arbor. These resources were rather quickly exhausted, but fortunately I was then able to form a solid collaborative arrangement, which lasted many years, with Wolf Zuelzer, the highly research-oriented pediatric hematologist at Children's Hospital in Detroit. Now there was an abundance of patients for study. Pursuing these children and their families in their home setting led me to become as familiar with the African-American districts of Detroit as I had been with the Italian-American districts of Rochester.

The first 42 parents of children with sickle-cell anemia I tested all sickled. At the same time, it became increasingly apparent that there was a clear hematological distinction between the sickle-cell trait and sickle-cell anemia with, by all the usual hematological parameters, no intermediate states (but see below). The red blood cells in persons with the anemia could be induced to sickle much more easily and assumed much more bizarre forms than those in persons with the trait. It was easy to visualize how the more severe sickling in the anemia could lead to red blood cell destruction. My hypothesis was substantiated, and, in 1949, I published a note in the journal *Science* to that effect.[6] I would like to believe that nailing the hypothesis down depended at the time as much on meticulous hematology as on genetic principles. By following children with the diagnosis of sickle-cell anemia into their homes, when they were doing as well as they ever would, and demonstrating that at all times there was really no overlap between the trait and the anemia, I had established that the sickle-cell trait was essentially benign but sickle-cell anemia was always serious.

Four months later, there appeared an article, also in *Science,* which I read with more intensity than I had ever before (or have since) accorded a scientific note.[7] At the California Institute of Technology, Linus Pauling and three of his postdoctoral associates, H. A. Itano, S. J. Singer, and I. C. Wells, had examined the hemoglobin of persons with the sickle-cell trait and sickle-cell anemia by the relatively new technique of electrophoresis. This technique involves placing a solution of a protein—in this case, hemoglobin—in an electric field. Because protein molecules carry a charge (except when at their neutral point), they will migrate in this electric field. Each protein has a characteristic rate of migration, and mixtures of proteins can be separated on this basis. In individuals with the sickle-cell trait, these investigators observed an abnormal hemoglobin component in addition to the normal one. In sickle-cell anemia, all the hemoglobin was of the abnormal type. Their biochemical studies and my genetic studies dovetailed perfectly: in the

43

presence of one abnormal (sickle-cell) gene, roughly half the hemo-globin was abnormal; with two such genes, it was all abnormal.

Many years earlier (1908) an English physician, Sir Archibald Garrod, had devoted a series of Croonian lectures to what he termed the "Inborn Errors of Metabolism."[8] In these he considered four quite rare, recessively inherited conditions (albinism, alkaptonuria, cystinuria, and pentosuria), for all of which he postulated a genetically determined block in the normal course of the cell's biochemistry, due to an inborn deficiency of a specific enzyme. He then used these observations as the basis for speculation concerning a high degree of biochemical diversity between the members of our species. Subsequent findings have extended this speculation far beyond his possible imaginings.

I imagine that although Sir Archibald's scholarship was respected, he was considered something of an eccentric by most of his contemporaries, fiddling as he did with such rare entities at a time when influenza, pneumonia, tuberculosis, gastrointestinal disease, diphtheria, and typhoid fever were still among the principal causes of death. The lecture series had been largely forgotten until G. W. Beadle and E. Tatum, in the late 1940s and 1950s, developed the one-gene one-enzyme hypothesis on the basis of studies on the bread mold Neurospora. Rediscovering Garrod's lectures, they were quick to acknowledge how well their concept had been adumbrated in Garrod's writings. My studies and those of the Pauling group fitted clearly into this tradition. Now, however, we were dealing with a more common disease, in which the protein product of the abnormal gene could be readily isolated for study. (In retrospect, the discovery by Karl Landsteiner and colleagues of the ABO, MN, and Rh blood groups in the first 40 years of this century really defined common biochemical differences, but the understanding of the precise nature of these differences did not come until later.)

Just eight years after the papers by the Pauling group and myself, Vernon Ingram, working at Cambridge University, demonstrated that the basis of this abnormal electrophoretic behavior was the substitution of a single amino acid, valine, for the normal glutamic acid in one of the two kinds of chains of amino acids (polypeptides) of which hemoglobin is composed.[9] This single amino acid substitution makes the hemoglobin in its deoxygenated form much less soluble than normal. In the venous circulation, with low oxygen content, the sickle-cell hemoglobin literally precipitates out in the erythrocytes of affected persons, the rod-like paracrystals resulting in the strange deformations characteristic of the sickle cells. These cells cannot circulate normally, thus causing the various symptoms of the disease. The precision of genetic control implied by Ingram's demonstration and its implications for other inborn errors of metabolism were immediately apparent. Investigators, buoyed by the availability of government funding, flocked into the field. I had been attracted to the study of hematological disorders because of the ease and clarity with which the diseased cells could be visualized. Now another great advantage began to emerge: the protein of primary interest could be so easily obtained and manipulated. And when, in due time, the study of the hemoglobin diseases (including thalassemia) moved to the DNA level, the fact that developing erythrocytes are so heavily devoted to the synthesis of a single protein, hemoglobin, made it relatively easy to extend the study of hemoglobin variants to the ultimate genetic reference point, the DNA.

One Phenotype but Multiple Genetic Causations: Hemoglobin C

A little over a month after the publication of my paper in *Science* on the genetics of sickle-cell anemia, I received an interesting and unexpected letter. W. N. Powell of the Scott and White Clinic in Temple, Texas wrote to say that he and J. G. Rodarte of the same clinic had for some years had under observation a 38-year-old, white male of Sicilian origin who appeared to have mild sickle-cell anemia. One of his parents had the sickle-cell trait, the other thalassemia minor. Both of the patient's children had thalassemia minor. It was clear the patient owed his disease to the simultaneous presence of one gene for the sickle-cell trait and one gene for thalassemia. We put our heads together to produce a little paper pointing out that there was a second genetic basis for what clinically would pass as sickle-cell anemia and suggested this was probably the basis for most of the cases of sickle-cell anemia reported in Caucasians.[10] A year later, John Lawrence, with whom I had maintained a warm relationship ever since the hospitality of his laboratory mentioned earlier, wrote to say he had a Caucasian patient similar to the one Powell, Rodarte, and I had described earlier. We collaborated on the necessary family study, with essentially the same interpretation. About this same time, Italian investigators quite independently recognized the same entity.

The study of sickle-cell anemia and related disorders by the technique of electrophoresis was clearly a powerful approach. In those days it required the rather complicated and cumbersome Tiselius apparatus. Not having ready access to such equipment at Michigan, but blessed by now with a large number of families in which the anemia occurred, I wrote to Harvey Itano at Cal Tech shortly after the publication of that group's paper in *Science,* suggesting we make common cause on certain problems. The first was to see if the amount of sickle-cell hemoglobin in heterozygotes varied significantly by family. (It did.) While this study was in progress, I was confronted with two most atypical children in the sickle-cell patients we were studying. Their disease was unusually mild. (By now I had acquired a rather good feeling for the nuances of the disease.) They both had large spleens, whereas in typical sickle-cell anemia repeated episodes of clotting in the spleen (because the sickled erythrocytes tangle together and impede circulation) lead to the formation of a great deal of scar tissue and, by an early age, a spleen so small it cannot be palpated on physical examination. Equally provocative, in each instance one parent failed to sickle. I sent blood samples to Itano. It was a day of great excitement when he wrote that both these children had been found to have a mixture of sickle-cell hemoglobin and a new type of hemoglobin, migrating in the Tiselius apparatus quite differently from normal or sickle-cell hemoglobin. The clincher: in each family the parent whose blood failed to sickle possessed the new hemoglobin. The children in question clearly had inherited one sickle cell gene from one parent and one for this new type of hemoglobin from the other parent. In those days, before it was realized how many abnormal hemoglobins might be found, we proposed to assign them letters in the order of their discovery, "A" being reserved for the normal type. Since the new hemoglobin was the second type of abnormal hemoglobin to be recognized, we called it "C."[11]

What was emerging was what by now has become almost axiomatic in human genetics. Examine any inherited disease closely enough, and it begins to break down into subtypes, with these subtypes often accounting for the sometimes puzzling nuances of the disease. The sickle-cell anemias were the first and still the clearest example of this phenomenon in human genetics.[12]

African Journey

From very early on in the studies of thalassemia and, later, sickle-cell anemia, I was intrigued by the scientific issue presented by the relatively high frequency of thalassemia major and sickle-cell anemia. These two diseases were in effect genetic lethals, only rarely permitting those affected to reproduce. How in the world did such diseases attain the frequencies we were observing in selected ethnic groups in the United States? Since each child who dies prematurely with one of these diseases removes two of the responsible genes from the population, natural selection should have held the frequency of these diseases to the much lower frequency of most of the other recessively inherited genetic disorders that result in early death. The dilemma was all the greater because the average individual with thalassemia minor has a blood hemoglobin level about 13% less than normal; it was difficult to see this as an advantage.

During the 1950s, surveys of the frequency of the sickle-cell trait in the U.S., including some by ourselves, revealed an average frequency for the trait in African-Americans of about 8%. Concurrently, surveys in Africa were yielding trait frequencies in some tribes of 15% to 20% and even higher. The application of the Hardy–Weinberg formulation led me to suggest that in such tribes, approximately 5 to 10 in each 1000 newborn infants should be destined to develop sickle-cell anemia—a truly surprising frequency. Even in the U.S. the frequency should be about 2 per 1000 newborns; sickle-cell anemia was not nearly as rare an entity as the medical literature would suggest.

Once again, the application of a genetic principle to the prediction of the frequency of an inherited disorder elicited controversy. Some physicians with experience in Africa were quick to doubt that such frequencies could have gone undetected. As a way out of the dilemma, A. B. Raper and H. Lehmann suggested that the admixture between Negro and Caucasian that had occurred in the U.S. was somehow responsible for the high prevalence of the overt anemia.[13] The issue was resolved in two ways. In 1951 and 1952 a husband–wife team of Belgian physicians working in the Congo, the Lambotte-Legrands, showed that when one concentrated attention on young children, the frequency of the anemia really was as high as predicted.[14] In the meantime, in the U.S., I could demonstrate no relationship between the apparent degree of white admixture in African-Americans and the manifestations of the sickling gene,[15] thus casting doubt on the alternative explanation suggested above.

There were in principle two logical explanations of the high frequency of the sickle-cell gene. One was that the trait carrier, the heterozygote, was somehow favored over normal in the African environment, in the sense of surviving and/or reproducing better. The resulting transmission of sickle-cell genes by the heterozygote would offset the loss of the same kind of genes through the early death of the homozygote with sickle-cell anemia. The term for this situation is balanced polymorphism. The alternative explanation would be a high mutation rate, far higher than mutation rates were thought to be.

In 1949 the agile mind of J. B. S. Haldane has raised the possibility that a relative resistance to malaria during the recent past might explain the relatively high frequency of thalassemia major and minor that Bill

Valentine and I were finding so troublesome. This suggestion lay fallow for some years, but then in 1954 it was extended by several investigators, especially A. C. Allison, to cover the sickling phenomenon.[16] With respect to *Plasmodium falciparum,* which causes the most severe type of malaria in Africa, Allison developed three lines of evidence in favor of the hypothesis: 1) a lower rate of *P. falciparum* infection in sicklers than in nonsicklers, as judged by the presence of parasites in the blood stream; 2) a greater susceptibility of nonsicklers than sicklers to experimental malaria; and 3) a high correlation throughout Africa between the importance of malaria as a disease and the frequency of the sickle-cell trait, the correlation being interpreted as a positive response of the sickle-cell trait to natural selection. All this evidence was circumstantial; the hypothesis remained quite controversial until, some three years later, several different groups of investigators working in Africa supplied the "clincher": among children admitted to hospital with malaria so severe that they subsequently died, there were fewer with the sickle trait than expected on the basis of the prevalence of the trait.[17] Meanwhile, in family studies conducted in collaboration with J. Vandepitte, a Belgian physician working in the Congo, Zuelzer and I were unable to demonstrate a mutation rate high enough to account for the frequency of the sickle-cell gene.[18] The high frequency of the sickle-cell gene is now generally considered to be the result of a balanced polymorphism maintained by malaria, the most striking (and best documented) example of a balanced polymorphism in all human genetics.

The malaria hypothesis was for several years highly controversial, and we, of course, wanted a piece of the action. This was 1955; Africa was still colonial, and British, French, and Belgian investigators seemed to have the facilities most suitable for the requisite research well in hand. The only area in all Africa where there was what might be termed an American presence was little Liberia, roughly 53,000 square miles of mostly tropical lowland on the southern coast of West Africa. Beginning in 1822, Liberia had been colonized by freed American slaves who could be returned by the American Colonization Society to this undesirable and often swampy part of West Africa because none of the dominant European powers had showed much interest in this area in the nineteenth century when they were apportioning Africa amongst themselves. For a century these Americo-Liberians, as the "repatriated" former slaves termed themselves, were perched precariously along the coast, an ocean behind them and unenthusiastic tribes before them, over whom they had established an uneasy hegemony. Periodically, as troubles overtook the Republic of Liberia, the U.S. had quietly intervened, in a rare exercise of conscience. In 1926, the Firestone Tire and Rubber Co., stimulated by the shortage of rubber that developed during World War I, negotiated a lease for one million acres of land, to be developed into a rubber plantation. The company, in concert with the American Foundation for Tropical Medicine, had been instrumental in the construction and funding of the small Liberian Institute of Tropical Medicine, dedicated in 1952, situated just off the plantation. There had been no studies of the sickling phenomenon in this highly malarious region.

At this strategic moment in my thinking about how to establish an African presence, a graduate student in anthropology, Frank Livingstone, was working part time on the sickle-cell project. Hearing the validity of the malaria hypothesis and how to test it discussed on all possible occasions eventually was too much: he volunteered to go to Liberia to make this the subject of his thesis research. Wolf Zuelzer and I helped in the necessary arrangements with the Liberian Institute.

I then planned a trip to Africa that would take me to the laboratories of a number of the investigators active in the study of sickle-cell anemia and the malaria hypothesis: Jean Hiernaux at the Institute Recherche Scientifique en Afrique Centrale in the western Belgian Congo; A. B. Raper at the Medical Laboratory in Kampala, Uganda; J. Vandepitte at the Government Hospital in Luluabourg, the Belgian Congo; J. Colaert in Leopoldville, the Belgian Congo; and George Edington at the Medical Research Institute in Accra, the Gold Coast. The trip would finally bring me to Liberia in late September, several weeks after Frank had arrived.

At Jean Hiernaux' suggestion, I timed my trip so that the first stop would be Bukavu on Lake Kivu in Ruanda-Urundi (then a United Nations Trust Territory administered by Belgium), where, in a gorgeous setting, the Commission for Technical Cooperation in Africa South of the Sahara was sponsoring a Conference on Social Sciences, which included a Section on Anthropology. There was one complication Jean and I had not foreseen. This was a very protocolish conference. There was no provision for people like myself straggling in. Fortunately for me, one of the official American observers could not make the meeting; Jean could vouch for my credentials, and within a day the Secretariat designated me a substitute for the missing American observer, which let me participate in a series of workshops discussing the problems of emergent Africa. This was a year before Harold MacMillan, then British prime minister, made his famous statement in South Africa, to the effect that, "The winds of change are blowing across the continent." One could sense among the delegates a great deal of uncertainty regarding the future, although in retrospect I don't believe any of them suspected how rapid colonialism's collapse would be. The sociology of the meeting was to me every bit as fascinating as the subject matter of the conference. My most discriminating observations were made in the bar, where the approach of the delegates to the waiters varied from a softly-spoken *"S'il vous plait . . . "* to a snap of the fingers and a curt "boy" from the South Africans. Some aspects of the future were predictable even to a rank outsider like myself.

This was my first exposure to the tropics, of which much more later. Although earlier, as a Trailside Naturalist for the Cleveland Museum of Natural History, I had lived in splendid isolation in the closest approach to primeval forest to be found near Cleveland, the tropical forest simply dwarfed any previous experience. On a hot summer day, a climax beech–maple forest in the temperate U.S. is a cool retreat; in the tropics at midday the forest also is a cool retreat, but one can almost feel it pulsate under foot with its myriad, complex biological exchanges.

In due time, the Conference over, I worked my way through Uganda, the Belgian Congo, the Gold Coast, and was off for Liberia. The various visits filled me with enormous respect for the high level of intellectual curiosity maintained by my various hosts in the face of what I would call an enormous work load. All along the way, with the utmost of cooperation, I was collecting blood samples, examining unusual patients, and getting a cram course in tropical medicine.

My plane reached a very poorly lighted Roberts Field in Liberia at 2 a.m. As I started to disembark, I saw Frank at the foot of the stairs and, even at that distance and in that poor light, his bearing was that of a man overtaken by disaster. As soon as I got within greeting distance, he made the reason clear: "Dr. Neel, there's less sickling here than in Detroit—how can I do my thesis?" Frank had already been to the Firestone Plantation, whose labor force was recruited from all over Liberia, and had tested a small number of the workers for the sickle-cell trait, finding, indeed, less sickling than in Detroit African-Americans. During the next three weeks, we tested the blood of over a thousand laborers at the Firestone Plantation for the sickle-cell trait. The sickling frequency was 8.6%, in Detroit it was 9.1%. Our samples came from members of 14 tribes; there seemed to be tribal differences in the sickling frequencies, but the numbers weren't large enough for a firm opinion. After I left, Frank settled down to accumulate the numbers necessary for a tribal analysis, and a fascinating picture emerged. There are 16 distinct tribes within the boundaries of Liberia. In the northwest, among the Kissi, Gbande, and Loma, some 15% of those tested had the sickle-cell trait. As one proceeded across the country, the frequency of the trait fell, and, in the extreme southeast, adjacent to the Ivory Coast, only 1% of those sampled exhibited the trait. This was not the place to try to obtain critical data bearing on the malaria hypothesis, but Frank had a new and important problem: why this strange distribution of sickling-trait frequencies?

The eventual most plausible explanation of the finding, as developed in Frank's thesis,[19] was that the sickle-cell gene was only now penetrating this part of Africa. This was at first most surprising, since the anthropological evidence suggested that these tribes had been in West Africa for a long time. As, however, others filled in the map of Africa with respect to the frequency of the sickle-cell trait, it all began to make sense. The highest frequencies of the trait are encountered in the north and east of Nigeria. It is commonly accepted that some 2000 years ago there was a major population explosion in this region, the peoples involved sweeping both through the rain forest to the south and across the savannah to the west. Given the generally high frequency of the gene in this region and in these derivative tribes, all primarily Bantu-speakers, these people are thought to be the primary source of the gene in Africans. In the west of Africa, the tribes who could not compete with these migrants were forced into less desirable terrain, such as present-day Liberia represents. How long the gene had existed in northern Nigeria before this population explosion is uncertain, but, given the migratory nature of most tribes, such a localization of the gene suggests it had arisen in this area only a few thousand years previously. We geneticists had happened along just in time for a ring-side seat at a very exciting event—the origin and spread of a "new" human gene.[20]

In the meantime, studies in Africa of the distribution of hemoglobin C heterozygotes had revealed a situation as unexpected as the sickle-cell distribution. In West Africa, the Niger River executes a great loop embracing a portion of Mali and Upper Volta. Within and adjacent to that loop, 15%–20% of the inhabitants of the region possess the hemoglobin C trait. As for sickle cell anemia, this implies that 5–10 among every 1000 newborn infants should be homozygous for the gene. We now know that homozygosity for this gene results in a mild anemia, and its combination with a sickle-cell gene, as discussed earlier, results in a rather severe anemia. In all directions from this region, the frequency of the hemoglobin C trait falls off rapidly. In Liberia, only 0.7% of the population has the hemoglobin C trait. This is again the classical picture of the origin and spread of a locally advantageous gene, but, again, because of the deleterious results of homozygosity, it has to be a balanced polymorphism.[21]

We also made another unexpected finding in Liberia. E. B. and K. Olesen, a husband/wife physician team in the Firestone Plantation Medical Service, brought to our attention several children with unusual, severe anemias which, when studied by the available techniques, appeared to be thalassemia major—only once before observed in Africa, in the Belgian Congo.[22] Later, back in Ann Arbor, this led us to certain biochemical studies of our stored Liberian blood samples, from which we estimated that some 4%–5% of Liberians had thalassemia minor. [Since this estimate was not developed under the most favorable circumstances, I was greatly relieved when, in the early 1980s, M. C. Willcox, and U. Bienzle and colleagues, confirmed our findings by more detailed studies, reporting that some 9% of native Liberians had thalassemia minor.[23]]

As in the cases of thalassemia and hemoglobin S, hemoglobin C also achieves its highest frequency in areas where pressure from *P. falciparum* malaria has been or still is intense. Another hemoglobin abnormality, hemoglobin E, is present in as many as 70% of the population, either in the homozygous or heterozygous condition, in the areas of most severe malaria pressure in the Orient. There is a tendency to extend the umbrella of the malaria hypothesis to explain the high frequencies of hemoglobins C and E—and thalassemia—but, in fact, only for hemoglobin S is the evidence really satisfactory. Since the homozygous state for the alleles responsible for both hemoglobins C and E is accompanied by a mild anemia, the presence of these genes in a population imposes an impaired fitness on some individuals, although by no means as severe as the hemoglobin S gene. The hemoglobin genes thus illustrate rather dramatically the "bargains" sometimes struck in the evolution of biological fitness.[23]

The findings with respect to these four genetic traits have caused a great deal of discussion concerning the frequency of balanced genetic polymorphisms in humans, as well as the manner in which these polymorphisms confer their selective advantage. I suggest that we must be very careful in generalizing from the sickle-cell situation to the nature of genetic resistance to disease agents in general. *P. falciparum* is thought by some to be a relatively recent human parasite, having "escaped" only a few thousand years ago from a primate or avian reservoir. It produces much more severe disease than the other human malarial parasites. This in itself may be evidence that it is a recent parasite, still adapting to its new host, since a truly successful parasite is one that causes minimal disease in its host, as illustrated by many of the intestinal parasites or filarial infections. One can speculate that in evolving genetic mechanisms to counter *P. falciparum* malaria, our species "seized upon" the first useful mutations at hand, serious though the consequences are. This point of view would hold that when *P. falciparum* malaria first appeared upon the scene, there already existed (or arose shortly through mutation) one or more very low-frequency pockets of hemoglobins S and C and thalassemia, the carriers of which found themselves at advantage in confronting the disease. (There still is no agreement as to the precise basis for this selective advantage.) However, the advantage enjoyed by these trait carriers carried a price: as their frequency increased, so did the frequency of the disease caused by the gene when homozygous. The highest frequencies of the sickle-cell trait in tribal populations are some 30%–40%. This is presumably the point at which, in areas of extreme malaria pressure, gene loss from sickle-cell anemia balances the gain in malaria resistance in the heterozygote. This polymorphism, and possibly the other hemoglobin polymorphisms, must be viewed as dramatic testimony to the expediency of the evolutionary process.

Presumably, in time, mutations will occur that will confer protection against malaria without exacting so high a price. In fact, there may already be an example of this "better approach" to malaria. It has long been known that Negroes, especially West Africans, have a high "natural" immunity to another malaria parasite, *Plasmodium vivax*. It has also been known for some years that with respect to the Duffy blood group (one of the common human blood group systems), whereas virtually all Caucasians and Orientals test positive to either or both of two typing sera (anti-Duffy[a] and anti-Duffy[b]), 70% of all African Negroes and almost 100% of West Africans, test negatively to both these typing sera, i.e., are Duffy-negative. These are inherited traits. More recently it has been clearly shown that it is only Duffy-negative Negroes who are resistant to *P. vivax* malaria. Apparently the same erythrocyte membrane-bound protein which is the cause of the Duffy blood group (either a or b) is also a specific receptor to which the *P. vivax* parasite must bind during its free-living stage in the blood stream if it is to gain access to the erythrocyte.[24] In the genetically determined absence of this receptor, the parasite cannot enter the erythrocyte. There is no apparent ill effect due to the absence of this receptor, so this is a much more benign protective mechanism than the sickling trait. This "explanation" still leaves unanswered the question of why, if the blood group malaria receptor really is dispensable, it achieved such high frequencies in Caucasians.

A reasonable explanation of our unexpected finding of a high-frequency pocket of thalassemia in Liberia would be that when the (recent and spreading) parasite causing *P. falciparum* malaria reached Liberia, there were, as a result of the mutation process, a few families in which thalassemia minor occurred (but none with hemoglobins S or C). Individuals with the trait would have a relative resistance to *P. falciparum* malaria, and the frequency of the trait would in time increase up to the levels we had observed. We would presume that hemoglobins S and C are, by contrast, relatively recent introductions to this region, destined, if *P. falciparum* malaria persists in West Africa, to achieve higher frequencies under the selective pressure of this disease.

Time to Move On

By the early 1960s, there were hundreds of investigators all over the world working on hemoglobin and its diseases. Although the era for studies at the DNA level had not yet arrived, the important studies were increasingly biochemical. Competition between investigators was, in the best scientific spirit, intense. This is not the atmosphere in which I like to work. I find much more pleasure in identifying a significant area of opportunity which is amenable to my talents and, for one reason or the other, not overpopulated.

With respect to a continuing involvement on my part in the study of hemoglobin, there was an additional problem: I wasn't all that biochemically oriented, and although, of course, prepared to adopt new techniques, was not enthusiastic about becoming the bench biochemist that circumstances were demanding. The application of simple genetic principles and exacting hematological standards to a carefully chosen disease had yielded exciting initial results, but others were better qualified for the next steps. In human genetics, the hemoglobinopathies were to be the principal opening wedge into the reductionist approach that has culminated in the "DNA revolution." My interests were increasingly in the genetic systems that operate in populations. I had by then, by virtue of this and the program in Japan, discovered how much I liked the challenges of field (as opposed to laboratory) work. I was beginning to visualize some major studies on the surviving relatively unacculturated tribal populations. With Don Rucknagel, a physician whose Ph.D. I had supervised, I wrote a major review of the state of the hemoglobin diseases, published in 1961.[25] That and a few data papers published that year were the end of my personal involvement with the hemoglobin diseases.

Times do change. Several years ago as I walked across our campus, a student asked me to buy a raffle ticket "for sickle-cell anemia." I smiled inwardly, thinking back to that same campus 40 years earlier, when any familiarity with the disease was largely limited to physicians specializing in blood disorders. Its current celebrity status dates from 1971, when President Nixon, responding to increasing public recognition and interest in the disease, included in his annual message to Congress mention of the need for greater attention to sickle-cell anemia. In 1972, Congress passed the National Sickle-Cell Anemia Control Act, and since then large amounts of money have been spent on the disease. It remains stubbornly resistant to the medical axiom that once the basis of a disease is understood, an effective therapy can be devised. Thus far, no way has been found to safely alter to any substantial degree the abnormal behavior of sickle-cell hemoglobin, and the disease has become a prime target for the new breed of molecular therapists.

The devastated Suwa Shrine in Nagasaki (1946). The half-tori still standing at the former entrance to the shrine illustrates the vagaries of the bomb's destructiveness.

4

Organizing the
Atomic Bomb Studies

In Chapter 2, I recounted the circumstances that propelled me in a relatively few days from the routine of a post-war Army General Hospital to a member of a team charged with making recommendations to the National Academy of Sciences concerning the feasibility of appropriate follow-up studies in Hiroshima and Nagasaki. The other members of the team had similarly short notices. Although this was a "civilian" operation, because of the circumstances we operated under military orders. Our travel would be by military air transport, and our orders with respect to Japan involved a coordination between the top medical brass in the Pentagon and General MacArthur's headquarters in Japan, a coordination which was by no means automatic. Somehow it all came together and I departed Washington on November 21, 1946 to rendezvous with the other members of the team at Fairfield-Suison Air Base in California for our trip across the Pacific.

We Arrive

Our little group reached Japan early on a foggy November 25, 1946, some 15 months after the bombs were dropped. It consisted of two VIPs in the study of radiation effects, Austin Brues, a physician from the University of Chicago, and Paul S. Henshaw, a cell biologist from the Clinton Laboratory of the Manhattan District (later termed the Oak Ridge Laboratory), plus three very junior medical officers, Captain Melvin Block, Lieutenant (j.g.) Frederich Ulrich, and myself. I had learned of this assignment only 24 days prior to our departure—long by military standards but short in terms of the necessary preparations. In addition to subjecting myself to a cram course in radiation biology and medicine during that period, I had been attempting to understand the somewhat complex background of our mission. I had learned that immediately after the war all of the military services had been (understandably) highly motivated to conduct firsthand studies of the biomedical aftermath of the atomic bombings. Research teams were hurriedly assembled by the Army, Navy, and Air Force which, as they became aware of each other, made common cause. Once in Japan, the U.S. team also made common cause with a group of Japanese investigators who, despite many handicaps, were attempting a similar evaluation. The activities of what came to be called the Joint Commission resulted in an extensive document which was still top-secret and served as our bible. (Subsequently declassified and edited by A. W. Oughterson and Shields Warren, it was published in 1956).[1] The Joint Commission, which extensively documented the acute effects of exposure to atomic weapons, not surprisingly concluded that long-range follow-up studies regarding the possible delayed effects of these exposures, especially those related to radiation, would be highly desirable. At the conclusion of the Commission's work, its chairman, then Col. Oughterson, recommended to the Surgeon General of the Army that the National Academy of Sciences and its operating arm, the National Research Council (NAS-NRC) be requested to undertake an appropriate study, a recommendation endorsed by the Surgeon General and transmitted to the NAS-NRC. This organization convened an advisory group in June of 1946 that further recommended that the NAS-NRC attempt to obtain a Presidential Directive to undertake such a program, and, if the Directive was obtained, establish a standing committee to supervise the program. The group also recommended a fact-finding mission to Japan. Captain Shields Warren, then the ranking naval medical officer in the Joint Commission, drafted a one-page letter incorporating the advisory group's recommendations, which, over Secretary of the Navy Forrestal's signature, went to President Truman, who simply wrote "Approved, HST" at the bottom of the letter. Thus had come into being the Academy's mandate. Our little group was the fact-finding mission recommended at that June meeting.

I never have been able to "close the circle" on precisely how my assignment to this group came about and now the opportunity is gone, the principals dead. I presume that Col. Staff Warren, C.O. of the University of Rochester unit of the Manhattan Engineering District, because of his background was in touch with the developments that led to the dispatch of our mission and, at the right time, on the basis of the Rochester contact, had thrown my name in the hopper. I should have asked— there were many opportunities in the late 1940s and 1950s, but it simply never occurred to me. At any rate, my early contacts with the Academy left me with the impression that my assignment to the group was a bit of a fluke, and, while it was hoped I would be helpful in various ways, the real guns were, of course, our VIPs.

Our trip over had been uneventful enough save for a retrospectively amusing incident resulting from our stopover on Guam. The island was of course, in military parlance, secure when we passed through, but the debris of war was much in evidence. We stretched our legs during the layover, wandering about with the kind of curiosity you might expect. When we were back on the plane, revving up for the next lap of the trip, to Iwo Jima, Paul Henshaw, sitting next to me, pulled from his pocket a shiny bit of metal. It will make a nice paper-weight, he said. The nice paper-weight, with its conical base surmounted by a coil-enclosed plunger, could only be the percussion cap for some type of explosive. At this point it was too late for immediate action—we were bouncing and careening down the very rough runway, vibrating as only a much-used DC-4 could. As soon as we were aloft, I got the attention of the sergeant who was cabin crew. There was little doubt in his mind either. We wrapped it in everything soft we could find, and no bomb was ever more carefully aimed than the way we dropped our package through the open latrine of our nonpressurized plane. It was at this point I began to suspect that my judgment of situations was not always going to coincide with that of my VIPs.

Once in Japan, we were, for administrative purposes, attached to the Public Health and Welfare Section, General Headquarters, Supreme Command Allied Powers (PH&W, GHQ, SCAP). Every activity had to have a name—we called ourselves the Atomic Bomb Casualty Commission (ABCC), a term that came to apply to the entire subsequent activity as it unfolded. Our concern was to proceed to Hiroshima and Nagasaki as quickly as possible. Our mission fortunately had a high priority. Shortly after our arrival we learned that the 406th Medical Laboratory in Tokyo had fitted up three railroad cars to serve as a mobile laboratory and living quarters in case it was required to contend with an epidemic someplace in Japan. Ten days after we landed, we were on our way to Hiroshima and Nagasaki, this three-car unit our headquarters for the time being. In the interval, we had learned a great deal about the organization of the Occupation in Japan, as well as about Japanese efforts to understand the effects of the bombs and to initiate follow-up studies. Traveling with us was Masao Tsuzuki, Professor of Surgery at Tokyo Imperial University, who had served as the number one person in the Japanese participation in the Joint Commission. Because of his high military rank (admiral) during the war, he had been "purged" by the

The railroad car converted by the 406th Medical Laboratory into a mobile medical research unit, which served as our first headquarters. From left to right, Paul Henshaw, Austin Brues, and Masao Tsuzuki (1946).

Occupation, but temporarily given a clean bill to work with us. His know-how and great prestige among the Japanese were to facilitate our activities enormously.

The Initial Survey

During the next 18 days we visited Kyoto, Osaka, Kure, Hiroshima, Fukuoka, and Nagasaki, in a grueling round of consultations with Japanese scientists and, in Hiroshima and Nagasaki, public health officials. We needed to learn everything we could about what the Japanese had been doing with respect to studying A-bomb effects, as well as the general framework in which any long range study would be conducted.

The three strongest impressions of those first weeks were not science. First, as a Midwesterner, with typical American concepts of space and natural resources, I was amazed to see just how small this country with which we had waged a protracted and bitter war really was. Maps show the Japanese islands to be about the size of California, but they don't show that at best only some 16% of the area of this mountainous country is suitable for agriculture. More than 65% of the land area consists of hills and mountains whose average slope exceeds 15°. That so small a country had launched such conquests had required superb organization, an organization which I now was seeing firsthand.

Second, the devastation of the war was everywhere in the cities we visited. In my experience, Americans, whose country has never really been touched by an external force, talk much more freely of war—even atomic war—than those who have personally experienced war's impact on their homeland. Hiroshima and Nagasaki were, of course, special shocks. Familiar as I was at that point with the Joint Commission's Report, the reality had added dimensions. No picture had conveyed the extent of the devastation. As we visited various hospitals to obtain some idea of the medical facilities available, patients badly scarred from the flash burns caused by the heat of the explosion were often in evidence. The Japanese had fought to the point of exhaustion; their principal cities were a shambles. No one could have predicted at that time their phenomenal recovery at which, thinking back, I marvel each time I return to Japan. In subsequent years, because of my continuing involvement with the biomedical consequences of nuclear weapons, there were numerous opportunities to witness atomic bomb test shots in the Pacific, in connection with the "familiarization" program of the Atomic Energy Commission. I was never tempted to go. I had seen enough.

Third, this was indeed a different culture, fascinating but one not easily penetrated. At the moment, under the Occupation, it was extremely pliant: as we discussed possible follow-up studies, we were repeatedly assured by the Japanese side how easily each study could be accomplished. It was, in fact, difficult to get an objection as we explored possibilities. It quickly became evident that I would have to be quite circumspect in eliciting Japanese input and be certain I had not unknowingly implanted my views on a culture in which they were not a valid basis for action.

It was understood by the Academy and by our group that my responsibilities to the mission were primarily genetic in nature. The issue was clear: how to obtain accurate data on the characteristics of the children being born to the survivors of the bombings and to a suitable comparison group. Wherever we went, then, I was constantly probing to determine the circumstances under which any genetic study would have to be performed. Two points quickly became focal. First, the Japanese appeared to have excellent registration of births and deaths. Traditionally, the Japanese system for processing vital statistics was quite different from ours. Since the enactment of the Family Registration Law in 1871, early in the Meiji Restoration, their statistics were organized about the fact that each individual had an "ancestral home" (*honseki*) where a record of vital events for his family were kept in a special office (*koseki*). These were semipublic records, often, for instance, consulted by professional matchmakers. Although the Occupation had imposed the U.S. system of registration of vital events by place of occurrence, the Japanese had (fortunately for us, as it turned out) not abandoned the older procedures. Both systems were working well! Secondly, most Japanese births were at home, attended by midwives, or else in the small lying-in facilities maintained by the midwives. Only the pregnancies recognized in advance or during labor as likely to present complications were transferred to hospitals by the midwives. Technically, a child who lived for only a few hours required both a birth and a death certificate. It was a sobering moment when Tsuzuki matter-of-factly pointed out that in olden (and not quite so olden) times the stigma of bearing a malformed child was so great that if liveborn, it might be quietly smothered

A totally devastated area in the Urakami Valley, Nagasaki, facing toward the former Mitsubishi Arms Manufacturing Plant. The small, intact wooden buildings in the foreground represent the first attempts at reconstruction (1946).

The marker in the foreground of this devastated area was erected by the Japanese some months after the bombing to designate the hypocenter of the explosion in Nagasaki. The lone figure studying the marker is myself. We obtained a translation of what was written on its various sides. To our great amusement, the Japanese had deduced and inscribed on the marker a number of facts concerning the bomb—its height at detonation, its approximate explosive power and the number of casualties—which were still top secret at home (1946).

and quickly buried, entered into the *koseki* record as a normal-appearing stillborn.[2]

The figures on pages 56, 62 and 63 depict the destruction within 1.5 km of the hypocenter in Nagasaki as we encountered it in 1946. Previously this had been a densely populated region; now only the shells of the few reinforced concrete buildings were still standing with, here and there, a rare attempt at modest rebuilding. Hiroshima presented a similar picture. The survivors of the bombings, their homes destroyed, were, if still living in the cities, scattered throughout outlying areas, often living with relatives or friends. Since it was already obvious that a focal point of our study would be the frequency of congenital defects, it was an immediate concern that under these circumstances it would be difficult to document major defects in a child born at home at night. The attendant midwife would be reluctant to stigmatize the family who retained her.

Even if we could capture the loyalty of these midwives, they were not trained medical observers, although they delivered babies very well. Our data would receive scrutiny far beyond that accorded most studies. It was going to require a very major effort to obtain world-class data.

From the standpoint of the genetic studies, a salient development was the discovery that the Chief of Health for Hiroshima City, Ikuso Matsubayashi, difficult though the times were, had attempted to organize an inquiry into the types of pregnancy outcomes occurring in the city, requesting the attendant at all births to submit to his office a brief special report. It was clearly not an adequate approach—midwives were not trained to describe congenital defects, the radiation exposure history was inadequate, and there was no way to check on the completeness of reporting—but the resources to do more were lacking. At least, however, the Japanese side was prepared for my interest in the question.

An entirely unexpected development during this first reconnaissance, in Hiroshima, was the discovery that an enterprising young Navy Lieutenant j.g., Fred Snell, bored with a routine assignment at the Yokosuka Naval Hospital near Tokyo, had talked his commanding officer into detailing him to Hiroshima, where, billeted with an Australian unit, he was conducting a one-man follow-up study.

Back in Tokyo from this "familiarization tour" on December 22, it was a time for decision: what next? The "what next" quickly became obvious: our VIPs would return to the States, to report to the Academy, while we military types, it was decided (not by us), would remain in Japan, to "maintain contact with the situation," as it was put. On January 8, 1947, we bade farewell to our VIPs. Lt. (j.g.) Ulrich left the theater on January 12; never overenthusiastic about his assignment, he had suffered from a severe and relapsing conjunctivitis during most of his stay in Japan, which had prevented his participation in the majority of our activities. Mel Block (a nascent surgeon) and I, now loosely attached to PH&W, GHQ, SCAP, were as close to being free agents as can happen to very junior officers in a military setting but, fortunately, with the kind of A-bomb-associated credentials that commanded the requisite attention when we voiced our needs.

Five Months in Limbo

Neither Mel nor I had much taste for the life of a junior army officer in occupied Tokyo. We agreed to spend as little time there as possible and promptly took off for Hiroshima. There Mel embarked on a study of keloids, the exuberant scar tissue encountered not only in individuals burned by the flash of the atomic bombs, but also in persons burned following incendiary bombings. For my part, I had two objectives. The first and major one was to try to develop a real plan for the study of the potential genetic effects of the bombs. The second emerged from conversations with Fred Snell, now loosely attached to our "group," and in a sense, equally a free agent. We wanted to block out some worthwhile observations on the survivors that could be made with the limited means at our disposal. One of the salient observations of the Joint Commission on individuals receiving substantial amounts of radiation (as judged by the occurrence of epilation and petechiae following the bombings) had been a profound anemia some three months post exposure. We decided it would be appropriate now, a year and a half later, to determine the extent of the recovery, a project with which I felt comfortable because of the hematological proficiency I had gained in my earlier studies of thalassemia.

In Hiroshima we fitted up a small hematological laboratory in the Japanese Red Cross Hospital and conducted a simple study of the hematological indices of school children, contrasting the findings in those who had experienced epilation (with or without other symptoms of radiation sickness at the time of the bombings) with the findings in an age- and sex-matched group from a nearby control city. The former group could be assumed to have been quite anemic during the early months following the bombings. Our study revealed only very small differences between the two groups, differences which, because of the possible greater disruption of life for those exposed to the bombs, we were reluctant to attribute to a delayed effect of radiation.[3] Working conditions that winter were interesting. Japanese hospitals were unheated at this time because of fuel shortages. Our laboratory enjoyed the luxury of an electric heater to break the chill, but not only did our breath regularly condense on the microscopes when we read our slides, there was a day when the heater was out that it actually froze.

While these hematological studies were in progress, I was digging into the kinds of details that would determine the magnitude of any genetic study. To plan the scope of the study, we had to know the numbers of births occurring in Hiroshima and Nagasaki and the completeness with which these were being registered. Getting the numbers of *registered* births was relatively easy: in 1946 there were 6,389 in Hiroshima and 4,528 in Nagasaki, but how many went unregistered? L. V. Phelps, chief of vital statistics at PH&W, gave us a powerful assist in looking into this question, applying all the tricks of his trade. It was a milestone when one day he conceded that the completeness of birth registration in Japan was probably better than in the U.S. The reason?

Not only were the Japanese respectful of authority, but unless a child's birth was registered in the *koseki* books, he was a nonbeing legally.

The salient scientific development of this period was as unexpected as it was beautiful. I have earlier mentioned Matsubayashi's effort to mount a system of recording congenital defects among newborn infants in Hiroshima. With no authority as yet to initiate any active study, I could at least keep in touch with Matsubayashi. At an early meeting following my return to Hiroshima, I expressed concern over the possibility that abnormal terminations could be easily concealed and pointed out how desirable an early registration of all pregnancies would be, followed at the time of pregnancy termination by an examination of each and every outcome by a physician. His response snapped me to rigid attention. The Japanese Government, he said, had a special ration system for pregnant women, so that if they registered the fact of their pregnancy (with midwife or doctor certification) as soon as the diagnosis was certain, then at the completion of their fifth month of gestation they had privileged access to various items and foodstuffs important to pregnant women. Here was the potential key to a *prospective* study.

We quickly organized a little investigation into the completeness of pregnancy registration. It was very high, 93% of the births in the city being to women who had registered their pregnancies earlier. The 7% discrepancy was due almost entirely to women who having registered their pregnancy elsewhere, then came in to Hiroshima for delivery. Furthermore, there were some registered pregnancies whose outcome did not appear on the city's books; these, we found, were women who had gone elsewhere for delivery. Both the losses from the registered series and the gains of deliveries from unregistered pregnancies reflected the Japanese custom that a young wife return to her mother's home for her first one or two deliveries; this practice should have introduced no bias regarding outcome into our study. If we could develop a system to ensure that qualified observers examine the outcome of all registered pregnancies, we had the basis for a quality study.

The second salient development during this period was administrative. It was Occupation policy that any initiative in Japan be in tight collaboration with a Japanese counterpart, a policy calculated to promote the "Americanization" of Japan. Our nascent activity just didn't fit into the Occupation's tables of organization. In May of 1947, Col. C. F. Sams, the commanding officer of PH&W, the military unit to which we were attached, had an idea. Disturbed by the apparent lack of quality control of vaccines and other biologicals in Japan, he had instigated the organization of a Japanese National Institute of Health, within the Ministry of Health and Welfare. Its first function was the control of biologicals, but he envisioned a broader mission, comparable to our own National Institutes of Health. Why not create a Section on Atomic Bomb Effects within this new Institute, with which we would work?

This thought, logical enough in context, which quickly formed into policy, turned out to be the single most unfortunate development of those early days. This struggling new Institute, however good its intentions, was in no position to give us any real help. The problem, however, went far beyond this. The mission of the Japanese Ministry of Health and Welfare at that time did not call for a research orientation. All of the Japanese investigators with whom we had come in contact thus far were based in one of the former Imperial Universities, at that time very prestigious, all funded through the Ministry of Education. Despite the outward appearance of cooperation and coordination between Ministries, each Ministry seemed to function like a feudal domain. The gulf between Education and Welfare was especially great. Our assignment to Welfare created difficulties in working with and recruiting university-based Japanese scientists, difficulties that persisted down to the reorganization of the Atomic Bomb Casualty Commission, to which we come in due time. I was deeply enough into the Japanese scene to recognize the implications of this directive but could no nothing to prevent the action. This problem was greatly accentuated when, effective March 31, 1948, the Government decreed a "full-time" system, which in our situation meant that a primary employee of a university could not be a part-time employee of the Ministry of Welfare. As an example, F. Kida, one of the few Japanese I had been able to identify with an established professional interest in genetics and congenital defects, with a primary appointment at Kumamoto University, was forced to resign his consultantship with our organization because of the way the decree was interpreted.

Much of my physical exercise during this interregnum came in the form of solitary climbs in the mountains in the Kure region, the area near Hiroshima where many U.S. Army personnel assigned to the region were billeted. One gorgeous spring morning, as I was driving my jeep along a disreputable mountain road, looking for a suitable place from which to start a climb, I encountered a road repair crew. They waved; I waved; they returned the salute with great shouting and waving of arms. My, how much more enthusiastic the reception the further you get from headquarters, I thought. Suddenly the road just ahead rose up and the jeep trembled. They had obviously been trying to warn me of imminent dynamite blasting. It was then I learned a jeep will go about as fast backward as forward, albeit not quite so accurately. That was the closest I ever came to "hostile" enemy action.

During this interim period, it became apparent from correspondence that there was a substantial difference of opinion developing between myself and our VIPs, especially Henshaw, as to how the Academy should best proceed in Japan. By now, the Academy's operation was funded by the newly established Atomic Energy Commission,[4] and the NAS-NRC had established an oversight committee, the Committee on Atomic Casualties. The Committee held its first two meetings in March and May of 1947. At these meetings, Henshaw, impressed by the evidences of research interest and activity he had encountered at the various conferences during his tour in Japan, had suggested that the necessary research could be largely left to the Japanese, with U.S. funding. I felt that for many reasons, while indeed the best Japanese available should be recruited, a strong U.S. presence was necessary. I note in this connection that the development of the atomic bombs had spurred much more research in radiation effects in the U.S. than in Japan, with a correspondingly greater reservoir of interest and competence in the U.S. Furthermore, with respect to my own area of interest, it had simply been impossible to identify (except for Dr. Kida) any Japanese active in the field of human genetics who might be brought into the program.

Back to Washington

In May of 1947 I received several letters from Henshaw, indicating that there would be a third meeting of the Committee on Atomic Casualties in June, at which my presence was desired. The Committee had already at its earlier meetings, on the basis of a preliminary proposal I had prepared, endorsed in a general way the concept of a genetic study, and even created a Subcommittee on Genetics; this would be my opportunity not only to present the details but also to register my views on the conduct of the total program. I couldn't leave Japan without military orders. They finally reached PH&W in Tokyo on May 31, Memorial Day—a holiday. The meeting was set for June 6! I happened to be in Tokyo for some conferences I had scheduled. Unfortunately, the documents I needed for the meeting were all in Hiroshima. I caught the next Allied Limited south (there were only two trains a day, and scarcely the Bullets to which the present-day traveller to Japan is accustomed), "cleared the post," and caught the next available train north. Returning to PH&W, I found high priority orders from Washington. By then it was June 3. I started across the Pacific, plane hopping with the Military Air Transport System. This consisted in reaching an island on one plane and using my priorities to get on the next outgoing plane while the one I had come in on was being serviced. I reached a military airport in California on the 5th. There was a very early morning flight to Washington on the 6th. It would get me to Washington by early afternoon, in time for the latter half of the meeting. We were to make on refueling stop, at a military field in Kansas. Now, in the preceding several months there had been a number of serious crashes of DC-4s of the type in which I was flying. The cause had just—I think it was the day before—been determined to be defective maintenance of an aspect of the plane's tail. We landed in Kansas to the news that all planes of our type were grounded for tail inspection. Ours was inspected and sure enough, its tail was worn, too. We sat all day in Kansas while a replacement part was obtained. I arrived in Washington at midnight, having missed the meeting, in what state of cumulative frustration you can imagine.

In my absence, there had been no formal review of a genetic program at that third meeting of the Committee, but within the week of my return, it became clear that both the Committee and the Academy staff associated with the Japan project felt the situation in Japan warranted a major study on the offspring of the survivors. Lt. Neel was directed to organize that study forthwith. Furthermore, the Academy, desperate for someone to represent in Japan the activity they were sliding into, would attempt to secure my early release from the Army if I would return as Acting Director of their nascent activity. It was time for some difficult decisions, the need for which had been building for six months. I had "volunteered" for the initial experience in Japan as a professional exercise in an interesting country, which I thought should prove much more stimulating than the humdrum of a Military General Hospital, but without any thought of a long-term commitment. Once again, it was time to fish or cut bait. I was still a very junior Army medical officer, my rank not recognizing any of my genetic training. Should I not accept the Academy's proposal, I could be ordered back to Japan—or elsewhere. But, whether I returned to Japan in or out of the Army, should the return be with the idea of a long-term commitment to the genetic problem? Because of the slow pace with which the data would accumulate, a short-term commitment would yield very little professionally. Again some floor-pacing. It was not an obvious decision. The involvement would certainly alter my projections for Michigan, and it was still by no means clear that a proper genetic study could be mounted. With many concerns regarding the administrative complexities of the operation, I decided to take the gamble of a long-term commitment to the program (but not with any glimmer of the 48 years it has now become).

The plan I had evolved for the genetics study was straightforward. We simply had to learn as much as possible about the characteristics of the babies being born to the survivors of the bombings and a suitable comparison group. Observations should include the presence of congenital defects, birth weight, viability at birth, sex of child, and survival through the neonatal period. These are very "impure" indicators of a genetic effect; although an increased mutation rate should have some impact on these indicators, an impact whose magnitude was difficult to specify, a multitude of environmental factors also influence these endpoints, as well as such factors as maternal age and parity. We would have to collect detailed demographic and socioeconomic data on our population in the course of the study, to factor into our final analysis; we would have to conduct frequent checks on accuracy of diagnosis and opportunities for concealment of abnormal outcomes. We were still quite uncertain as to the amount of radiation to the gonads received by survivors, but by the standards of the experimentalist, it looked to be small. We could not accurately predict how many children would be born to the more heavily exposed, i.e., within 2000 meters of the hypocenter at the time of the bombings (ATB), but guessed that it might be 12,000 to 13,000 children over the next ten years.

Given the probable gonad doses and the "impure" nature of the indicators, it seemed to me most unlikely that this study, which by its very nature would be of considerable duration and very expensive, would demonstrate statistically significant differences between the children of parents who had received increased radiation at the time of the bombings and the children of (unirradiated) parents who had been on the outskirts of the cities, or had come to the cities after the bombings. Why proceed? There were two arguments on the pro side. First, we might be mistaken in our projection. The sensitivity of human genes to radiation was so ill-defined in these early days that as late as 1950 the official manual of the Department of Defense and the Atomic Energy Commission on atomic bomb effects contained the statement: "published estimates of the dose that might be expected to double the gene mutation rate in man range from as low as 3 r to as high as 300 r, and it is conceivable that the true value lies outside these limits."[5] (The units in which radiation exposure is presented are discussed in note 6.) Were the lower guess—which I did not accept—to prove correct, then we should indeed be able to demonstrate an effect on our indicators. Second, speculation and rumors concerning the monstrous children being born in the bombed cities were rampant; there was urgent need for facts.

There were, finally, two hidden inducements for me to proceed. During my preliminary survey, I had learned of the very high (by our standards) frequency of consanguineous marriage in Japan. This was another confounding variable in any study; we would have to ascertain whether the parents were related every time we registered a pregnancy. With a study of any magnitude, we should in time accumulate a superb body of data on consanguinity effects. The second inducement was that there was really very little in the medical literature about the pattern of congenital malformations in non-Western (i.e., non-Caucasian) cultures; we would in the course of any proper study accumulate a unique set of data on congenital defects among the Japanese.

One thing became very clear as, back in Washington, I contemplated the situation: I needed a lot of moral support; it was time for the first meeting of that Subcommittee on Genetics. It was arranged, on short notice, for June 24, 1947. In attendance were G. W. Beadle, D. R. Charles, C. H. Danforth, H. J. Muller, L. H., Snyder, and myself. It was a distinguished group. One of my disappointments in missing that first organizational meeting was that I had not been able to put on the record just how dicey a genetic study in Japan would be. Now I did it in spades.

Although Don Charles, from the University of Rochester, was the least known of these Committee members, he was in this setting most influential to my thinking. Don was not only a very competent mouse geneticist but an excellent statistician. Whatever statistical sophistication my earlier Ph.D. thesis on *Drosophila* had displayed was due to the Charles influence. During the war, while Stern and Spencer had, within the framework of the Manhattan Engineering District, been demonstrating with *Drosophila* the mutagenic effects of X-rays at doses as low as 25 r and 50 r, Don had been directing a project designed to determine the effects of chronic radiation of male mice on congenital defect and survival in their offspring. This work, a preliminary analysis of which was available to us at the time of the meeting, has never been properly written up; Don, always a procrastinator at publishing, was found to have Hodgkin's disease in 1949 and quietly took his life in November, 1955. The studies of his group suggested that preweaning mortality, rare morphological abnormalities, and mutations reflected in litter size, taken all together, increased by about 0.01% per r unit of chronic radiation to the gonad. A variety of calculations on Don's and my part suggested that to the extent that mice could be used as a guide, the probability of demonstrating an unequivocal effect of the bombs seemed small indeed, but who could be sure the human response could be predicted from the data on mice?

Muller was greatly concerned that failure to demonstrate a statistically significant effect would give the world a false impression, but agreed that a study had to be conducted. A study was endorsed. The meeting over, its action supported by the parent committee, I moved to prepare a statement to be published in *Science,* a statement which would make it clear that there were no false illusions about what a major study might produce. As published several months later, the critical section read:[7]

> . . . the Conference on Genetics voted unanimously to record the following expression of its attitude toward the genetic program: "Although there is every reason to infer that genetic effects can be produced and have been produced in man by atomic radiation, nevertheless the conference wishes to make it clear that it cannot guarantee significant results from this or any other study on the Japanese material. In contrast to laboratory data, this material is too much influenced by extraneous variables and too little adapted to disclosing genetic effects. In spite of these facts, the conference feels that this unique possibility for demonstrating genetic effects caused by atomic radiation should not be lost." (p. 333)

My interest in getting this statement out was practical as well as scientific: already it was very clear that if this study proceeded it would be large and very expensive; it would be best if "Neel's Folly" had the appropriate endorsements.

The rest of the summer was spent thinking through the details of the study, developing the necessary forms, and recruiting. By summer's end, I had a team of three: Masuo Kodani, a second generation Japanese-American (*nisei*) cytologist–geneticist, fluent in Japanese; Richard Brewer, a vital statistician who would keep our records; and Ray Anderson, a physician who had just begun his compulsory military service, who like myself had a Ph.D. in genetics. It was a good beginning. I also was released from the Army. Another critical point was now emerging. As I have earlier emphasized, the National Academy is the nation's highest scientific body. When, during the brief period prior to my initial departure for Japan, I had first begun to circulate through its offices, it was with awe and reverence. Now I was viewing it as a support base for a far-away operation. The Academy, through its National Research Council, was without doubt the most important nongovernmental source of scientific advice there was. It had done yeoman service during the war (see, for instance, Kevles' "The Physicists").[8] Its mode of operation was through advisory committees, who told others what to do. It conducted no primary research itself, despite the name. Now it was being asked to assume operating responsibility for a complex, long-range piece of research in a remote and foreign culture, a type of research for which there was no real precedent. This was not really its forté. Never did a more difficult assignment fall on less prepared soil, illustrious though the soil was.

Our family life that summer deserves a comment. My military orders to Japan had not permitted me to take my wife and daughter to Japan, where housing was in short supply and priority went to those with the most years of active duty. While I was in Japan, Priscilla had joined her mother in the Boston area. We very much wanted to pick up the threads of our life together. A few days after I returned from Japan, I was casting about at the Academy for advice on how, on the very limited budget of a first lieutenant, my wife and our young daughter could now join me for my indefinite stay in Washington. A new acquaintance offered me a month's sublet, as he was going on vacation. Just when that was due to expire, we found another month's sublet, then, miraculously, another—and then it was time to return to Japan, where I again could not take my family because my brief overseas service did not qualify me for Occupation housing.

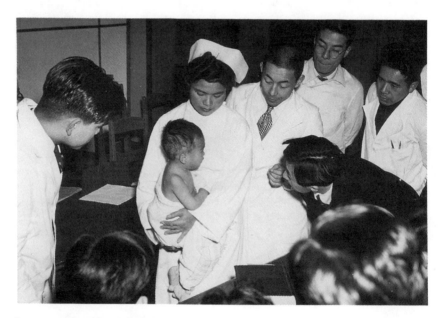

Japanese consultant and staff examining a child seen during the "nine-months examination" (Hiroshima, 1949).

5

A First Appraisal of the Bombs' Genetic Effects

Returning to Japan in September of 1947, after a summer in Washington working on shaping up a genetics program, I developed a split personality. Half of me was trying to bring the genetics program up, the other half, a director more by circumstance and necessity than by conviction, trying to lay the groundwork for the larger organization. The program of this larger organization would be essentially tripartite. One major component would be concerned with studies of the effect of the exposure to the A-bombs on the survivors themselves. A principal concern in this study was to search for an increase in a variety of cancers, 5, 10, 20 or even more years following the exposure, as well as any other delayed effects. The second major component was the genetic study, concerned with the detection of any radiation-induced increase in the frequency of germ-line mutations in these survivors, an increase which at that time could only be detected by the kinds of studies of their children described in the last chapter. A third, but minor, component would be a study of the children who had been *in utero* at the time of the bombings. In the early years of radiology a number of unfortunate experiences, all over the world, had indicated that the developing fetus was especially sensitive to the effects of radiation; it would be important both to the parents and medical science to determine how these *in utero* infants were faring. A major question at the outset was how to divide the organization's efforts between Hiroshima and Nagasaki. At this point we must consider in some detail the differing natures of the two cities on which the bombs were dropped, since these differences came to shape the program in many ways.

Background on Hiroshima and Nagasaki

Hiroshima, located on Japan's main island, Honshu, is seldom mentioned in Japanese history. The area was poor, its people not known for any particular activity or talent. It is undoubtedly because of this low position in the Japanese hierarchy that Hiroshima Prefecture, together with neighboring Yamaguchi Prefecture, contributed so many of the early Japanese immigrants to Hawaii and the U.S. Most of the great events in Japanese history occurred on the island of Kyushu, to the southwest, or higher up on Honshu, in the Kyoto-Osaka-Tokyo area. Situated on the delta of the river Ohta, the city was unusually level by Japanese standards. Like every other Japanese city, it housed, during the war years, a variety of enterprises supporting the war effort, but more importantly, it was the headquarters of the Second Imperial Army, a very legitimate military target. Its population during the war years, including a fluctuating military contingent, has been estimated at 330,000, but a considerable but indeterminate number of young and old had been evacuated to the country at the time of the bombing.

By contrast, the history of Nagasaki, on the southwestern aspect of Kyushu, is rich and colorful. The first Europeans, Portuguese, reached Japan about 1542, passengers on a Chinese junk driven ashore in a storm. Some 28 years later, both the Portuguese traders and the relatively large contingent of Portuguese Jesuit missionaries who had quickly followed up on this initial contact, were induced to locate their headquarters in a newly founded village, Nagasaki, at the head of a superb harbor. Over the next 300 years Nagasaki was Japan's principal eye to the Western World. Even during Japan's self-imposed period of seclusion, from 1641 to 1870, Dutch traders were permitted to maintain a trading post on a small, man-made island in the harbor. With the reopening of Japan in 1870, Nagasaki was one of the "special treaty" ports, authorized to engage in foreign trade, and, because of its past, a center for Christian activities. During the nineteenth century, 11 consulates were established there, and a handsome foreign settlement came into being on the eastern bank of the harbor. A mountain rises rather abruptly at the head of the harbor. As the city grew, it extended along the two sides of the harbor and also up into the valleys on either side of the mountain, thus assuming the shape of an X, the lower arms of the X straddling the harbor, the upper arms extending along valleys on either side of the mountain. The principal potential military targets were the Mitsubishi shipyards, situated midway on the left lower leg of the X, and the Mitsubishi Arms Manufacturing Plant, at the edge of the city, in the Urakami valley which constituted the upper left arm of the X. Nagasaki's population during the war years was approximately 250,000, but, again, with a considerable number of its young and old evacuated to the country.

On the original list of possible targets for the two atomic bombs drawn up by the office of General Leslie Groves, in charge of the Manhattan District, priority had been assigned to Kyoto, Kokura, Niigata, and Hiroshima, all cities relatively lightly bombed and so suitable as demonstrations of the bomb's power. Secretary of War Stimson, knowing of the cultural heritage of Kyoto, had ruled that city out; Nagasaki was substituted. Hiroshima, bombed on August 5, 1945, was the primary target of that run; the bomb detonated some 580 meters above the city, at almost precisely the planned point. The primary target for the second run, on August 9, 1945, was a very major munitions plant at Kokura, some 170 kms northeast of Nagasaki. The drop was to be visual, not instrumental. The Kokura target was obscured by both clouds and smoke from fires still burning from previous air raids. Short on gas after prolonged circling waiting for the smoke to clear, the plane carrying the atomic bomb made for Nagasaki, the secondary target.

There is some confusion in the various accounts of the Nagasaki bombing as to the precise target in that city. The "official" history of the Air Force's activities in the Pacific leaves open the issue of which (if either) of the two Mitsubishi plants was the primary objective.[1] On the plane's first pass over Nagasaki, both complexes were obscured by clouds. On the second and, because of the gas shortage, last pass, the representative of the Manhattan Project on board authorized an instrument drop, but at the last possible moment, a hole opened in the clouds, and the bombardier took over. The bomb detonated over the Urakami valley, at a height of 500 meters. J. Beser, the radar counter-measures specialist of the flight crew, states in his account that the target was the Mitsubishi shipyards, situated as described above, and states flatly that "we . . . missed our target, the city of Nagasaki by at least a mile and a half."[2] F. W. Chinnock[3] and J. L. Marx,[4] in well-researched studies, agree that the shipyards were the primary Nagasaki objective, but identify the Mitsubishi Arms Manufacturing Plant as the alternative objective, and suggest this was what the bombardier visualized and aimed for. Since this was at the edge of Nagasaki, not surrounded by densely built up city, it was a much dicier objective than the shipyards, in terms of potential for damage. In this case, the hypocenter was within 0.5 km of the target. Given that the Hiroshima bomb was dropped in a way to maximize casualties, and given the imperatives of war, I have little doubt the primary objective was the Mitsubishi shipyards, if not the center of the city itself (the intersection of the X). One point is accordingly certain, had it not been for the clouds, the casualties, assuming a hit on the primary Nagasaki target, would have been much greater. The lesser casualties in Nagasaki than in Hiroshima were not by design. Even so, the Urakami valley was cleaned out. The Urakami Cathedral, surrounded by the Catholic community of Nagasaki, was within 500 meters of the hypocenter. The Nagasaki Medical School and Hospital, one of the oldest and most distinguished in Japan, was within 600 meters of the hypocenter. Both were demolished. For what it was like to have been in the Nagasaki Medical School at that incredible moment, the reader might like to consult a little known work, "The Bells of Urakami," by Takashi Nagai.[5]

The true number of casualties will never be known. The number usually given for Hiroshima is between 65,000 and 90,000 immediate and delayed casualties, the latter term implying death before January, 1946, and a similar number significantly injured but surviving. For Nagasaki, the corresponding number is some 40,000 to 50,000 killed and 30,000 to 40,000 injured who survived. Some estimates of the death toll are considerably higher.[6] The Nagasaki bomb was more powerful, and, as noted, had the bomb detonated over the primary target, the casualties would have been much higher. As it was, much of the force of the explosion was contained within the Urakami valley. Hiroshima has been given (and sought) much more publicity from this event than Nagasaki. It's true the casualties and damage were greater. I've often wondered, however, whether the psychological basis for Hiroshima's attitude was that the bombing was the first really noteworthy event in its history, whereas for Nagasaki, this was only the latest in a long series of historical happenings, to some more of which we will return in the next chapter. Some of the Nagasaki Christians, a group that suffered so disproportionately, would view the bombing simply as another of their trials and tribulations, the experience assuming mystical overtones to be shared quietly.

Aware, because of my Rochester background, of the major research program on the biomedical effects of radiation in the Manhattan Engineering Project, I had always assumed that the prospect of A-bomb survivors suffering the aftereffects of their radiation exposures was very much in the minds of those responsible for developing the atomic bombs. It came as a great surprise in 1984 to encounter in Wyden's excellent report on the development and effects of the atomic bombs,[7] the thesis that the physicists who had developed the atomic bombs were taken completely by surprise by the occurrence of radiation sickness in the survivors. He quotes a principal and highly respected figure in the development of the bombs, Norman Ramsey, as follows:

The people who made the decision to drop the bomb made it on the assumption that all casualties would be standard explosion casualties. . . . The region over which there would have been radiation injury was to be a much smaller one than the region of so-called 100% blast kill. . . . Any person with radiation damage would have been killed with a brick first. (p. 16)

Now, it is a fact that anyone *completely in the open* within 2000 meters of the hypocenter would very probably be killed by flash burns and/or flying debris. (The records of the ABCC classify several thousand of the survivors within the zone of significant radiation from the atomic bombs as "in the open," but most of these were partially shielded, if only by other humans. For instance, a group of school children were lined up on school grounds about 1500 meters from the hypocenter at the time of the blast. Those in the front rows were killed at once, but those in the back rows, while sustaining serious flash burns to their face and upper body, were not burned on their lower bodies, and some survived.) It was difficult for me to believe, however, that those contemplating the use of the weapons overlooked the possibility that there would be survivors shielded from the heat and blast by buildings but who would still receive relatively large amounts of radiation. Furthermore, if there was no fore-thought about radiation effects, why then did the Manhattan District mount such a large program of research on the biomedical effects of radiation, at the University of Rochester, as well as at its Oak Ridge Laboratory, the University of Chicago, and University of California at Berkeley?

I have now satisfied myself that Wyden is absolutely correct. The inquiries of the Manhattan District were undertaken with worker's health in mind (i.e., to set permissible occupational exposures). For instance, the early experimental work on the induction of mutations in *Drosophila* by X-rays had all been conducted at quite high doses of radiation. The oh-so-hush-hush experiments of my early mentors, Spencer and Stern, within the Manhattan Project, were intended to determine whether ionizing radiation was proportionately effective in inducting mutations at lower doses, of 25 r and 50 r. These were doses which at that time it was thought might characterize occupational or military exposures in connection with the deployment of nuclear weap-ons. The Japanese reports of radiation sickness among survivors were at first regarded as a ploy to gain world-wide sympathy. The hastily assem-bled Joint Commission that was in Hiroshima and Nagasaki within two months of the bombings was quite unprepared by anything stateside for the extent of the radiation casualties. It is a tribute to the adaptability to the unexpected of our species that several decades later the U.S. military floated the discussion of a "neutron bomb" whose desirable attribute was that it would be highly lethal to people but leave installations/buildings relatively intact, for the convenience of the victor! When, in the 1980s, President Reagan so actively promulgated the Strategic Defense Initiative which was to provide the United States with an invulnerable shield of new defensive weapons, I could not help but wonder what biological fall-out from these proposed new weapons the physicists had overlooked this time.

First (Acting) Director

Logic dictated that our major effort be invested in the area in which the majority of the survivors receiving significant amounts of radiation were to be found. The work of the Joint Commission had suggested that because of the topography in Hiroshima, there were between two and three times more survivors who had received relatively large amounts of radiation there than in Nagasaki. As it developed, there was funding for only one set of major research laboratories, so that the effort in Nagasaki could not even be proportional to the casualties. (I must confess that through the years, fascinated by the unique and improbable history of the city, I've slipped down to Nagasaki somewhat more often than the strict work load demanded.) With these considerations in mind, I concentrated on getting the Hiroshima operation going first. We needed to bring up a temporary clinic–laboratory as soon as possible but also at the same time to plan permanent quarters. Far enough from the hypocenter to be apparently structurally intact was the gutted Asano Library, named in honor of the clan that had so long dominated the region. Military and Japanese consultants said it could be rehabilitated. We began to develop plans. With respect to the more permanent quarters, we needed not only a site but also a "staging area" at which we could assemble building materials and equipment, much of it to be shipped from the States. When I explained the need to the city officials, my attention was directed to a large ex-military installation down on the waterfront, far enough from the hypocenter to be intact. Ironically called the Gai-sen-kan (literally, Triumphal Return Building), amongst other functions it had been the place where the formal send-off parties for troops of the Second Army en route to the South Pacific were held. It had a large, now unused auditorium—excellent for storage—plus a variety of office space. Access was easy—by rail, road, and water. I took steps to procure it.

The site for a permanent building took a little more time. There were many areas formerly occupied by the military that now had no function. One in particular seemed attractive, adjacent to the projected 100-meter-wide "peace" boulevard transecting Hiroshima, which was so prominent a feature of the reconstruction planning. I negotiated to hold this and several other sites until appropriate consultants from the U.S. could make a final decision. Everyone concerned—military and Japanese—was very helpful, but during the occupation an inch of progress seemed to require a foot of memoranda. I have long since lost count of how many times, when the red tape was too much to bear, I produced that letter of authorization for the study, with "Approved, HST" scrawled at the bottom, and hinted that higher powers were keeping an eye on us.

Meanwhile, back in Washington, the Academy had been busily recruiting. The two greatest needs—in addition to professional personnel for the other, nongenetic, research aspects of the program—were for administrative staff and engineering/architectural competence. A chief accountant was found—Ken Brewer—who reported during February of 1948. His first official request was to take over my books. This being a military theater, all my requisitioning was done on the basis of a "memorandum of receipt." I still recall his look of disbelief when I passed him the box into which my military memoranda of receipt had been tossed— my "books." The architect, Sig Pfeiffer, also arrived in February. Securing the Gai-sen-kan was seen as a great move, but the rehabilitation of the Asano Library was quickly ruled out, on the grounds that it would be unwise to attempt, in this earthquake-prone region, to rehabilitate a building that might have hidden damage. I couldn't argue with that; we could make do temporarily with the Gai-sen-kan, peripherally located though it was. Then came the major decision. At first the sites I had selected for our permanent building found favor. Then Sig became aware of Hijiyama, a small "mountain" (oversized hill) rising out of the Ohta delta, within two km of the hypocenter. It mattered not that Hijiyama was a public park with strong Shinto overtones, with a large military cemetery, part of which would have to be moved. The site had the prominence that seems so irresistible to the architectural soul. In researching, at Sig's request, my choice for a primary site, the Japanese found that it had been flooded 50 years earlier. That was the coup de grace to my plan. (Since then the dike structure at the site has been immeasurably improved and there has been no flooding since WW II.)[8]

So the permanent facility was eventually sited on Hijiyama, a collection of prefabricated domed Quonset huts which, because of their resemblance to longitudinally split Japanese fish cakes, were soon dubbed *kamaboko-jo,* "fish cake palace," by the Japanese. The choice of the Quonset hut motif was predicated on the assumption that such building materials could be acquired throughout the Pacific Theater as "war surplus," but eventually all the necessary supplies had to be imported from the United States. The location was not a wise decision. Access was difficult, at a time when transportation was strained. It was an unnecessary psychological affront—as soon as the occupation ended, the city began to make representation for the return of the area. (The representations have finally had the desired effect; the operation is scheduled to move into new quarters down in the city in 1996, situated not very far from my original choice of a location.) Yet I must confess that from time to time, as Priscilla and I have sat in one of the small ABCC apartments built atop Hijiyama, looking out over the city at day's end as the sun, following its accustomed path, works its way through layer on layer of cloud mountains, finally arriving at and then disappearing behind the real mountains encircling the city, the city lights meanwhile emerging in the foreground, I have been inclined to understand why the site was so architecturally irresistible. The issue of working space in Nagasaki was solved much more prosaically by leasing an undamaged building.

Bringing up a Genetics Program

Concurrently with my administrative responsibilities, I was attempting to activate the genetics program. Here, too, there were also the proverbial thousand things to do. We put the final touches on a questionnaire, the first half of which was to be completed by a clerk when a woman registered her pregnancy, the second half by the attendant when the child was born. The first half dealt with the radiation exposure history of the woman and her husband: distance from the hypocenter at the time of the bombings, shielding, and history of radiation sickness, as well as the couple's reproductive history prior to this pregnancy.[9] The second half dealt with the attributes of the child. There were conferences with midwives, to gain their cooperation, to assure them that confidentiality would be maintained, to instruct them in the use of the questionnaire, and to supply each of them with a standard scale with which to weigh the newborn infants. We arranged that all newborns, alive or dead, would be examined by a physician, just as soon as possible if the midwife thought the child was abnormal, on a more leisurely schedule if the child appeared normal. A staff of eight to ten Japanese physicians had to be recruited in each city and instructed in a standard examination of the newborn. To move them about the town for the necessary physical examinations of newborn infants, we maintained a fleet of about ten jeeps in each city.

A second questionnaire was developed—our Long Form—to be used if a child was abnormal. It included questions concerning the parents' socioeconomic status. In addition, this Long Form was to be completed on every tenth pregnancy, to provide data for comparison with that collected for the abnormal terminations. We attempted, so far as possible, to keep the physicians unaware of the radiation exposure histories of the parents of any child they were examining, thus controlling a potential source of bias. Whenever the report of a termination had not been received within three weeks of the woman's expected date of confinement, we would initiate a special follow-up. It developed that such events were rare; we saw little opportunity to conceal abnormal pregnancy terminations. In all of this planning, I was fortunate to have recruited the services of Koji Takashima, a Hawaiian-born *nisei* who had obtained his medical education in Japan and served not only as interpreter but a constant source of insight into how best to conduct a study on the delicate subject of congenital defect within the framework of Japanese culture.

The full program began in February of 1948, in Hiroshima. At the outset, we were so uncertain concerning the extent of the radiation exposures in the two cities and the composition of the population that we gave serious consideration to the use of control cities. Kure, a former naval base city some 18 km from Hiroshima, was selected as a control on Hiroshima, and Sasebo, another former naval base some 80 km from Nagasaki, as a control on Nagasaki. The program in Kure, initiated in March, 1948, was abandoned after two-and-a-half years (and the examination of 8,391 babies), when it became clear that with parents who had moved into Hiroshima after the bombings and parents so far from the hypocenter that they had received no radiation, there were adequate "controls" within the city. The situation in Nagasaki was found to be similar, and studies were never initiated in Sasebo.

I returned to the U.S. in March, 1948, the issues clearly joined, but the success of the program by no means assured. The first children to be conceived following the bombs would have been born in April–May of 1946. We thus had no data on congenital defects in children conceived during the first 18 months following the bombings. However, through the official vital statistics of the two cities, we were later able to reconstruct listings of these children, and they enter into many of the studies to be described in Chapter 13.

My replacement as Director was Lt. Col. Carl Tessmer, a pathologist who within the framework of the Army had developed a special interest in radiation pathology. Carl was certainly well qualified for the position, but given my repeated efforts in those early days to dissociate the ABCC from the Occupation and to establish our civilian image with the Japanese Government, scientific community, and people, his uniform did not send the right message. A further mixed message in those early days was the funding of the operation by the Atomic Energy Commission. Although on the U.S. side this was in fact "the" agency designated to fund research on radiation effects, with no alternative source of funding of this magnitude available, the association presented an easy target for those Japanese and U.S. critics who wished to impute military overtones to the studies.

Some Ethical Issues

I have until now treated as matter-of-factly as possible some troublesome emotional and ethical issues. I have found it very difficult to take a firm philosophical stand on the use of the bombs. There is first the morality of the bomb itself, which in this context translates into my personal participation in any activity related to it. War is for the most part the killing of good people who had nothing to do with initiating the killing process. This killing is immoral, the result of human failure to settle complex issues rationally. Once begun, the escalation of savagery suggests how thin the veneer of civilization is; throughout history, military technology has done what technology can. The 100,000 civilians who died in Tokyo the night of March 9, 1944, as 325 of our B-29 bombers destroyed 16 square miles of that city with a methodical, crisscross pattern of fire-bombing, died just as unpleasant deaths as the victims of the atomic bombs. Had the atomic bomb not been developed, the U.S. was prepared to expand this type of indiscriminate slaughter.

While I am a strong proponent of nuclear disarmament (to a level consistent with the possibility of the emergence of "wild card" nuclear national powers) and consider nuclear war as it might now be practiced a threat to civilization, a disaster of unimaginable proportions, I believe that fewer lives were lost because the bombs were dropped than if they had not been dropped. They provided the Japanese a way out of the war.[10] The Japanese would not have hesitated to use similar weapons had they possessed them, and understood why we used them. During the early years of the program in Hiroshima, I was billeted near Kure, the former great Japanese naval base. As already mentioned, I found exploring the mountains behind that base good exercise (and therapy for input overload). At every turn, or so it seemed, there were former gun emplacements with supporting tunnels and trenches. Storming these islands would have been very costly. It is true we might not face our present range of nuclear problems if the weapons had not been developed, but this is a commentary on human failure to cope with technological advance rather than on the morality of the weapons.

On the other hand, the delicacies inherent in our follow-up studies were very much on my mind. I tried to imagine the converse—the Japanese organizing studies in the U.S. after dropping bombs on Chicago and San Francisco. How would we respond? I concluded the Japanese were much, much more cooperative than we would have been under similar conditions. It was inescapable that in the early days we were cloaked in some of the authority of the occupation, a fact undoubtedly bearing on our reception, but, except for a small minority of Japanese, this cooperation has extended at all levels down to the present. We were dealing with a very sensitive matter—congenital defect. The lurid and feckless speculations in both the Japanese and American press concerning the genetic effects of exposure to the bombs had certainly prepared the Japanese for our interest. These speculations stigmatized the children born to survivors following the bombings; in a country where arranged marriages were still common, they were regarded as less desirable mates. We attempted in all our dealings to observe Japanese proprieties, but I do not doubt that occasionally our teams gave offense. This raises a question to which I have no answer. Since, as we shall see, we found little in the way of a genetic effect of the A-bombs, the principal tangible result of our study has been to dispel the nasty rumors, to provide reassurance. Is it ethical to offend a few in the course of providing reassurance to the many?

One of the most frequent Japanese complaints has been that we (the ABCC) only examined them (like guinea pigs), but did not offer treatment in the event of findings of medical significance. The fact is that the terms under which the ABCC operated did not permit treatment, but any finding, whether on a child or an adult, was not only explained carefully to the patient (or parents), with the recommendation to see his/her physician, but also the patient's personal physician received a detailed letter describing the findings. The amazing cooperation which all the programs of the ABCC have enjoyed down through the years suggest that the complaints concerning the ABCC emanated more from a politically oriented, vociferous few than from the "silent majority."

The First Six Years of the Genetics Program

The program as just described continued for six years, during which I returned to Japan annually. Early in 1949, it was a stroke of great good luck to recruit W. J. (Jack) Schull for two years in Japan. Jack brought to the program the solid background in statistics the program required. When his tour finished, he joined me in Ann Arbor; as will become apparent, off and on we have worked together ever since.[11]

Very early in the study, as personnel became available, the quality of the genetic study was upgraded. An autopsy program was initiated, in a search for congenital defects not apparent in a clinical examination. Between 1948 and 1953, 717 children who were stillborn or died during the neonatal period were autopsied. In addition, since even severe defects in vision or hearing, or severe mental retardation, may not be detected at birth, in January of 1950, the ABCC instituted a "nine-months examination," bringing to the clinic for reexamination as large a sample as the facilities permitted—ultimately some 19,000 infants. This was possible only because during these early years the Academy recruited a very able set of pediatricians interested in the physical and mental development of both the directly exposed children and the F_1 of the genetics program.[12] At the time of reexamination, the child's physical development was also recorded. Over the years, these follow-up examinations and special studies drew on the efforts of a team of some 25 Japanese and American professionals.

One of our concerns from the beginning had been that the adverse effects of the bombs on pregnancy termination might be expressed early in gestation, resulting in increased rates of abortion. The study of abortion in any culture, under any circumstances, has always been difficult. This is especially true with reference to very early fetal losses. We initiated an effort to study early spontaneous abortions in the fall of 1949. At the same time the Japanese Government was promulgating a program of abortion literally on demand as one means of controlling the size of a population which, as the post-war repatriation of Japanese from Korea, Manchuria, and Formosa proceeded, was perceived as becoming too great for the home islands to support. The frequency of induced abortion so complicated our efforts to study spontaneous abortion that after several years we abandoned the undertaking. Our data pertain only to children who survived through the fifth month of gestation.

In the early years, on each visit I would "spot check" the accuracy of as many of the descriptions of congenital defect in our records as possible. Periodically, Neel's jeep was seen all over Hiroshima and Nagasaki. To this day, I am convinced that the birth rate was unusually high in Maruyama, the traditional area in Nagasaki for masculine relaxation and populated by women of the demimonde. This area required at least its fair share of these spot checks. One day the C.O. of the U.S. Military Government team stationed in Nagasaki, who never really had understood our study, called me in and made it very clear that my so-frequent and obvious presence in Maruyama—and in broad daylight, too—was really not good for the image of the occupation.

There is a wealth of problems in the conduct of a long-range program such as this. On the practical side, the Academy had to develop the infrastructure and the experienced personnel at the Washington level, for an interface with the Atomic Energy Commission, for the difficult task of recruiting for a distant study, and to respond to the needs of the organization in Japan. This took several years, but in due time a smoothly running office was in place.

A second and very major problem was to define program content in Japan. This was relatively easy for the genetics program, but much more difficult in the search for radiation-related diseases (somatic effects) in exposed children and adults. The early years of these latter studies were characterized by a series of disjointed probes, as investigators pursued their special interests (such as cataracts, retarded physical development, blood dyscrasias). The AEC staff who served as "project officers" for this undertaking at times were quite critical of its conduct. Relations between the Academy and the AEC reached a nadir in 1951, when the AEC expressed its displeasure in the conduct of the program by reducing its budget, and the Academy responded that it would rather shut the program down that limp along on such inadequate funding.

I remember this confrontation so well because in 1950, anticipating the possibility of this action, the AEC had sent Ernest Goodpasture, a distinguished pathologist, Merle Eisenbud, director of the New York City-based Health and Safety Laboratory of the AEC, and Willard Machle, a consulting industrial physician, to Japan to advise on the most appropriate orderly phase out should this become necessary. The qualifications of Machle for this rather exacting responsibility were never quite clear to me. I happened to be on one of my periodic visits to ABCC at the time this mission was active, and reacted in total horror and indignation when Machle told me that if the activity was phased out, the first program to go would be the genetics program. Why? He was very pleasant about it. Nothing personal, but since the genetics program had by now accumulated considerable data, it was only efficiency and good management to shut it down and use the remaining money during the phase out to ensure that the other programs had more to show for their efforts. This scarcely seemed an appropriate reward for past performance; the reader will have no difficulty envisioning the lack of cordiality that quickly characterized the relationship between Machle and myself.

Eisenbud[13] relates that Goodpasture and Machle were of the opinion that the programs should be discontinued. However, on his way back to the U.S., Goodpasture met in Tokyo with General MacArthur, who, hearing Goodpasture's doubts about the wisdom of extending the program, strongly urged its continuation, on the grounds that discontinuation of the program would create a scientific vacuum into which investigators of uncertain credibility would be drawn. The group's report took this into consideration, and, after a tense year or so, the relationship between the NAS and the AEC got back on track, as both sides realized the constraints upon the other. One can only speculate concerning the fate of the ABCC had MacArthur's busy schedule not permitted him to see Goodpasture on that particular day.

It was a programmatic landmark when, in 1955, an NAS team headed by a distinguished virologist–epidemiologist, Thomas Francis, after surveying the situation, strongly recommended that study cohorts of exposed and nonexposed persons be defined in the two cities, to be examined on a biennial schedule, with various special studies to be superimposed on the basic examination in a carefully coordinated fashion. Since the genetics program was in the course of building up a similar cohort of children born to survivors, as well as a control cohort (both later also to be subsampled for special studies), this recommendation brought the somatic studies in line with the genetic, creating cohorts whose study has been pursued down to the present.

The major administrative issue in those early days was that of programmatic continuity. Directors and staff were usually recruited on a two-year contract; just when an individual really had gotten his teeth into the program, it was time to leave. It was not difficult to recruit able persons who desired a two-year experience in Japan but very difficult to persuade good people to make a long-term commitment to a slowly moving epidemiological study when so many exciting advances in medicine were occurring at home. I had solved the conflict by making frequent visits from a base in Ann Arbor, but this created a somewhat delicate relationship between a long-time consultant such as myself and the directors, each of whom had his own ideas about the program. There was always some ambiguity about my role as "consultant." I felt keenly responsible for what I had initiated, bearing in mind the words of Maréchal Joffre, who, when asked who was responsible for victory in the battle of the Marne in World War I, in which he was the commanding officer, replied, "Of that I am not certain—but I do know who would have been blamed were the battle lost."

Once a procedure is introduced into a study such as the genetics program, it should remain the same, otherwise the early data will not be comparable with the later data. This was especially important in this study, since the proportion of all births to relatively heavily exposed parents was decreasing year by year. To study the later births with better procedures than were available for the earlier births would be to introduce a possible bias. I well remember discovering on one of my periodic trips to Japan that the current director, a pediatrician by training, now that improved facilities were available, had, with the best of intentions, just introduced a series of radiological examinations into our "standard" nine-months check to facilitate the detection of skeletal abnormalities. When I pointed out the potential for bias and requested that the procedures by discontinued, there ensued a somewhat spirited exchange. His final comment was, "alright, if you want *inferior* examinations I'll see that you get them," to which I replied, "yes, but *uniformly inferior*, please."

Not unexpectedly, the funding of the follow-up studies in Japan by the Atomic Energy Commission has on occasion led to charges that our activities and writings were slanted by association with an organization devoted to the development of atomic energy, an organization that might be expected to take steps to minimize reports of the unfavorable consequences of exposure to radiation. I myself feel that on several occasions, spokespersons for the AEC have been less than candid on some of the health issues stemming from the AEC's activities. Such obfuscations as have occurred (concerning, for example, fall-out at weapons testing sites) have emanated largely from the Weapons Division of the AEC; the Division of Biology and Medicine, through which the follow-up studies in Japan were funded, has been much more forthcoming. The National Academy of Sciences, keenly aware of the delicacy of these studies, has been almost paranoid in sniffing out efforts to influence its direction of these activities. I can state categorically that at no time over the past 48 years have I been aware of any improper pressures with regard to the content or analysis of the genetic studies. Exchanges with the AEC concerning program there were, but it was science, not politics, that prevailed.

An Analysis of the Results of the First Six Years

By mid-1953, data had accumulated on some 70,000 pregnancy outcomes. We had been summarizing our results year by year. Now Jack and I, in collaboration with Newton Morton and Duncan MacDonald, the two "resident geneticists" then in Japan, undertook a preliminary evaluation of all the data. It was a complicated set of data, for whose analysis there were no good precedents. Communications flew back and forth across the Pacific, as each of the four of us pushed his opinion of how the data should be analyzed. In the end, there was a reasonable meeting of minds. There was very little if any difference between the children born to parents who were judged to have received significant amounts of radiation at the time of the bombings and the children of parents judged to have received little or (if not in the city at the time of the bombings) no radiation. From the knowledge that the Atomic Bomb Casualty Commission had acquired of the demographic structure of the two cities, we could estimate that approximately 70% of the children ever to be born to the survivors had by now been born. The birth rate had declined rapidly with the Japanese Government's program of population control. The remaining 30% of all births to survivors within the zone of radiation, to occur over approximately the next 35 years, were not apt to change the picture substantially. The genetics program as it then existed was large and expensive, dominating the activities of the ABCC. Recruiting for the program and maintaining standards had not been easy. It seemed to Jack and myself that the Academy had met its charge in this respect. We arranged a meeting of an NAS-appointed committee in Ann Arbor on July 10–11, 1953, at which we presented the results of this preliminary analysis. Participating were G. W. Beadle, Don Charles, C. C. Craig, L. H. Snyder, and Curt Stern, Jack and myself.

Our consultants concurred in our appraisal of the situation. It was recommended that the large clinical program be discontinued at the end of the year. There were, however, two points on which it seemed desirable to continue to collect data, namely, mortality among the children being born to survivors, and the sex ratio of their children. Our data on mortality covered with the necessary completeness only the neonatal period; it seemed important to follow (and enlarge) the cohort of children the study had defined with respect to survival through childhood and adolescence. The situation with respect to sex ratio was more complicated. Because of the crisscross nature of sex-linked inheritance, deleterious mutations induced on the X-chromosome of radiated mothers should decrease the proportion of male births, whereas such mutations induced in fathers should decrease the proportion of female births. Our data reflected this expectation, just at the edge of statistical significance. Additional data on both survival and sex ratio could be collected inexpensively; it was merely a matter of abstracting information being registered with the city and relating it to the radiation experiences of the parents, now on file with the ABCC. We decided to continue to collect these data. Continuing data collection had an additional function: it maintained a "genetic presence" in the total program of the ABCC, so that if new approaches to the genetic effects of the bombs ever became available, it would be a matter of expanding an ongoing program. As we shall see, such developments did materialize.

Our definitive analysis of the effects of the atomic bombs on the manner in which pregnancies terminated in Hiroshima and Nagasaki, covering 76,626 registered infants, appeared in 1956 as a book, jointly authored by Jack and myself, with the assistance of some 13 colleagues.[14] This was the second book we wrote together, since we had coauthored a little textbook on human genetics in 1954.[15] The analysis dealt with six endpoints: sex ratio, congenital malformations, stillbirths, birth weight, death during the first nine months, and physical development at age nine months. Each of these indicators was influenced by many factors in addition to the potential for a genetic effect from parental exposure to the atomic bombs, and rather elaborate statistical procedures were necessary to ensure that these factors were also taken into account in our analysis. In the end, the analysis confirmed the earlier, tentative findings, of small but far, far from significant differences between the children of the radiation-exposed and unexposed. These differences were, however, more often in the direction of a radiation effect than not.

Since the ABCC still had not developed its working estimates of the amount of radiation received by the survivors, we were forced to generate our own, based on the distance of survivors from the hypocenter, their shielding from direct exposure, and their symptoms. In "The Effects of Atomic Weapons," published by the Department of Defense and the Atomic Energy Commission in 1950,[16] there was a distance–dose curve for a "nominal" atomic bomb. This revealed very little radiation beyond 2000–2500 meters. For those closer to the hypocenter, we estimated that the average surface (skin) doses received by survivors might be the equivalent of about 120 r, with individual doses ranging from a few r to the maximum compatible with survival, perhaps 500 r. There were not yet available the results of studies on models of the human body that would reveal how much of this surface dose might be expected to reach the germ cells. It was thus impossible to undertake a rigorous comparison of these "findings" with those of the experimental geneticist.

Again I stress that we accept the fact that since ionizing radiation has produced mutations in all the life forms in which its effects have been properly studied, some mutations were produced in the survivors in these two cities by the bombs; the problem has been to generate a best estimate of the magnitude of the effect and place bounds on the estimate. At this point in the study, in view of the many wild rumors and surmises that had circulated in the months and years just following these explosions, it seemed that the most appropriate use of the data was to consider what kinds of effects the study excluded with high probability. We therefore attempted to calculate upper limits to the effects of the bombs. We found that the data excluded, with the usual statistical provisos, an increase in the malformation rate of 100% or more over the value in the controls, or an increase in the stillbirth or neonatal death rate of 80% or more over the controls. This in our opinion meant that we could reject the suggestion of one very prominent geneticist, J. B. S. Haldane, that human genes were so sensitive to the effects of radiation that as little as 10 r, or even 3 r, would double the present mutation rate. On the other hand, the findings were consistent not only with the data of Don Charles, to which I previously referred, but also with newer reports, primarily from the Oak Ridge National Laboratory in the U.S. and the Medical Research Council Radiobiology Unit in the U.K., on the sensitivity to the genetic effects of radiation of another mammal, the mouse. It was to be some 30 years before a variety of additional studies on these F_1 and better dosimetry, plus growing knowledge of human genetics, would permit us to evolve from a treatment limited to what end-point changes the data could exclude, to an estimate of the genetic doubling dose of radiation most consistent with the data.

In Chapter 4, I mentioned that there were two additional professional incentives to a major commitment of time to the "Japanese endeavor." One was the fact that in the course of the study of A-bomb effects, we would register a large number of infants born of consanguineous marriages. What this led to is the subject of the next chapter. The other incidental outcome of this study was the accumulation of an unusually complete set of data on congenital defects in Japanese newborns. Before proceeding to the consanguinity studies described in the next chapter, I paused to analyze these data. The most striking finding was that despite the enormous differences, genetic and cultural, between Japanese Mongoloids and American Caucasoids, the total frequency of major malformations was quite similar in the two groups. However, this total frequency was achieved in different ways in the two groups, Americans having a relatively higher frequency of defects of the brain and spinal cord, Japanese a higher frequency of cleft lip and palate. It was as if in the course of evolution the various human populations had all "accepted" that the interaction of genetics and environmental factors would result in a similar impact from congenital defect, but had in the details balanced their books differently. We still do not understand how this comes about.[17]

A *"kakure kirishitan"* (hidden Christian) village, Shishi, on the far side of Hirado Island (1965).

6

When Cousins Marry

---

arly in the course of planning the studies on the genetic effects of
the atomic bombs, I had become aware of the relatively high
frequency of cousin marriages in Japan and introduced a special
item into our study questionnaire to identify such marriages. An early
tally of the response to this item revealed that 7.45% of the children
included in the study were born of consanguineous parents.[1] When in
1954–1955 we undertook the analysis of A-bomb effects described in the
last chapter, we treated the results of the consanguineous and nonconsanguineous marriages separately (there was no obvious difference). By
this time, we had registered 5,163 children born to consanguineous
parents. These ranged from children who because of multiple loops of
relationship were more closely related than first cousins, to a few children more remotely related than second cousins. It was certainly the
largest prospectively ascertained (i.e., registered before birth) sample of
consanguineous children in genetic history. Clearly the situation called
for a special study.

Why Study Cousin Marriages

Geneticists have been drawn to the study of the outcome of consanguineous marriages for many years. The reasons are simple. Most of the
genetic diseases that have been described in humans are recessive in
inheritance. The genes for these diseases are usually quite rare. Many
family lines carry one or more of these genes, but in the heterozygous
condition, so they cause no obvious ill effect. Supposing, now, you
choose to marry your first cousin, with whom you share a set of grandparents. With respect to any specific gene, there are two copies in the
grandfather and two in the grandmother. With respect to any one of

95

these four genes, there is one chance in four that you carry that gene, and a similar chance that your first cousin carries that gene. The probability that a child of this first-cousin marriage will receive that specific gene from both parents (i.e., by homozygous for that gene) is thus 1/64. Since there are, as noted, altogether four representatives of any particular gene in your paternal and maternal grandparents, the probability of being homozygous for any one of these genes is 4 × 1/64 = 1/16. Since this holds true for all genetic loci (we ignore the complications introduced by genetic linkage), then the child of first cousins should be homozygous at 1/16 of all its genetic loci, above and beyond the homozygosity that may result from the marriage of unrelated persons. If you marry an unrelated individual, the probability that he or she carries the same deleterious recessive gene as you do is much less than if you marry your first cousin, as is, then, the likelihood that your child will be homozygous for any deleterious gene present in the population. To the extent that there are deleterious recessive genes in the population's gene pool, this child of a first-cousin marriage has an increased risk of manifesting their effects. Thus, a study of such children yields data on aspects of the gene pool that cannot be approached any other way. These data can be used empirically, as in genetic counseling, or manipulated in an effort to understand the forces responsible for maintaining genetic variation—although, as we will see, there are some serious limitations to our interpretation of inbreeding effects.

It is important to recognize that consanguineous marriage does not of itself create any new genetic problems. It only reveals what deleterious alleles already exist in the gene pool. The medical literature abounds with case reports of children with rare, recessively inherited diseases resulting from consanguineous marriages. This results in a very biased view of the outcome of such marriages, since those marriages in which the children are all normal don't make the material of medical case reports. The situation which we had encountered and recorded in Japan in the course of the studies on atomic bomb effects provided an almost unique opportunity to obtain a balanced view of the results of consanguineous marriage. Accordingly, in 1956, Jack and I set about planning what we hoped, because of favorable circumstances, would be the most comprehensive study of this subject ever performed.

The Child Health Study

Any proper study of consanguinity effects demands that a sample of children from the marriages of nonrelated parents be examined with precisely the same care as the children of consanguinity. As we thought about our study, it became clear that these latter children (the "controls") would be of almost as much scientific interest as the children of con-sanguinity, since they would provide a baseline for disease in Japanese children, especially genetic disease, a baseline to be compared with the findings in other ethnic groups. These findings would in effect result in an extension of the earlier study of congenital defect in the Japanese. We therefore decided to call our program the Child Health Study.

Stage one in the planning involved generating the necessary fund-ing (from such disparate sources as the Atomic Energy Commission, the Rockefeller Foundation, the National Institutes of Health, and the Asso-ciation for the Aid of Crippled Children), developing a protocol, nego-tiating with the Atomic Bomb Casualty Commission for the necessary space in which to examine the children, and assembling a team. In developing a protocol, we were fortunate that the funding permitted bringing to Ann Arbor for one year three Japanese physicians who had developed a strong interest in human genetics, namely Toshiyuke Yanase, Norio Fujiki, and Koji Ohkura.

The study would require that our subjects would miss some school-ing; good school attendance is deemed much more important in Japan than in the U.S. We decided that half a day away from school was the most we could ask for, into which half day we would pack a physical examination, anthropometric measurements, a set of X-rays designed to help evaluate physical growth and development, a dental examination, a test of neuromuscular and mental status (the latter through a series of maze tests), and laboratory studies based on urine and a sample of blood obtained by finger prick. The examinations would be scheduled in such a way that no examiner knew whether he was examining a child of inbreeding or a control child. Since the average child would be age nine, this all took a bit of doing. Only in Japan could such a schedule be crowded into a half day! Fortunately, the mothers often came along (and later received a report of any medically significant findings).

The introduction of a program like this to a Japanese community is not something to be taken for granted. There would be no weight of authority behind the study. Since the children would have something to say about it—and word does get around—we had to omit unpleasant procedures, such as venipunctures, as well as certain diagnostic studies that would be a normal follow-up to the various findings on physical examination. Guided by our Japanese colleagues, in the best Japanese tradition we explained our program to every Japanese group in either Hiroshima or Nagasaki that might conceivably feel it had a legitimate interest in the study, and might in the end influence the decision of a set of parents to approve their child's participation. These included, in both cities, the appropriate representatives of the medical schools, city and prefectural health departments, mayor's office, Medical Association, Board of Education, School Principals Association, and PTA!

The study got underway in September 1958 and lasted until August 1960. We had recruited four other Americans, who also assisted in the planning: J. N. Spuhler, an anthropologist, R. W. Miller, a pediatrician, J. D. Niswander, a dentist, and A. L. Drew, a neurologist. Seven young Japanese physicians were involved; we received continuing help as well from the three Japanese who had assisted in planning the study.

In content, the study had two major subdivisions. In technical terms, we wished to establish the fate (alive or dead) and the present physical condition of a cohort of children born between 1948 and 1953 and registered in the genetics program. One of the truly unusual aspects of this study was that we would have identified this cohort at birth, then a subset at nine months, and then all surviving at an average age of nine, so that we would have a cumulative picture. Accordingly, one subdivision of the study was essentially a records operations, tracing through the *koseki* records every child in the cohort with respect to survival, and establishing cause of death for the deceased. The other subdivision was an effort to learn as much as possible in the time available about the physical status of all the living children in the cohort.

We eliminated from the sample all children whose parents fell into our relatively heavily irradiated categories, so that the suspicion of a radiation effect could not detract from our findings. With this, and after a few exclusions for technical reasons, there were left with 1789 births of consanguineous parentage to be followed up in Hiroshima, and 2686 in Nagasaki, the larger sample in Nagasaki due to the higher birth rate and the higher rate of consanguineous marriage in that city (8.0% vs 6.0%). We then matched this sample from among the other births registered in the previous genetic study by the random selection of 1970 children of nonconsanguineous parentage in Hiroshima, and 2847 in Nagasaki. In the end, we had a 97.8% participation rate among living children in Hiroshima, and an amazing 99.4% in Nagasaki. We would like to believe in our super-salesmanship, but local rumor had it that the fact that the children were transported to and from the examination in the fleet of jeeps maintained by the ABCC—the first jeep ride for many— had something to do with our success.

One of the potential pitfalls of a study like this is a socioeconomic difference between the groups being contrasted. The little questionnaire which we asked the parents to fill out before the children came in for examination contained a number of items that permitted us to compute a socioeconomic score. When we analyzed the data, it became clear that the consanguineous marriages had slightly lower socioeconomic scores than the nonconsanguineous. The chief apparent reason was that a disproportionate number of the consanguineously married were relatively recent immigrants to the two cities from rural areas (where consanguinity rates are higher), who tended to have lower occupational ratings than the long-established city dwellers. We had to make allowance for this fact in our analysis.

The findings from the study may be considered in three ways: 1) as a set of data, from the children of unrelated parents, defining the impact of genetic disease in childhood, 2) as a basis for objectively describing the effects of consanguinity and providing empirical counseling regarding the risks of consanguineous marriage, and 3) as an insight into the nature of the human gene pool. With respect to the first of these questions, we found that in the children of unrelated parents the expectations of some significant medical problem appearing in a child, from birth to an average age of nine, was 8.5%. These problems extended all the way from congenital defects incompatible with survival to severe but fully correctable defects of vision or hearing. This sounds high, but McIntosh and his colleagues in a study of New York City children roughly comparable to our own obtained very similar results (7.5% defective).[2] It was not easy to estimate just how much of this disability was due to simple recessive inheritance, but, from a variety of approaches, we estimated the contribution to be between 10 and 20% of the total. This implies that between 0.8 and 1.6% of a cohort of Japanese children assembled at birth would exhibit some recessively inherited disability by age nine, much of this not in the form of well-known genetic syndromes.

This estimate, based primarily on the occurrence of similar medical findings in siblings, is substantially higher than the later estimate of 0.2% by Baird and colleagues[3] for a Caucasian population, based on an enumeration of *recognized* recessive entities. For technical reasons, I suspect ours may be a bit of an overestimate but Baird's an underestimate, a figure of between 0.5 and 1.0% being appropriate. More recessively inherited disease can manifest itself later in life, and our figures do not include the dominantly inherited diseases, many also of later onset, nor some disease due to chromosome abnormalities. When all of these additions are taken into consideration, we estimate that roughly 1.5–2.0% of a general population (Japanese or Caucasian) will at some time exhibit disease with a simple genetic basis,[4] and this figure does not make allowance for the many diseases of later life in which genetic predisposition is a contributory factor.

Turning now to the second aspect of the study, we must divide the effect of inbreeding into two categories: the effect on child survival and the effect on the characteristics of the children we examined. It is perhaps most convenient to compare the findings in the group of children not resulting from consanguineous marriage to those in the group resulting from first-cousin marriages. With respect to survival, there had been 9.0% deaths (up to an average age of nine) in the noninbred children but 10.5% in the children of first-cousin marriages. With respect to health status, we found, as already noted, 8.5% of the outbred children to have some meaningful physical defect, versus 11.7% of the children of first cousins. This "inbreeding depression" extended to the anthropometric measurements—the children of first cousins were slightly smaller (about 0.5%) in all dimensions measured. Our medical histories suggested that the inbred child was somewhat more susceptible to a variety of infections than his outbred counterpart; some of this lag in physical development might be due to the effect of disease.

One of the most interesting findings came in the field of neuromuscular and mental status. We consider "age when walked" and "age when talked" very imprecise indicators of development, in a scientific sense, since they are so colored by maternal recall and impression. Nevertheless, our statistical analysis showed the child of a first-cousin marriage to be slightly retarded in these respects, by about half a month. We employed two tests of strength and coordination, namely, grip strength as measured on a dynamometer and tapping rate as measured with a simple apparatus. Again the average child of first cousins was slightly inferior (1–2%). Our "intelligence tests" consisted of ability to follow a color–shape trail in a simple field and ability to master a simple maze test. Here also the child of first cousins was slightly inferior (3–4% slower in executing the task). Finally, we were fortunate to gain access to the school-performance record of the preceding year. Schooling in Japan is taken more seriously than in the U.S. We found it striking that there was an inbreeding depression in every sphere of activity: language, social studies, mathematics, science, music, fine arts, and physical education. The grading system in Japan is normalized, given proportions of children receiving each grade, a "1" being highest, "3" average, and "5" lowest. The average grade of the inbred child was 3–4% below the noninbred child.

In Hiroshima, through the cooperation of the Department of Psychology of Hiroshima University, we were able to obtain a series of 2111 individually administered psychometric examinations, a much more exacting test of mental status than the color–shape trail and maze tests. The inbreeding depression was again about 4%. We were amazed at how well the results of this more sophisticated approach agreed with the results of our simpler tests and school performance.

Except for cumulative mortality and frequency of handicap, all of the above data were adjusted for the lower socioeconomic status of the consanguineous parents. Failure to have made this adjustment would have inflated the consanguinity effect by about 16% for the physical measurements, about 18% for the neuromuscular tests, 26% for school performance, and 18% for the individualized I.Q. tests. For technical reasons the data on mortality and defect could not be so corrected. Relatively few other studies of consanguinity effects have attempted to factor out the influence of socioeconomic factors on the data; the inaccuracies this has introduced into the results and conclusions of these other studies are simply indeterminate. The data we collected in Japan are now widely used in genetic counselling when the question is the outcome of consanguineous marriage.

What can we infer from these studies concerning the human gene pool (the third way of looking at the data mentioned earlier)? The effect of inbreeding may be partitioned into an effect on mortality and morbidity and an effect on fertility. What we have generated through these studies is a partial picture of the mortality and morbidity effect (we will come to the fertility later). The view of mortality is partial because we have no data on early fetal loss, prior to the fifth month of pregnancy. This is of great theoretical significance, although in terms of psychological and sociological impact, these early losses are of much less significance than later losses. Our view of the mortality effect is also partial because our average child was aged nine; there could be later inbreeding effects on survival. Since, however, approximately 80% of the mortality prior to the age of 20 has occurred by age nine, we do have data on most of the postnatal, prereproductive mortality.

The inbreeding effect that we observed was lower than reported in many previous studies but was similar to that reported by a Japanese group conducting a comparable major study at Shizuoka about the same time.[5] It is also similar to the effect encountered in other recent studies. It is commonly estimated that humans have about 50,000 different genes in an egg or sperm. A standard question for geneticists is, in the average egg or sperm, how many of these genes are so defective that when a zygote is homozygous for these genes, its survival is jeopardized. In a first-cousin marriage, the fertilized egg should be homozygous for approximately $1/16 \times 50,000 = 3125$ different gene pairs beyond normal expectation. From the effect of inbreeding on survival that we observed, it is easy to calculate how many of the allelic forms of these genes are so defective that when two such alleles are brought together, the individual cannot survive, i.e., the allele is a genetic lethal.

Our calculation was 1.06 such genes in Hiroshima, and in Nagasaki, where the inbreeding effect was lower, only 0.33. In other words, less than 1 of the 50,000 genes in the average sperm or egg of the parents producing these children was so defective that if a child had two copies of this gene, it could not survive. One lethal gene effect can be produced by one lethal gene, or two half-lethal genes, or four quarter-lethal genes, etc. As suggested, this estimate does not include early fetal deaths; their inclusion (could they be studied) might increase this figure substantially. There are some gross theoretical oversimplifications in this calculation, but it does supply a sort of "ball-park" figure. Our estimate is lower than most, but since our data set is much larger and our subjects so carefully defined, we prefer our estimate. To me, this estimate of the number of concealed lethals has always been surprisingly low. Here let me remark that, given the complexity we now know the genetic material to have (see Chapters 14–15) and the high input of random error into the system from mutation (which will be discussed in Chapter 12), isn't it remarkable how "clean" the system seems to be?

These studies, too, resulted in a book, but this time Jack had clearly borne the brunt of the work, especially the complicated statistical analysis, and became the senior author.[6] Even as the book was being written, however, it was clear to us that we were not done with the study of consanguinity. We were missing two components of the total picture, known to be important from the results of inbreeding domestic and experimental animals. One was the relative fertility of the consanguineous marriage. The other was somewhat more subtle: to what extent was the reproductive performance of the individual who was the product of inbreeding depressed, even though the individual might be married to an unrelated person? We began to think about further studies.

Coming back from Japan in 1959, at the completion of this study, Priscilla and I (and our three children) stopped off in India at the request of the Rockefeller Foundation, to visit Kerala State, site of the radioactive sands of Travencore. The Foundation wished to know if the situation was favorable to the study of radiation effects. In Trivandrum, the capital of Kerala State, we stayed in the almost palatial former residence of the British Commissioner for that area, now converted into a hostel for transients like ourselves. At dinner our first night there was a quiet young African American couple, with whom we struck up a pleasant conversation. At that time the man's name, Martin Luther King, meant nothing to me. He had come to study the nonviolent Indian tactics for civil resistance.

In the course of the next ten days, I became very certain—especially by comparison with the Japanese situation—that the radiation exposures at Travencore were so low (only five to ten times the background rate), the exposed population so mobile, and the registration of vital events so poor, that I really could not recommend that Rockefeller funds be invested in a study. Despite the various problems involved in the genetic studies of those exposed to the atomic bombs, the study of the children of Hiroshima and Nagasaki has offered far and away the most realistic opportunity in the world today to determine the genetic effects of radiation on humans.[7]

Consanguineous Islands

Although by the standards of the West consanguineous marriages were unusually high in Hiroshima and Nagasaki, in isolated areas in Japan, and especially in some of the smaller offshore islands, consanguinity rates were known from the efforts of Japanese investigators to be much higher. In our new role of connoisseurs of consanguinity, we were interested in studying a few such islands at first hand. They would never supply the kind of data base emerging from Hiroshima and Nagasaki, but they could provide important insights into the extremes possible in a society with Japan's structure.

Since most of my time in Japan in connection with the Child Health Study involved the Hiroshima phase, it was appropriate I would look about for a "case study" in the Hiroshima area. With help from our Japanese colleagues and a city official, Norio Ishikuni, who was aware of our interest in consanguinity because of our incessant use of the *koseki* records, I selected for study the population of Hosojima, an irregularly shaped island in the Inland Sea, 60 km east of Hiroshima, only 1.64 km in length and .74 km wide, on which in 1959 there lived 175 persons. To the north, approximately 2.0 km away, is the main Japanese island of Honshu, while 0.6 km to the south is Innoshima island (population 41,164 in 1958), famous in local history as a stronghold from which pirates sallied forth to plague the China Sea—and perhaps occasionally the more remote portions of feudal Japan itself. A steady stream of small coastal shipping passes Hosojima on all sides. From the geographical standpoint the island is really less isolated than hundreds of similarly sized islands dotting the Inland Sea. Yet, as we were to find, it represented one of those juxtapositions of the old and new which make Japanese culture so fascinating.

Although the desired degree of documentation is lacking, the islanders regard themselves as descendants of a samurai of the Matsumoto clan who, having the misfortune to have been allied with the losing Hosokawa clan in the Battle of Funaoka-yama in Eishō 8 (1511), was forced to flee the site of action. It is stated that the first inhabitants of Hosojima, some 470 years prior to our study, were this samurai (who had become a "mountain priest" in the tradition of vanquished samurai) and his bride, whom he acquired in the small settlement of Shigei on nearby Innoshima. The number of children born to this couple is unknown, but in the following generation the eldest son is said to have returned to his mother's birthplace, while at least one younger son remained behind on the island. The eldest son established the dominant branch of the family in Innoshima, while the younger son established a subsidiary branch on Hosojima.

Some 300 years before our study, a second family, from the Hamamoto clan, came to the island. The head of this family is believed also to have been a fugitive samurai, from what is now Shimane Prefecture, but details are lacking. For the first several hundred years of its history, apparently only a handful of people lived on Hosojima. It is alleged that because of the smallness of the island, there was great reluctance of outsiders to settle there, in consequence of which there resulted over the years an extremely close degree of inbreeding, at times amounting to incest. In 1877, when in the course of the Meiji Restoration Japanese commoners acquired surnames, all of the inhabitants of Hosojima were given the surname Nishihara. In 1872, an additional family, later given the surname Sakurai, came to the island, while following World War II, a Nishihara woman who had married off the island and acquired the surname of Suzue returned with her husband. At the time of our study, in 26 households, the head bore the surname Nishihara; in one, Suzue; and in one, Sakurai.

In feudal Japan, the rights of primogeniture were rather strictly observed. This and other feudal institutions underwent rapid changes the latter part of the nineteenth century. However, the Hosojima islanders did not follow the general trend. The island is devoted entirely to agriculture. Literally all of its farming activities are undertaken in cooperative form. To this day, by tradition, the eldest son inherits the farm—younger sons usually leave the island to find employment elsewhere. There are no hired laborers on the island. The eldest son finds his bride on the island if possible, a tradition apparently reinforced by the reluctance of "outside" women to share the isolation of the island. One tangible result of this policy has been that farm holdings on the island have not been fragmented to the degree encountered elsewhere in Japan; the average farm on Hosojima has an area about twice the present average for Hiroshima Prefecture. The island is relatively prosperous. Its inhabitants are not obliged to supplement their livelihood with part-time fishing, as is so often the case with Japanese farm families dwelling on small islands. A further unusual feature of the island's economy is that no rice is raised there. The principal crops are wheat, sweet potatoes, tobacco, and pyrethrum. The latter two, which occupy approximately 30% of the acreage, are the "cash crops" that make possible the purchase of the necessary rice and fish. Each time now that I fly over Japan, with its many offshore islands, I wonder how many of these islands are also "genetic jewels" in the Hosojima tradition, inconspicuously spotted within the proverbial stone's throw of booming centers of Japanese postwar technology.

By virtue of two trips to the island, plus extensive recourse to the local *koseki* records, we were able to establish that consanguinity existed for no less than 29 of the 45 marriages represented on the island. This was almost certainly an underestimate, since the *koseki* records, on which we relied in part for establishing remote degrees of relationship, only came into being in 1871, following the Meiji Restoration. At that time, all individuals recorded the identity of their parents, and some- times their grandparents, but, since as commoners these ancestors did not have surnames, the *koseki* records were great for establishing rela- tionships as remote as second cousins, but became very shaky for third cousins and beyond. We surmised that there was—would have to be—a great deal of remote consanguinity we could not document. The best we could do was find one marriage in which the participants were related as first cousins, second cousins, second cousins once removed in three ways, and third cousins in two ways.

We found the sociology of this "extended family" fascinating. One of the outstanding evidences of the persistence on Hosojima of the old way of life is the survival of the institution of the "young people's group" (*seinendan*), which serves in time of fire or disaster, polices the beach, etc. The group had its own dormitory (*yado*) where the young bachelor males slept every night. These youths carried on their individual work and ate their meals at their own farm household, and then gathered at the dormitory each evening. The dormitory serves as a center for social recreation and a meeting place for young mixed company. Traditionally, the group included only young men, but, since 1936, all single males and females of the village 16 to 25 years of age are automatically mem- bers. When we were there, the group included seven males and three females. The grammar school children gathered in the early evening at the dormitory to be tutored by the young men and women. Within the recollection of the present inhabitants there had been no crime or sui- cide on Hosojima.

Although the circumstances did not permit a thorough physical ex- amination of all the residents, we encountered only two manifestations of this degree of inbreeding: Eleven persons, in eight sibships, exhibited a congenital or early onset nerve-type deafness, so severe that eight of them were mute; the parents of affected children were usually normal. The mutes were completely integrated into the life of the island. Eleven persons also exhibited a rare, essentially benign pigmentary anomaly of the skin, *dyschromatosis symmetrica hereditaria*; again the familial pattern was of recessive inheritance. We may presume that among the early ancestors of this group there were one or two carriers for these alleles, the relatively high frequency of the condition being a good index to how few ancestors had contributed to this (private) gene pool.[8]

Most of Jack's time in connection with the Child Health Study was spent in Nagasaki. He accordingly looked about for a case study in that region, and found it in Kuroshima, an island some 17 km off the coast of Kyushu, almost due west of the city of Sasebo. Larger than Hosojima (4.5 km long and 2.1 km wide), it had a population of 2262 persons, the majority Catholic, the minority Buddhist. It was a good opportunity to determine to what extent the Catholic proscriptions against consanguineous marriage were observed in a quite isolated population, or, put otherwise, what happens when religious edict encounters geographical complications. The answer: taking the frequency of consanguineous marriage among the Buddhists as the norm for this situation (44%), the frequency in the Catholic group was reduced (31%), and the marriages tended to involve lesser degrees of consanguinity. Clearly the church made its views known, but equally clearly, was prepared to issue a dispensation for good cause, an adaptability that has accounted for much of the success of Catholicism in the past.

It was in the course of these studies on Kuroshima that Jack became aware of Hirado, some 10 km to the northwest of Kuroshima. Considerably larger than Kuroshima (33 km in length and 9 km wide), with a population of some 40,000 persons, it should provide a sufficient data base for the study of those remaining questions concerning consanguinity effects mentioned on p. 102. Back in Ann Arbor, at the termination of the Child Health Study, we began to plan our study of this island.

Hirado

Separated from the Kyushu mainland by a strait only 100 meters at its narrowest point, Hirado is not your ordinary offshore Japanese island. As mentioned earlier, the first Europeans to reach Japan were some Portuguese traders aboard a Chinese junk driven ashore by a storm about 1542. In a few years, Portuguese ships were visiting Japan regularly. Hirado, headquarters of the Matsuura clan, became a frequent port of call and the site of a trading post. When the Portuguese were induced to make Nagasaki their principal trading headquarters, about 1570, they ceased to call at Hirado so regularly, but, in 1609, the Dutch East India Company, wishing to share in the profits of the Japan trade, gained permission to open a post at Hirado, and a few years later, in 1613, English traders gained the same privilege.

The Portuguese now began to find themselves in increasing difficulties in Japan. The Portuguese Jesuits had been extremely successful in gaining converts, including prominent local feudal lords. Along with Christianity had come the European arts of warfare, sufficient, in the hands of the local feudal lords, to threaten the balance of power in Japan. In 1614 the Shogun, Tokugawa Ieyasu, reacted, banning Christianity. Historians generally agree the ban was much more on political than religious grounds. At first the ban was only half-heartedly enforced, but a series of indiscreet acts by the Portuguese finally led to a ruthless and savage repression of Christianity, the expulsion of all Portuguese from Japan, and the cessation of trade, in 1639.

The Japanese then embarked on their famous, self-imposed isolation, but still desired a very limited exchange with the outside world. The Dutch had strictly avoided missionary activity. In 1641 they were instructed by the government that if they wished to continue trading, they should leave Hirado and settle on a small artificial island in the Nagasaki harbor, to be constructed by the local merchants. Hopeful of profiting by the departure of the Portuguese, the Dutch had little choice but to accede to this strange arrangement. The English had already abandoned their post on Hirado in 1623. Hirado's day in the international sun was over. So far as we are aware, when in 1964, first the Schulls and then the Neels moved in for the study I am about to describe, we were the first foreigners to live on the island for 323 years.

Jack's interest in Hirado had also been stirred by an unusual aftermath of what is sometimes called the "Christian Century" in Japan. After the Meiji Restoration, when Catholic missionaries returned to Japan, it quickly became apparent that despite the persecution, Catholicism had not been exterminated but had survived in many parts of southwestern Kyushu, especially on the small islands, in a secret and syncretic form which incorporated elements of Catholicism, Buddhism, and Shintoism. These groups surfaced with the reopening of Japan, and were commonly referred to as *kakure kirishitan* (hidden Christians).

Almost immediately after the Restoration most of these groups realigned themselves with the Catholic Church, but a sizable minority chose to retain their autonomy. The far side of Hirado, it developed, had been largely populated by *kakure*. Although most of these, joined by other *kakure* from more remote islands, had recognized the second coming of the mother church and embraced Catholicism in the 1870s, it was known that there were still perhaps a dozen villages on the far side of the island whose inhabitants practiced *kakure* Christianity. Until recently, these villages had been highly isolated, the roads being very poor. Thus Hirado not only had the population base and, we anticipated, the level of consanguinity appropriate to the study we had in mind, but an additional sociological twist. The ban of the Catholic Church on consanguineous marriage issued from the Council of Trent in the sixteenth century, and presumably was known to the Jesuit missionaries. To what extent had the *kakure* been able to observe these bans, given their isolation and the necessity to maintain secrecy?

Our studies of Hirado covered the period 1964–1965. Funding was from the Atomic Energy Commission. This may sound strange, but with the Commission's interest in the question of the consequences of an increased mutation rate had gradually come the recognition that predictions of the effect of radiation required a much better understanding of the gene pool than was then (and still is now) the case. The studies involved a close collaboration with personnel from four Japanese universities: Juntendo University, Kyushu University, Kyoto Prefectural University School of Medicine, and Tokyo Medical and Dental University. The study had two phases. The first, under Jack's direction, was a survey of the total island, household by household, to identify the consanguineous marriages plus those in which either spouse was a product of consanguinity, and to obtain an extensive reproductive history, as well as socioeconomic data, on all households. When a spot check revealed about 10% underreporting of consanguinity, a large scale study of the consanguinity based on the *koseki* records was initiated, resulting in a data set as accurate as possible. The second phase, under my supervision, was to entail physical examinations of a subset of the island's population. We were anxious with respect to the physical examinations to have as many points of cross-reference with the Hiroshima–Nagasaki study as possible.

The study rather quickly settled one point. There were 8,228 households on the island, in which were represented 10,457 marriages with one or both spouses alive. Of these, 8209 involved Buddhists, 1226 Catholics, 761 *kakure*, and 261 a miscellany that we termed "Other." Where the data on consanguinity were satisfactory: 14.6% of the Buddhist marriages were consanguineous, 8.3% of the Catholic, and 27.2% of the *kakure*, the latter figure a commentary on how the habit of endogamy to preserve secrecy had overcome whatever scruples there might have been against consanguinity. Further, for about 10% of the marriages, it could be shown that the mother was the product of consanguinity, and for another 10%, the father. The true figure was probably higher, but the facts were obscured by their remoteness in time. Since it had been something of a gamble that the rates would be this high, we were delighted.

Priscilla and I arrived on the island in the late fall of 1965, having left our daughter at the International Christian University in Tokyo and our two sons at the Canadian Academy at Kobe. (A second son had been born in 1954.) Because I was also intermittently still working on atomic bomb effects, the ABCC made a house available to us in Hiroshima, at which we held frequent family reunions throughout the year. Housing arrangements on Hirado were to be a bit unusual. The Japanese world is very orderly, with carefully defined privileges and obligations. There was nothing in the Hirado social dictionary about renting living accommodations to foreigners, and for want of finding something, Jack and his wife, Vicki, had at first been forced to live in a hotel. However, the director of the largest private hospital on the island, Shinobu Kakizoe, was a graduate of Kyushu University, and through the good offices of those of our medical associates who were Kyushu graduates, it was arranged that the Schulls would live in his hospital, rent free. "Atop" the hospital would be a better word than "in," for what was made available was a single large room sitting on the roof which, just incidentally, also housed the motor for the elevator. It was being used as a "ward" for ambulatory patients, who were moved out when the Schulls moved in. This is what we inherited for living quarters.

We shared toilet facilities with the patients, on the top floor of the hospital. But not the *o-furo*, that wonderful big hot tub with which a traditional Japanese family, communally or in succession by rank, finished off the day whenever possible. For this we were indebted to Dr. Kakizoe, whose private home articulated with the hospital. Our turn came about 6 p.m., sandwiched between the baths of his three children and of the Kakizoe adults. Access to the Kakizoe home from our penthouse was only through the hospital lobby. One of the high social points of the hospital routine came to be the passage of the kimono-clad Neels through that lobby, *o-furo* bound, returning to the right and left the bows of the large number of visitors and patients who always seemed to turn up about that time.

The entire island was administratively a single city, Hirado-shi, but almost a third of the population was clustered around the principal harbor, located well toward the northern aspects of the island, facing Kyushu. This had been the stronghold of the Matsuura clan, and a small castle and museum still showed this influence. There were still Matsu-uras on the island, of which a bit more later. Once out of this area, and especially on the far side of the island, one was in very rural Japan indeed. Outer Hirado, in fact, and especially the *kakure* hamlets on the far side of the island, seen through half-closed eyes, was almost the Japan of Lafcadio Hearn, of old Japanese prints—a contrast with Tokyo as great as any contrast I have encountered in any country. But it was to change quickly. Already there were two new tourist hotels, catering to the Kyushu trade. Some years after our stay there, in 1969, the Empe-ror's visit to the island, his first, prompted a rushed black-topping of roads. A bridge spanning the straits at their narrows was opened in 1977. Hirado has now become popular with Japanese tourists and the island's population is much more mobile. We were fortunate to be there at the right time for our studies.

Arranging for the physical examinations required a bit of country-style diplomacy. Our team could not possibly examine all of the island-ers. In general, the further one progressed down the island from the city proper, the higher the consanguinity rates. We decided not to include the people of the town area in our examinations. On the other hand, the roads and accommodations were so poor in the far end of the island that attempting to work there would have been very inefficient. As a com-promise, we elected to examine all the children attending the two middle schools (grades 7–9) located in the central part of the island, and as many of their parents as possible. The children would be 13–16 years old; 21.4% of them were the product of a consanguineous marriage. All 1,555 of them could be examined at school, but that would clearly be inconvenient for the parents. At length we hit upon the idea of examin-ing the parents in *kōminkan*, small community centers scattered over the island, that in many respects were similar in function to the grange halls of the rural United States.

Jack had arranged for a little office in the city hall. As in the Hiroshima and Nagasaki studies, from this base I made the rounds of every organization whose understanding and approval might help enlist the cooperation we needed. There was now, however, an additional group on this island whose blessing was absolutely indispensable. In feudal times, the administrative unit for the rural holdings of the feudal lords was the *mura* or *buraku*, a hamlet and the immediately surround-ing area, each of which had a chief, a *kucho*, elected by the other peasants. It was through him that their lord's agents made his wants known. The *buraku* system was still very functional on Hirado; we had to negotiate with the *kucho* of each *buraku*, first for the cooperation of that district and then for the use of the *kōminkan*.

Our pointman in these negotiations was Manabu Yamamoto of the Kyushu Prefectural University School of Medicine. The consensus of our Japanese professional team was that these negotiations would somehow be dignified by my (unspeaking) presence. The *kucho* were an interesting lot, obviously chosen in part for force of personality. There was another prerequisite. No negotiation, it seemed, no matter what the time of day, could be conducted without liberal potions of *shōchū,* a distillate of rice wine not unlike a weak vodka. A clear prerequisite for the position of *kucho* was a capacity for *shōchū* that often left us so far behind that by the time we got around to discussing the needs of our study, we were not quite sure why we had come.

The Japanese social structure is often characterized as rigidly stratified. This is not quite the best way to describe it. Each individual's social relationships and loyalties are primarily determined by a series of what the mathematician would call "intersecting sets." There is one set from primary school, one from high school, one from university, one from marriage, and one from business. These are fixed and enduring, but there may also be temporary sets. If I am taken to an exclusive bar by Japanese friends, I become a member of a temporary set, with jocular exchanges not otherwise considered good form. If on an early morning walk around Hijiyama I encounter Japanese doing the same, I am readily greeted. I have become a member of the "walking set"—but these same Japanese would not dream of greeting me on the street. In case of doubt, "greet, not greet," you will find Japanese eyes studiedly averted.

These alliances form early. One day when Yamamoto and I called on a *kucho,* we were informed by his wife that he was at a local school graduation. Having come a long way, we decided to go to the school and speak to him when graduation was finished. When we arrived, it was to a thoroughly alcoholic scene following a kindergarten graduation! Two sets were being established, the children who would go through school together, and perhaps equally meaningful, the parents of these children, who would henceforth be interacting in many ways.

In due time we were so fortunate as to meet the second son of the current Matsuura generation, who still resided on the island, and to join the family for tea on several Sunday afternoons. Learning of my interest in heredity, cousin marriages, and the *koseki* system, which maintained a pedigree for each individual, he volunteered to show me the scroll on which the Matsuura genealogy was recorded. Unrolled, it extended across the room, covering some seven centuries, beginning with the emergence of the Matsuuras as the dominant local clan in the thirteenth century. I was aware of the custom in these feudal Japanese lineages

111

that if there was no male heir, or if one's own son was not really competent to further the interests of the family, of marrying one's daughter to a gifted commoner and then legally adopting him. If there was no marriageable daughter, simple adoption of a "son" would do. The Japanese term for such sons is *yoshi*. Such a practice was greatly facilitated by the concept in feudal Japan of *ie* (stem family) in which the lines between kin and nonkin are not as sharply demarcated as in the Western World. This was an opportunity to probe a bit into how biologically meaningful these apparent male lineages are. Rather casually, I asked him how many were *yoshi* sons. It was impressive to me, as a geneticist, how, without hesitation, he walked through that pedigree from earliest times, identifying about one *yoshi* each century. (I was of course in no position to judge his accuracy.)

The younger Matsuura daughter was then undergoing the traditional prenuptial instruction in the Japanese arts, and Priscilla was privileged to join her weekly "classes." Preeminent among these was guidance in the art of the tea ceremony, the session held in the Matsuura garden tea house with, as befitted her background, the old masters of the region presiding. At first acquaintance, we Westerners have difficulty understanding this apparent fuss over a cup of tea. In the hands of the Japanese, it is a momentary detachment from life's pressures, a ritual exercise in studied elegance and symbolism that breaks the tensions of the moment—a release device of a type with which we Americans are not very familiar.

All our staff found lodgings in the port area. With the site for the examinations regularly at least 10 km away, a problem arose. Our "company" station wagon couldn't possibly move our team of seven professionals and technicians plus our equipment about. Fortunately, one of the new tourist hotels had just purchased a minibus, to meet and transport tourists from the ferry landing. We put it under contract, and in succession set up our medical shop at two middle schools and a dozen *kōminkan,* the latter, unheated and in various states of disrepair—and this was winter time.

I've mentioned those rare moments of insight, of fulfillment, for which I have such clear recall. They're not major speeches at scientific gatherings nor the receipt of an honor. One of these moments occurred in Hirado. At the end of our round of physical examinations, in March, I returned to Hiroshima, in connection with the ongoing studies of atomic bomb effects, while some of the clerical staff remained on the island, completing various aspects of the records. Then came May, and it was time to transport the records to Hiroshima for shipment to Ann Arbor. So concerned was I over some possible loss along the way that I arranged for an ABCC photographer to accompany me down to Hirado, to photograph the irreplaceable part of the data over an estimated five-day period. I had been warned that this particular (excellent) photographer was known to drink too much, and the further he got from home, the more he drank. Hirado could be regarded as a long way from home. We lodged in a very old-fashioned inn fronting on the harbor. I practically lived with that man for five days, intermittently referring to a great party that would take place when the work was done. He stayed sober, we finished on schedule, and celebrated with a grand bash at the inn. Our study had come off without major problems. I still recall waking the next morning, stretched out on the *tatami* of my room in that old inn, the sun and early morning cries of the first fish peddlers on the dock area just beginning to penetrate the room, not sure whether I was in the Tokugawa Era or the Twentieth Century, but all my senses in tune with whichever it was.

Back in Ann Arbor, the analysis of the data, which extended over a three-year period, fulfilled all of our expectations with respect to rounding out the story of consanguinity effects.[9] Considering, first, the effects of a consanguineous marriage on the child, the excess prereproductive mortality (up to age 21) in the children of first-cousin marriages could be calculated to be 4.6%. This was somewhat higher than the findings in Hiroshima and Nagasaki, but allowance must be made for the fact that on Hirado, the data covered the full first 21 years of life expectancy, and the time span for the births on which this calculation was based, 1920–1939, was a period when living conditions were somewhat harsher. The physical examinations had included the indicators found to be most useful in the earlier study: anthropometrics, tests of vision and hearing, and tapping rate. There was also an examination of the heart and lungs, which included blood pressure, more for public relations than because these would provide useful information. Finally, we had access to school-performance records for the preceding year, and also to the results of a Tanaka-Binet intelligence test administered to all children on admission to middle school. For these indicators, the effects of consanguinity were again similar to those encountered in Hiroshima and Nagasaki, albeit generally even slightly smaller.

The Hirado study permitted us to explore an aspect of consanguineous marriage not included in the earlier study, namely, the total number of pregnancies and the total number of live births to such marriages. Here was a surprise. Both pregnancies and livebirths were significantly increased in the consanguineous marriage, although, because of the higher death rate in the liveborn children of such marriages, the "net fertility" was about the same as for nonconsanguineous marriages. This

113

finding could be explained either of two ways. One was that for some unknown psychological reasons the consanguineously married couple was motivated to have more children. The other, which we favored, was "reproductive compensation," i.e., the replacement of a child who had died early by shortening the interval between successive pregnancies. Several other studies in Japan have yielded similar findings. Whichever the explanation, this reproductive compensation for the higher death rate of children in a consanguineous marriage, if it is widespread, implies that consanguinity is not as effective in eliminating deleterious genes as if there were no such compensation.

With respect to the effect on the child if the mother or father was inbred (but not consanguineously married), our results were, as might be expected, rather different for fathers and mothers. Mortality was increased in the children of inbred mothers; the mortality in the children of mothers who were the product of first-cousin marriage being increased about half as much as the mortality in children themselves the product of a first-cousin marriage. No explanation for this was immediately apparent. On the other hand, although not statistically significant, in all our analyses the effect of fathers' inbreeding was actually to improve the survival of his children slightly. We put this down to a statistical fluke, concluding only that there was no evidence to suggest that a father's being inbred adversely influenced the survival of his children. The only other study of Japanese for the effect of parental inbreeding, by Yanase, has yielded a similar result. One possible explanation—if it is not a statistical fluke—is that the inbred fathers were more often first sons of a consanguineous marriage, a marriage arranged to keep the farm in the family, and, having inherited the farm, were slightly better off economically.

Earlier we expressed the results of consanguinity in terms of "lethal equivalents." In view of the results from Hirado, I now need to amend our earlier estimate. Although difficult to do so with total accuracy, I suggest that in view of the results from maternal inbreeding, we now place the frequency of lethal equivalents expressing themselves in Japanese children born at or near term at about two. We return in Chapter 12 to the problem of reconciling this (still small) number with what we know about mutation rates and the extensive genetic variation in our genes.

And what of Priscilla while all this was going on? During her first extended stay in Japan, in 1958, Priscilla, already accomplished in water-color painting, had become fascinated by the very demanding Japanese ink-brush painting on rice paper called *sumié*. The lessons, begun when we lived near Hiroshima, continued even on Hirado (although they involved commuting to Nagasaki). In a corner of the room in the Kakizoe Hospital where we lived, the Schulls had installed three Japanese *tatami* mats for sleeping purposes, which arrangement we took over. These three straw mats, each approximately 1 × 2 meters, gave us a king-sized sleeping surface. But these mats also provided the ideal (and traditional) surface on which one did *sumié*. Returning from the field in the evening, I would find our *tatami* covered with the day's output, and, if her genius was really flowing, it was not clear whether the paintings or I would occupy the mats that evening. Several years after starting *sumié* she had also begun the study of calligraphy (*shōdō*), and her paintings combine both arts. These interests have continued right down to the present; each time we return to Japan, there are more lessons. Her work has been exhibited in Japan, and her paintings adorn both our home and my office.

It was later in this stay in Japan that our family climbed Mt. Fuji. There were four of us, Priscilla, sons James and Alex, and myself (daughter Frances had returned to the U.S.). We did it in "classical" style, beginning in late morning near the base and reaching stage $8\frac{1}{2}$ by about 10 in the evening. There we all slept for a few hours in a hostel that specialized in soup and bare-board bunks (but with pillows) and resumed the climb about 2 A.M. We reached the summit about 15 minutes before the sunrise, and as a glorious sunrise unfolded, we could easily understand why the Japanese value the experience so highly. We then came down the "back way," to Gotemba, where, at his summer home, Taku Komai, one of the truly pioneer figures in Japanese genetics, whose welfare I had ascertained my very first trip to Japan, had a watermelon cooling in the well. Luxury can be variously defined.

Most of our fellow climbers had been young people, presumably members of student associations of one kind or another, but there was a lacing of people of Priscilla's or my age or older. Some of the latter greeted the sun in the traditional *shinto* fashion, a bow and a clap of the hands, even though, because of its patronage by the military, *Shintoism* had been discouraged since WWII. This complex fabric of the old and new, surfacing in all sorts of places, was very much part of the fascination of Japan.

115

Shavante men in the village of São Domingos (1962). The middle male is carrying two baskets.

7

Amerindian Studies: Testing the Water

The decision to devote a major share of the last 31 years to the study of some of the least acculturated Amerindians still to be found in South America was made gradually, driven by my increasing preoccupation with "gene pool genetics." Pursuing the sickle-cell gene in Africa and the studies of atomic bomb effects in Japan had, to be sure, involved "gene pool genetics," but a considerable part of my work was still clinical and medical genetics. By the late 1950s I was backing off from the study of the hemoglobin diseases and the program in Japan was enjoying a (temporary) lull. There was time to think about new directions. The term "gene pool" will be unfamiliar to many. It is a descriptive synonym for the collective genetic material of a population. It is a convenient term which, however, cannot be taken too literally. The genes of a population are of course not free-floating in this pool, but packaged in individuals. Their independent assortment from generation to generation is limited by the fact that specific subsets of genes are associated on chromosomes. Finally, family and ethnic considerations often channel the distribution of genes within this pool. Despite these limitations, the term is so evocative that it enjoys a wide currency, one that even lends itself to the title of a book such as this.

The Nature of "Gene Pool Genetics"

The question of the genetic effects of radiation brings gene pool considerations to the forefront. Any successful species is thought to owe its existence over any prolonged time period to a "reasonable" balance between mutation and selection, the latter serving to eliminate the mutations with unfavorable effects and conserving those conferring benefit to the species. As the possible threat of an increased mutation rate from exposures to radiation and the release into the environment of mutagenic chemicals began to emerge, many prominent geneticists, especially H. J. Muller, expressed concern that modern human populations were so delicately balanced between the input of deleterious genes through mutation and their elimination through selection that any increase in mutation rates would tip the scales, so to speak, and initiate a genetic decline in our species.[1] It was presumed that even without an increase in mutation rates, modern medicine, vaccination, sanitation, etc. had already seriously obtunded the action of the selective forces that had shaped us. This was an important thesis, one that picked up on some of the valid eugenic concerns of earlier years. The studies in Japan would provide some data on radiation-induced mutation rates but could not address the issue of how selective forces had changed.

In the late 1950s I began to devote considerable thought to the question of just how to study the operation of natural selection on our species. In 1957 I wrote a paper entitled "The study of natural selection in primitive and civilized populations" for a symposium[2]. Of course I discussed the hemoglobin genes—they were and remain the best examples of a genetic system under reasonably well understood selective pressures. The bulk of the paper, however, was devoted to speculations concerning the extent to which differential fertility had genetic significance (rather than being solely a socioeconomic phenomenon), and the extent to which the dietary changes that usually came with civilization (so much more salt and fat in the diet) might be stressing adaptive genotypes which had evolved under different circumstances. I found it impressive and depressing how little solid data there were relating specific genes to specific traits of selective advantage or disadvantage. (I exclude a small group of very rare and usually fatal genetic diseases from that statement.) As regards fertility differentials, it was clear that with respect to the prospects for mounting a significant study in civilized countries of genetic factors in differential fertility, human control of fertility for personal and socioeconomic reasons had reached the point at which it might largely obscure evolutionarily significant genetic differences in reproductive potential.

As I continued to think about how poorly we geneticists really understood the natural selection which we all endorsed, a further consideration emerged. Even if appropriate studies on biological selection could be organized on civilized populations, the findings would probably not be relevant to the forces previously driving human evolution: modern humans had arisen under very different conditions. To understand ourselves, and how the conditions regulating survival and reproduction had changed, we must understand the biology of precivilized man much better. I realized we would probably never assemble from studies of existing tribal populations the numbers of observations necessary to relate specific genes to specific selective advantages, but at least we could take steps to define the range of population structures within which the evolutionary forces shaping humans had to operate.

These concerns about the operation of natural selection very quickly merged with the fascination with the question of human origins and evolution that most humans share in one form or another. To the geneticist, evolution consists of changes in gene frequencies: the appearance and spread of "new" genes, the disappearance of some "old" genes, and the persistence of some genes already present, but usually with changes in their frequency in the population. These changes in the gene pool are modulated through the interplay of three factors, namely, the rate of entry of "new" genes into the population through mutation, the rate of elimination of "old" genes through the operation of natural selection and chance, and, finally, what we term population structure. This latter term encompasses such aspects of a population as the frequency of inbreeding, marriage patterns, the extent of differential fertility, the age pyramid, survival curves, and the size of the basic population breeding unit. All these latter factors determine how the phenotypes conditioned by genes are presented to the action of natural selection, as I shall be demonstrating later. Note that when I write of mutation introducing "new" genes, I put new in quotes. Mutation is repetitive, i.e., over the aeons of time, the same mutation probably occurs again and again; what is a favorable mutation in one setting may be an unfavorable mutation in another setting, as the data on the hemoglobin diseases indicated. If the appropriate populations could be identified and studied, they might shed some light on the dynamics of these evolutionary forces. The opportunity for such studies—in contrast to the opportunity to study DNA—was vanishing rapidly.

Backing into the Amerindian Studies

During 1960–1961, my thoughts as to what could and should be done marinated in a rich stew of very mixed anthropological reading. But while the stew was flavorful, it very quickly became apparent that the data collected by the social and physical anthropologist had simply not been the data the geneticist requires for any very precise approach to the subject of the dynamics of human evolution. It also became clear that with respect to a significant study, it was now or never; the relatively few remaining primitive populations of the world were so rapidly being disrupted that ours was almost surely the last generation to encounter any of them in a *relatively* undisturbed condition.

I began to accumulate a file on the reasonably intact tribal populations whose study might provide some significant insights. (In various publications I have referred to these groups as "primitive," applying the term "primitive" in a technical sense: preliterate, aggregating in tribes, social organization primarily determined by a kinship system, depending for livelihood on hunting-and-gathering and/or very simple agricultural practices. Although "primitive" is an accepted term for such groups, I have slowly come to feel not only that it is unduly pejorative but that I am having increasing difficulty defining the societal dividing line between primitive and nonprimitive behavior. I need only mention the recent atrocities in the wake of Yugoslavia's disintegration. Likewise, I now dislike the term "unacculturated," which implies that before they encountered Eastern or Western civilization, these groups had no culture. Accordingly, throughout this book I shall use the more noncommittal term "tribal" in referring to the groups we studied.) Within two years, the file I was accumulating on tribal groups grew to some nine different populations, among whom the New Guinea natives and some of the Indian tribes of the interior of South America appeared especially fascinating. The situation called for a pilot study, to determine if the sort of undertaking I had in mind was feasible.

At that point, two developments made the choice easier. During 1956–1957 Francisco Salzano of the University of Rio Grande do Sul in Pôrto Alegre, Brazil had been in the Department of Human Genetics at Michigan, on a Rockefeller Foundation Scholarship. As his fellowship drew to a close, we had begun to discuss what type of research he might pursue when he returned to Brazil. As was my custom, I suggested that he consider what unusual genetic resources his country might possess, rather than pressing the automatic pilot button that might lead to an effort to duplicate some of what he had experienced in Ann Arbor. As the discussion continued, it was clear he was drawn to genetic studies of some of the Indian tribes of Brazil. I encouraged him to move in that direction, which he did. One result, as my thinking matured, was a strong potential on-scene collaboration, the importance of which in field studies cannot be overstated. A second but unplanned development was an invitation from the World Health Organization in 1961 to attend a meeting in Rio de Janeiro devoted to consideration of studies in areas of high natural radiation. This meeting was prompted by representations to WHO by certain Brazilian scientists, regarding the possible feasibility and desirability of biomedical studies in the areas of high natural radioactivity in the state of Espirito Santo. The meeting, which occurred in December of 1961, revealed that the levels of exposure to ionizing radiation were so low—less than in Kerala, India—that even an extensive and carefully controlled study was not likely to yield important insights. But the meeting—my first in Brazil—was an opportunity to get a feeling for the climate of acceptance. It seemed good, and would definitely be better if I could learn a little Portuguese. Although Salzano was not part of that conference, we arranged a meeting just after the conference, at which we began a serious effort to identify the most suitable Brazilian Indian populations for study.

There were two additional reasons for this interest in Amerindians as contrasted with other populations. The first was their relative proximity to the laboratory facilities of the Department of Human Genetics in Ann Arbor. This was important, because in conjunction with the fieldwork we would be sending a stream of biological samples back to Ann Arbor for analysis. The other was a feature shared by only two other populations, the Australian aborigines and the Polynesians: the place and approximate time of entry of the forebears of the Amerindians into an extensive area devoid of other humans was known: the Bering Land Bridge, some 15,000 to 40,000 years ago.[3] Assuming reasonable homogeneity in the founding stock, an added bonus to studies of the Indians would thus be documentation of how much diversity could arise in this time span, i.e., documentation of the pace of evolution.

I was projecting rather comprehensive studies of some of the least acculturated tribes in existence, some contacted and "pacified" only in recent years. The master plan required that we would go in as a multidisciplinary team—geneticist, physician, linguist, social and physical anthropologist—do physical examinations, get reproductive histories and genealogical data, and finally, collect and return to the laboratory in good condition blood, urine, and stool samples. The studies on these samples which I had in mind would not only draw upon all the resources I had been building up within the Department, but many other specialty laboratories at the University of Michigan and elsewhere. This would be a very different sort of undertaking from the traditional fieldwork of the single cultural or physical anthropologist, or the dash of a geneticist to a remote area to obtain some blood samples. Central to this plan was a close-quarters interaction between diverse disciplines out of which would either emerge an exciting intellectual interaction and synthesis or mutual rejection, which in the field is not so easily handled as at home.[4]

It wasn't at all clear that an operation of this complexity and scale was feasible on the kind of tight timetable that working with perishable blood samples requires. It certainly wasn't clear we could find the necessary funding. This brings me in all candor to the final reasons why, at the relatively advanced age of 45, I began to think of such an undertaking. Behind our elaborate rationalizations as to why we do this or that are often rather simple motivations. From time to time men are driven, for reasons often difficult to articulate, to measure ourselves, each in our own way, according to our background and opportunities. Mine had thus far been a rather safe life. This study presented a situation in which I could test myself. I had thus far functioned in a setting where help in case of miscalculation was close at hand. This would be different. Even as a boy, I had empathized with the tragic fate of the American Indian. Now I would glimpse—and perhaps even ameliorate a bit—a vanishing world.

The third reason for this late conversion was pragmatic. In the intensely competitive atmosphere of the modern biomedical sciences, one cannot afford to choose too many unyielding problems. It would be a gamble whether we could obtain the data I hoped for and whether we could synthesize anything of real scientific value from our experience. At this point, my scientific career was reasonably well launched. I felt I could afford to take that gamble, a gamble which, if it proved scientifically productive, could result in the luxury of working in a relatively uncrowded field of endeavor. Furthermore, my thrust to escape from Wooster to a situation where one could do major science had been a grind in which vacations played a small part. This would not be a vacation, but it might be some compensation for the many times I'd reined myself in in order to get on with the job.

Organizing a Pilot Study on the Shavante

By early 1962, a concrete plan for a pilot study of a Brazilian Indian village began to emerge. We would take to the field in July. The team would involve Salzano, Friedreich Keiter (a German physical anthropologist who would be a visiting professor in Brazil at that time), myself, and Pedro Junqueira, a Brazilian hematologist directing the Blood Bank in Rio de Janeiro, who was not only interested in typing any blood samples we obtained, but who apparently knew everyone in Rio who might possibly be of help to us. We would try to pick an Indian village maintaining intermittent contacts with one of the outposts of the Brazilian Indian Protective Service. It would have to be one of the outposts with a landing strip, so we could move quickly when we had our perishable samples. First priority would go to any such villages in which a cultural anthropologist had already worked; we would try to persuade him to join us in the field.

We needed funding and credentials. I had been reading as much as I could on travel, exploration, and anthropology in the interior of Brazil; almost without exception every book devoted the first chapter to the two months spent in Rio getting equipment through customs and obtaining the necessary permits for the interior. We had to do better. A first step in that direction involved an application to the World Health Organization to fund our activities in Brazil. We were successful. The Organization also agreed that I could take the equipment I would bring from Ann Arbor into Brazil under a shipping document to be issued by them. The Atomic Energy Commission, supporting my studies on population genetics, agreed to finance the laboratory aspects of the work.

Letters were now flying in all directions. I was also taking a cram course in Portuguese. In April, Francisco and Pedro wrote jointly, announcing they had decided that our target population would be the Shavantes living near the isolated Post Pimental Barbosa of the Indian Protective Service, located in the Mato Grosso, on the Rio das Mortes. The prehistory of the Shavante, like that of all Indian tribes, was murky. They may at one time have extended considerably further to the east, and have been driven into the more inhospitable Mato Grosso by encroaching civilization. Their contacts with neo-Brazilians had been largely hostile until some 20 years previously. As late as 1941, the inhabitants of the village selected for study had formed part of a band that had killed an agent of the Indian Protective Service, Pimental Barbosa, and five of his assistants, the six having been entrusted with the mission of establishing friendly contacts with the Shavante and setting up a "Post of Attraction" for them.

This had been the last such massacre involving that group of dedicated men who took such risks in the peaceful pacification of the Indians of the interior; true to the motto of the service (*Morror, se presico for; matar nunca!* "Die if necessary but never kill!"), they had not used firearms when the fatal attack came. The present chief of the village— whom we were to meet shortly—took part in that attack. Subsequent efforts at making friendly contacts had been somewhat more successful, and an Indian Post had established intermittent but peaceful contacts with the Shavante in 1951. The Post had been managed almost from its inception by an exceptionally able agent, who of course by now spoke the language and could help in our contacts. Ten years earlier we probably could not have executed the studies I will describe, whereas by now, some 30 years later, increasing contact with the outside world has already grossly altered many aspects of what we encountered.

Half-way into our planning, Francisco wrote that he had just learned of a cultural anthropologist, David Maybury-Lewis, now at Harvard, who had spent much of 1958 with the Shavante. As soon as I learned this, I called David. He quickly came to Ann Arbor for two days of steady conversations, myself emphasizing the concept of a multidisciplinary approach, himself trying to imagine the feasibility of all this in that setting. A critical point in the discussions was the possibility of obtaining meaningful (i.e., reasonably accurate) pedigree information. Cultural anthropologists are great on defining systems of kinship and formal marriage patterns, but in all my reading I had not encountered a single village pedigree which met exacting genetic standards. By great good fortune, David had been planning to return to the Shavante briefly that summer, to check out some details for the monograph he was writing.[5] We agreed to join forces. The team was complete. (It also included Salzano's long-time technical assistant, Girlay Simões.)

Now it was time to assemble our medical and collecting equipment. There would be no corner drug stores or scientific supply houses in the Mato Grosso. Nor could I count on rectifying all of our oversights once I reached Rio. The list grew longer and longer. This was the beginning of what became a standard check list, but the first time around was difficult, since weight and space were also considerations. In the midst of all this I served a brief apprenticeship in the art of dental extractions, since my readings had led me to believe that the most common medical "emergency" in the interior was a toothache, and it was a poor sort of doctor who couldn't handle that.

I reached Rio the evening of July 9, 1962. Waiting to meet me were Pedro and a representative of the World Health Organization. With the latter's presence and my papers, I cleared Customs without opening a single footlocker or bag! The first test was met. The next several days were spent purchasing our "trade goods" and food for the interior. The trade goods would be the quid pro quo in our dealings with the Indians, and the choice was important: another set of expeditionary horror stories involves groups who brought articles in which the Indians were uninterested, and therefore less than cooperative. Our Brazilian colleagues had already obtained the absolutely essential permits from the Indian Protective Service to proceed to the village we had chosen.

Now came the diciest part: transportation to the Post. There were two alternatives. One was to go commercial as far as possible, then try to find a bush pilot to ferry us the rest of the way. This would not only kill our budget; in Brazil stories of the adventures (and misadventures) of bush pilots were the equivalent of our nineteenth century frontier legends. The other alternative was to fly with the Brazilian Air Force, one of whose functions was to service the isolated Posts of the Indian Service. One of their scheduled bimonthly flights to the interior included a

124

stop at the Post Pimental Barbosa, and my arrival had been timed to take advantage of the next flight in, even though we knew observance of the schedule to be somewhat erratic. We took our gear to the Air Force Cargo Depot on 10 July, only to learn that tomorrow's flight had been canceled—indefinitely. An unexpected need had diverted the planes normally available for this function to the south. Now came the crucial test of our organization and my first real insight into the very personal way that Latin America functions. Pedro "remembered" his friend the Brigadier General in the Air Force (I'm sure he had never been far from his thoughts). We paid a formal call. Such of the conversation as I followed seemed, in addition to amenities, to involve the honor of Brazil in meeting a commitment to the World Health Organization. The next day we had transportation, a DC-3.

The final link in the sequence being forged was completely imponderable. Maybury-Lewis had found the Shavante to be taking long and frequent hunting-and-gathering treks, leaving only a few people behind in the village. There was absolutely no way that from Rio we could determine the whereabouts of the Indians. Each of the Indian Posts was supposed to be in radio contact with headquarters in Brasilia, but the agent at Post Pimental Barbosa was out on leave (and when we got there, we found his transmitter was not functioning). If these Shavante were away, there were no substitute nearby villages. On this we simply took our chances.

I will not go into such detail again, as I describe our subsequent trips to the field. This, our first go at it, required more preparation than the others, but something of this nature preceded each of our expeditions. There is, however, a very basic point to be made by such a description. I suspect that most of my readers with a scientific background are much more familiar with laboratory-based research than with fieldwork, well aware of what must be done to pursue a laboratory problem successfully but less knowledgeable as regards field studies. Only by virtue of preparations such as I have described, only with the collaboration of strong, knowledgeable local colleagues, could fieldwork of the type I will be describing be carried out successfully and on any reasonable schedule. One also needs a bit of luck.

The First Round of Fieldwork

Flying in to the Shavante from Rio, the DC-3 overflowing with equipment, we stopped only in Goiana, to pick up the Indian agent to the Post, Ismael Leitão, who had been out on leave. Ismael took one look at David

Maybury-Lewis, and declared the plane was not large enough for the two of them. I never really fathomed the enmity, but suspect he felt challenged when an anthropologist (with different viewpoints) arrived to study the people, prolonged contact with whom was his chief claim to fame. The situation demanded the full range of Pedro's diplomacy, but at length all was well—the seating arrangement on the plane was designed to maximize the distance between David and Ismael.

Arriving, our concerns about finding the Indians at home were immediately justified—about one third were out on trek. Happiness was redefined: it is finding *some* Indians in the village (and more came straggling in the next two weeks). The village was located about a half mile from the little cluster of buildings that defined the Post. David moved in with the Indians, minimizing further contacts with Ismael, while we established ourselves at the Post, where we would examine, measure and bleed the Indians. First business, the first day, was to explain what we wished to do—and what gifts we had brought. But there was also time to visit the grave of Pimental Barbosa, in silent tribute to a kind of courage and dedication best glimpsed standing in the lonely, remote spot where he had given his life for a cause.

The very next day we began our examinations, beginning, at their insistence, with the males (since the Shavante were not yet sure of our intentions toward their women). We'll come to the findings in subsequent chapters. Let me say only that the males were collectively the most superb physical specimens I had ever seen. Our view of health among Amerindians has been unduly influenced by the findings on Indians huddled around Missions or Indian Services, or herded into reservations, 50 to several hundred years after first contact, when alcoholism, tuberculosis, and venereal disease have already begun to erode health. To obtain the overview we desired, we needed to examine a total village sample, and we negotiated to that end. Meantime, David was assembling a total village pedigree, a difficult, difficult job given the enormously complicated kinship system and terminology and the lack of any written records. This would provide the first assessment of the genetic structure of the village; a second would come later when we analyzed the results of our blood typings, and could determine how often a child's blood type was genetically inconsistent with the types of its nominal parents. Such a finding would either indicate a deficiency in taking the pedigree, or that, just as in our own culture, the matrimonial bonds do not always channel gene flow. For our subsequent analyses, we simply had to know how valid the pedigrees were.

It has proved enormously difficult to project to friends and colleagues the essence of an experience such as this, and those to come later, and its impact on one's orientation. For now, three episodes drawn from that first encounter must suffice. Two days after we had established ourselves. David asked me to come to the village to examine a sick Shavante woman. There is not much light in the thatched, windowless, dome-shaped Xavante dwellings. Even so, it was clear she was very pregnant. The history was of fever, chills, malaise, backache, and earlier that day, a profuse nose-bleed. Examining her abdomen as best I could under the circumstances, I felt in addition to a near-term uterus, a large spleen. Of the various possible diagnoses, there were two for which I had effective therapy: pyelitis of pregnancy and malaria. I left medication for both, with appropriate instructions.

Three days later, this same woman appeared where I was conducting physical examinations of the men, carrying her newborn infant. She had gone into precipitous labor shortly after my visit; the child had the massive extravasation of blood under the scalp (cephalohematoma) one sometimes encounters following a very short labor in which the child's head has been forced to dilate the birth canal too quickly. (Medical note: I had not administered quinine for the malaria, since quinine can induce strong uterine contractions.) The child was not nursing properly. It undoubtedly had hemorrhaged into the brain. In hospital, such a child would be fed by nasogastric tube and probably saved, perhaps only to exhibit the symptoms of severe brain damage later. Here, I could only suggest measures I knew to be ineffectual. The child died two days later. As mentioned, we had agreed to examine the Shavante males first. The child died the day we were to begin examining the women. Our program came to an abrupt halt. I was clearly responsible for the child's death, having laid examining hands on the mother's abdomen and administered medication just before her labor. After all, such a child had not been seen in the village for many years. For two days the Indians deliberated. Fortunately, the rest of my practice of jungle medicine had gone somewhat better, and on the third day we were allowed to proceed with our examinations. Ever afterward, in treating an Indian with an illness of any severity, I was careful to explain that malevolent spirits were clearly at work, and although I would do my best, it would be well if the village Shaman was also consulted!

The second episode began a few days later when in the early evening a young man rode into the Post. Although this was Indian country, there were already some small "ranches" scattered along the Rio das Mortes, whose owners ran a few head of cattle on the very poor grazing provided by the sandy soil of the Mato Grosso, and practiced subsistence agriculture on the occasional pockets of fertile soil near the river. The rider was a hand at the nearest ranch, where the owner's wife, known to be carrying twins, had delivered the first, but after two hours, not the second. A "midwife" was there. Would the doctor—news of whose presence had been carried by an infrequent river trader—come? There is often an interval of some hours between the births of twins. I sent him back, requesting a report first thing in the morning, and spent the next several hours assembling a makeshift obstetrical kit. Another rider was there shortly after dawn. The second twin had been born, but the placentas had not been expelled. This was a true emergency, a retained placenta being an open invitation to life-threatening infection or hemorrhage. We left at once, by water, in a leaky dugout canoe. After several hours on the river we reached a landing and a collection of half a dozen buildings, one adobe, the others thatched on sides and top, the latter of flimsier construction than Shavante homes. The circumstances, I would come to appreciate later, were more or less typical of the lot of the impoverished "caboclo" of the interior of South America.

My patient was in bed in the adobe house—exhausted, pulse of 140. Obstetrics had not figured largely in my medical training or thoughts in recent years, basic though it may be considered to genetics, but I am forever amazed at what one can recall under stress. First a generous shot of morphine—what I had to do would be painful. While I waited for it to take effect, that wonderful Brazilian touch: would the doctor care for a cup of coffee? Then the standard Credé maneuver, clasping through the relaxed abdominal wall the upper uterus between cupped hands and intermittently squeezing it with all the strength possible, Salzano exerting gentle traction on the two protruding umbilical cords. Five minutes, ten minutes, fifteen minutes, my shoulders began to ache and beads of perspiration gather. Then the desired result—I felt the uterus become smaller, and shortly the placentas were expelled. More massage of the uterus and an injection of pitressin, both to induce the uterus to contract (why at the last moment in Ann Arbor had I tossed the pitressin in?), a generous supply of antibiotics, and we were shortly on our way—but not before I had examined another pregnant woman—who badly needed some reassurance—and looked around a bit. One did not have to be a trained agronomist to appreciate the marginality of this existence—and how it might fuel social unrest. Going back upriver, to the Post, I found myself reflecting on how unbelievably the practice of medicine had changed in the past century—and the mindset of a life without medical care as we know it. (Follow-up: one week later the patient and the twins were doing well.)

The third episode was toward the end of our stay. Salzano and I had been working almost daily with David, to check the position of the persons we had examined in the master pedigree he was assembling.

The tensions being what they were, it seemed better that we go to the village for this purpose, where David was living, rather than that David come to the Post. We were always back at the Post by nightfall, from which, in that brief interval before I succumbed to exhaustion, we often heard the Indians chanting. One evening, at David's suggestion, I stayed on in the village for the nightly council. Beneath a fantastic canopy of stars, the pungent wood smoke evoking a hypothalamic response, I listened uncomprehendingly, as the mature males, gathered in a group, discussed, as was their fashion each night, the day's events, and planned for the next day. In the background, the young males began to discharge their nightly function of chanting before each house.

Suddenly the thought came to me that I was witness to a scene which, in one variation or another, had characterized our ancestors for the past several million years. The sudden realization of this contact with the thread of evolution resulted in another of those very emotional professional moments; this time I could feel the hair on the nape of my neck stirring, in a manner more often characteristic of physical fright than intellectual delight. Here was the basic unit of human evolution— the band or village—considering its interaction with other similar units and the environment. We were as close as modern man can come to the circumstances under which our species had evolved, under which our present attributed had arisen. What insights into the process could we, or any other group of investigators, hope to gain? I was momentarily encased in a temporary capsule of bygone time. As I stumbled back to the Post in the dark, I was very conscious of the presumption in what we were undertaking. The fascination of this setting was ultimately to take me to a total of 90 more Indian villages, distributed among 13 tribes.

This pilot study resulted in a monstrous, 88-page paper in the *American Journal of Human Genetics,* our first effort at a reasonably comprehensive approach to the biology of an entire village.[6] The publication included a pedigree of the village—David had done his job well—that the results of the blood typings suggested had the requisite accuracy. The paper ended with a series of specific hypotheses we proposed to try to test in further studies. The next step was to try to ensure that we were not being misled in our view of things by a single atypical village. To this end, in 1964 we made our second trip to the Shavante, in the immediate wake of the Brazilian Revolution of that year. We needed confirmation of some of the findings from our pilot study. The uncertainties surrounding our first round of fieldwork were almost insignificant compared with those of fieldwork on the heels of the revolution, but eventually the resourcefulness of Pedro and Francisco paid off and we were back in the

Mato Grosso, first at a Shavante village some eight miles from the Indian Service Post of Simões Lopés, then to another village, at a place called Saõ Marcos, now in close juxtaposition to a Salesian Mission (run by a charming and urbane Italian priest). Both villages were more ac-culturated than that at Pimental Barbosa, but the Indians were geneti-cally intact. The observations made at Post Pimental Barbosa—which I will be discussing in the next several chapters—were in general verified, and we had confidence we had not been misled by a single, atypical village.[7]

It was in connection with these sorties that we first established a relationship with the Missionary Aviation Fellowship, an organization of pilots and planes dedicated to support of the remote Protestant mis-sionary posts maintained by the New Tribes Mission and the Unevan-gelized Field Mission. Once we established our credentials, the orga-nization, with its little Cessnas, provided (for this and later trips) a type of reliable and careful transportation that removed a major source of concern. Since our blood samples were, of course, perishable (our access to local refrigeration was uncertain) a recurrent worry had been the possibility of a stranding at some remote airstrip, our hard-won samples slowly deteriorating. With the MAF willing to work with us, a major concern was met.

Meanwhile there were some significant administrative develop-ments. In mid-1962 the International Council of Scientific Unions blessed a proposal from one of its constituents, the International Union of Biological Sciences (IUBS), that there be an International Biological Program (IBP) of several years duration, one aspect of which would embrace studies of human adaptability. I became involved in the plan-ning at both the international and national levels. For the six years of the IBP our Amerindian program was considered one of the principal Amer-ican components. I consider it a tribute to the then Atomic Energy Commission that their view of the population genetics we needed to know to understand the effects of radiation was broad enough to support our Amerindian study. Further funding came from the National Science Foundation, the lead agency for the IBP.

Also in 1962, the World Health Organization, stimulated by our pilot study among the Shavante (which they had funded), convened in Gene-va a Scientific Group on Research in the Population Genetics of Primit-ive Groups, which I chaired. Our 11-person group produced a report, published by the World Health Organization, which laid out the ele-ments of a significant study, in an effort to ensure cross-comparability in the future between the findings of various investigators. For me, an important by-product of this was an ongoing consultantship to WHO which, as our fieldwork continued, was of great help in moving our scientific equipment across national boundaries.

With this general background, I now made a major decision. Together with the appropriate colleagues, I would mount an all-out effort to document in all the detail possible the genetic dynamics of one or more Amerindian tribes. With luck, we should emerge with a better approximation to the *biological* conditions under which man evolved than now existed. Not only a subject of interest in its own right, the study should provide a perspective against which to evaluate some of the implications of our current departures from those conditions, the subject of the last chapters of this book.

Yanomama children playing at night (1966).

8

On to the Yanomama

My involvement with the second Amerindian tribe we studied had elements of the same serendipity that resulted in the Japanese experience. In 1964, en route to Brazil for the second round of fieldwork among the Shavante, I had stopped off at the Venezuelan Institute for Scientific Investigations (IVIC), near Caracas, to meet Miguel Layrisse and Tulio Arends, two people who had been quite active in genetic studies of the Indians of Venezuela. This was in part an effort to familiarize myself with the findings of these two very able investigators, but, looking ahead, I was also concerned whether there might be in Venezuela tribes more suitable than the Shavante for the studies I was projecting. The day after my arrival in Venezuela, the papers headlined the Brazilian revolution of 1964. Our field team for that year had been scheduled to join me in Rio in another four days. All plans were placed on hold. Civil war seemed a distinct possibility. International flights were canceled. The crisis passed relatively quickly, and with cables and short-wave exchanges flying, we managed to coordinate a team arrival in Rio delayed by only four days.

Enter the Yanomama

Meanwhile, during the layover in Venezuela, I had talked with as many people with Indian experience as possible. Increasingly, my attention was drawn to the Yanomama, a group occupying the remote Amazonas Territory of Venezuela, in the area drained by the headwaters of the Orinoco River. Their distribution was known to extend over the border, into adjacent Brazil, although details were very sketchy. There were

133

thought to be at least 10,000 of them, living in villages of 100–150 persons. The Shavante, on the other hand, numbered only 1500–2000. There was no reason to suspect recent displacement; they should be even more culturally intact than the Shavante. A group as large and undisturbed as the Yanomama would be much more appropriate than the Shavante for the study of internal tribal dynamics. Their tribal distribution straddled the Brazilian–Venezuelan border; given the politics of Latin America, there was much to be said for having at least two strings on one's fiddle. There was an excellent natural airstrip on the Orinoco at a place called Esmeralda,[1] just about 35 miles down river from Yanomama territory; the Venezuelan Air Force flew in there on occasion, and our potential collaborators at IVIC had the official standing that would let them call on the Air Force for transportation. Finally, IVIC, one of the outstanding research centers in Latin America, would be a strong base from which to stage field operations.

After the appropriate collaborative arrangements, our fieldwork among the Yanomama began in 1966. For these studies, the indispensable cultural anthropologist became Napoleon Chagnon. Nap had sought me out in Ann Arbor several years earlier, having heard of our developing program. By virtue of the contacts I had already made, I could facilitate his entry into the field; he, for his part, in addition to pursuing his own interests, could put together the village pedigrees so basic to our work. For this purpose, we went through the same indoctrination concerning the nuances of genetic (as contrasted to classificatory) pedigrees that had been necessary in the collaboration with Maybury-Lewis. Those familiar with Nap's writings concerning the Yanomama know how well the lessons took. Our preliminary discussions involved his doing fieldwork in Brazil, probably among one of the Gé-speaking tribes of the Mato Grosso, such as the Cayapo, with us to come in later for the same type of collaborative effort we had enjoyed among the Shavante with David Maybury-Lewis. During my enforced layover in Venezuela during the Brazilian Revolution, I had come to feel that the Yanomama, not yet studied in depth by a cultural anthropologist, offered greater opportunities, and had so advised Nap. He had agreed, and by 1966, when we first visited the tribe, had spent 13 months among the tribe.

Characteristics of the Yanomama

Given the various ethnographic accounts of the Yanomama now available,[1] I am embarrassed to attempt to characterize them in a few paragraphs, but enough has to be said to provide a setting for the data resulting from 11 years of intermittent fieldwork. This tribe is distinguished from other tribes of the American tropical rain forest belt in many ways. Its language is not closely related to any other Amerindian language. Their material culture is meager; their hammocks (of bark strips), basketry, and clay pots are of very simple design and workmanship, in comparison with the artifacts of the surrounding tribes. Hunting is done with 7-foot bows of palmwood and arrows of equal length, made from arrow cane; the men lavish great care on the manufacture of their bows and arrows. Depending on the purpose, the arrows are variously tipped with curare points, barbed points (the barb an agouti canine), or broad palmwood points. The younger women sometimes wear a brief "skirt" of cotton fringe, older women nothing. Men wear a cotton belt, to which the penis is tied by a cotton string.

The "village" consists of a more or less circular lean-to, perhaps 50 meters in diameter, open at the center. Each family has its own area within this *shabano*. The principal cultivar is the plantain (*Musa* spp.), in contrast to the heavy reliance of most of the surrounding tribes on manioc (*Manihot* spp.). Until recently the Yanomama did not manufacture canoes and tended *not* to locate their villages near major waterways. They grow tobacco (*Nicotiana* spp.), whose ultimate destination is as a large wad between the lower lip and gum of both men and women. There is ritual endocannibalism of the bones of the dead. Hallucinogens, derived from a variety of plants and administered by nasal insufflation, are freely used, especially by the younger men. (Try to imagine the experimentation through which this practice evolved.) A stated reason for their use is to induce contact with the omnipresent spirit world.

In contrast to this material simplicity, in common with other Indian tribes, they have a highly complex patrilineal kinship system that provides the basis for social organization. Marriage must be contracted outside one's lineage; a woman joins the lineage of her husband. The preferred form of marriage entails a man marrying his mother's brother's daughter or his father's sister's daughter (i.e., a first-cousin marriage), but partners meeting these specifications are frequently unavailable, in which case a variety of marriage types are satisfactory as long as lineage exogamy is observed.

The violence which is an integral and formalized aspect of the Yanomama life-style has been much—perhaps unduly—emphasized (see note 1, Chagnon reference). Much of it is precipitated by the undue attention of one man to another man's wife, or by the capture of women from an adjacent village, which of course calls for reprisal. The lowest level of violence consists of chest-pounding duels, in which the men take turns (the challenger second) in striking each other with full force over the heart, until one of the two can no longer continue. The second level consists of duels with long poles, in which each person (challenger again second) strikes the other on the head. Every adult male bears the scars of such encounters, and some have healed depressed skull fractures in which one can lay a forefinger. A variant of this involves the spear-like use of sharpened sticks. Finally, the ultimate is a raid on a neighboring village, to avenge past wrongs. These are not pitched battles but hit-and-run affairs, in which one or two men may be killed or injured by arrows and some women captured. Although the casualties of any single encounter are seldom high, Chagnon estimates that approximately one-quarter of males reaching adulthood will die in this fashion.

Domestic violence is common; how many times, in the middle of the night, lying in my hammock slung in an Indian village, have I heard the shriek of an abused wife? But as statistics become available for countries such as the U.S., where the single largest cause of injury to women is abuse by the men they live with, this abuse causing more injury than automobile accidents, muggings, and rapes combined, it is not clear how gentling an influence civilization has exercised. Furthermore, the net of kinship within a village provides an abused woman with an important system for immediate support should a husband become excessively abusive, a support system largely lost in industrialized, urbanized populations. Having spoken of the violence, I am compelled to record the many manifestations of affection and respect I have witnessed, between husbands and wives, parents and children. There is a low flash-point for violence, but the event that most often initiates this violence—disputes over women—is a prime source of violence (or a socially acceptable substitute) all over the world. On the other hand, violence toward children—a cancer in the U.S.—seems almost nonexistent among the Yanomama.

The area currently occupied by the Yanomama is shown in the map on page 137. As a consequence of their relative inaccessibility, their substantial contacts with the "outside" world were initiated later than those of most Amerindian cultures. From the fragmentary data available, it seems clear that the Yanomama have in the last century expanded their tribal area, more or less centrifugally, by a factor of two or perhaps three, as shown in the map, with a proportional increase in population number. This geographical expansion is most readily explained by an increasing population at a time when many of the surrounding tribes—who had earlier buffered the Yanomama from contacts with civilization—were rapidly declining in numbers and influence. To what extent the population expansion was in part in response to the availability of new territory, because of the decimation of the surrounding tribes, and

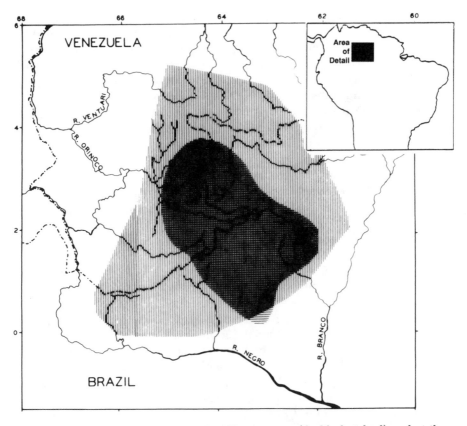

Yanomama territory, approximately 100 years ago (darkly hatched) and at the present time (lightly hatched). The area shown constitutes the extreme southern tip of Venezuela and the northern parts of the State of Amazonas, Brazil.

to what extent in response to the acquisition of new "technology" such as steel tools and, possibly, the plantain (*Musa paradisica*) as a principal cultivar, must remain moot points.

The tribal "heartland" prior to this expansion appears to have been the Parima Mountain range, which defines the border between Brazil and Venezuela between latitudes 2°30' and 4°30'North. The distribution of the Yanomama thus straddles one aspect of the watershed of the Orinoco and Amazon Rivers. The entire area they now occupy is tropical rain forest in ecology, ranging in elevation from approximately 130 to 1200 meters, throughout which are scattered patches of savannah, espe-

137

cially in the highlands. The number of Yanomama dispersed throughout this area can now, from our fieldwork, be estimated as 12,000–15,000 persons distributed among some 120 villages. Since the tribal area currently encompasses approximately 173,000 square kilometers, average population density is about 1 person per 13 square kilometers, a somewhat misleading statistic in view of the nonuniform distribution of people throughout the area. Prior to their expansion, to judge by the distribution of villages in the tribal heartland, the density was undoubtedly higher, more like one person per eight or nine square kilometers.

We have made a total of nine expeditions to the Yanomama. Layrisse was somehow always able to arrange that at the critical time the Venezuelan Air Force would be sending a "flying boxcar" (C-4) to Esmeralda. The night before the flight we'd work until midnight getting all our gear in order, then at 3:00 or 4:00 A.M., leave IVIC with our supplies for the military air base at Maracay, arriving at the base at dawn and taking off shortly thereafter. Often we shared the plane with missionaries and their supplies. The C-4 was slow and the distance such that it was an all-day affair for the crew to load, fly in, unload, and fly out. There was of course no weather service at Esmeralda and, more than once, arriving (approximately) overhead of Esmeralda, we had to turn back because the area was obscured by clouds. The most memorable of these failed landings was in 1969, when a brave pilot was trying to feel his way into a cloud-obscured Esmeralda. We knew we were in trouble when he summoned into the cockpit Velasquez, a Makiritare chief who knew the region well, returning to his village after business in Caracas with the Indian Agency. Suddenly, above the din of the interior of the plane, we heard Velasquez shout, "Duida!" the name of a mountain massif cresting some ten miles north of Esmeralda, to which we were uncomfortably close. It would never have occurred to me that so large a plane could heel over so quickly. We proceeded straight back to Maracay, to try another day.

Upon arrival at Esmeralda, we then took to the Orinoco River. Each time we flew in I could not but contrast the ease of our passage with that experienced by the great Alexander Humbolt who in his explorations of the interior of South America had in 1801 gotten as high on the Orinoco as Esmeralda, of which he left a most unflattering picture.[2] But to each his own: we too were enjoying a unique window in time, our studies depending as they did on the rapid transport of highly perishable samples by plane to a base laboratory. A generation earlier we could not have hoped for success, while a generation from now, the disruption of these societies will have been so devastating that our studies would be much less insightful.

On to the Yanomama

The Nature of the Fieldwork

The studies on the Yanomama were to be very similar to those on the Shavante, but on a more ambitious scale. Again we would evaluate general health by physical examination, make anthropometric and dental measurements, and investigate disease pressures by studying blood samples for antibodies against various agents of disease and stool samples for parasites. Of greater genetic interest would be an effort to work out the vital statistics of a group at this cultural level, determine the level of inbreeding, and evaluate patterns of exchange of members between villages (i.e., determine the migration matrix). Finally, we would type each individual with reference to some 30 genetic traits, not only to characterize the tribe but also to define the extent of village genetic differences. These latter observations would permit us to begin to define the population structure within which evolutionary forces operated.

The various findings will be the substance of the next three chapters. Before I proceed further, however, an important statement, which I put in capital letters not only for emphasis, but so it will be easy to refer to from time to time. THE YANOMAMA (OR ANY OTHER SURVIVING TRIBAL POPULA-TION) SHOULD NEVER BE CONFUSED WITH A TRUE STONE AGE CULTURE. They now derive some 70% of their calories from the plantain; there are reasons to believe this is a post-Columbian import. Even the most remote villages have at least pieces of steel machetes, enormously more effective in their slash-and-burn agriculture than the small stone axes still to be found throughout their territory. The European epidemic diseases reached many tribes well ahead of the Europeans, with massive loss of life necessitating village realignments. These epidemics often altered tribal relations and resulted in the absorption of small tribal remnants by more robust groups.[3] The Yanomama were less touched by such developments than any other of the tribal groups we studied. WE BELIEVE THAT IN MANY DETAILS OF THEIR MATERIAL CULTURE AND SOCIAL ORGANIZATION, THE YANOMA-MA ARE PROBABLY AS CLOSE TO THE SOCIETIES OF EARLY, EVOLVING HUMANS AS CAN BE FOUND TODAY. They present an imperfect mirror of the later stages of human evolution, the mirror cracked and dusty, but as accurate as any we have. Other current populations of this type are even more disturbed than we found the Yanomama to be in 1966.

The early expeditions to the Yanomama were all male affairs, but, in 1971, Priscilla joined that year's expedition (as well as several to follow), not as a bystander but as a working member: a major chore in the field was the accurate labeling of our blood, urine, saliva, and stool samples— and this became her job. On occasion, when our casual masculine approach to cooking became too much to bear, she took over. On that 1971 trip, as usual, we flew into Esmeralda with the Venezuelan Air Force, then took to the river, and by dusk had reached the Salesian Mission at the junction of the Ocamo and Orinoco rivers, where we overnighted. Early the next day, we started up the Ocamo, and by late afternoon had reached our destination, a quite unacculturated village whose women had never before (to my knowledge) seen a white woman at close hand. We approached the village, to negotiate slinging our hammocks in whatever vacant space the *shabano* offered. Priscilla, wearing a wide-brimmed hat and slacks, was immediately surrounded by a group of completely naked women who suspected she was one of them but were not quite sure. The next few minutes were a classic exercise in nonverbal communication. First, several, including some of the oldest, demonstrated their ability to express milk from their breasts, and pointed to her chest. She shook her head. There was a momentary pause, then one dashed off, to return with a male baby. Pointing to his penis, she then pointed to the appropriate part of Priscilla's anatomy. Again Priscilla shook her head "no," and now she was classified.

That night we hung our hammocks in a row, with Priscilla's on the far end, next to mine, this in an effort to provide a semblance of privacy for the other members of our group. All through the night (as we learned the next morning) a steady succession of mostly young boys and men were poking and prodding her hammock—how does a white woman feel? The following night we made the obvious adjustment—Priscilla's hammock was encircled by those of the rest of the team.

Indian villages seem to swarm with small, scrawny dogs who never really accepted us and were at best a noisy nuisance and at worst nipping at our heels and occasionally making contact. They especially bedeviled Priscilla. What with dogs and Indians, it seemed best that when she went off into the bush to answer the call of nature, I go along, swinging a stick at the dogs. Nothing escapes Indian eyes in this setting. We would leave the village to cheers and whistles from the Indians. In a few days Napoleon informed me that since any time an Indian male walks a woman into the jungle it is for quite a different purpose, I was rapidly establishing a reputation for virility. Shades of Nagasaki.

Studies on Other Tribes

The bulk of our fieldwork among Amerindians has involved the Shavante, the Yanomama, and a tribe immediately to the northwest of the Yanomama, the Makiritare. However, we have had lesser contacts with nine other tribes of Brazil, Venezuela, and Guiana, namely, the Baniwa,

Cayapo, Kanamari, Kraho, Macushi, Panoa, Piaroa, Ticuna, and Wapi-shana. We have also studied in lesser detail some six Chibcha-speaking tribes of Costa Rica and Panama, of whom the Guaymi have received the most attention. The locations of these tribes are shown in the map on this page. The sample is sufficiently varied to permit (later) some cautious generalizations about tropical-dwelling Amerindians. It is important to emphasize that although the time spent among each of these latter tribes was disgracefully short by the standards of the traditional cultural anthropologist, usually 4–6 weeks, by proper organization we could in this interval collect sufficient samples to occupy the laboratory for the remainder of the year.

Wherever we worked, the first step in our transportation to the field was almost always by air. Sometimes the airstrips were close to the Indian villages in which we were interested, but often we next took to

The locations in Central and South America of the 13 Amerindian tribes that have received the bulk of our attention.

the rivers. Local transportation was usually by bongo, a canoe shaped from a single log, whose durability compensated for its unsteadiness. The stern accommodated an outboard motor; our outboards were our prime pieces of equipment. Finally, since the Yanomama are not a riverine people (only now establishing their villages by rivers for trade purposes), we often ultimately took to jungle trails, clumsily attempting to keep up with our Indian guides and carriers who seemed to skip along while we floundered.

My inability to communicate to our professional colleagues the flavor of this fieldwork proved so frustrating that on the expedition of 1968 to the Yanomama we took with us a professional ethnophotographer, Tim Asch; the resulting film entitled, "Yanomama: A Multi-disciplinary Study," earned Tim, Nap, and myself a blue ribbon at the American Film Festival in 1972 but, more importantly, seems to have caught much of the flavor of the work. I unabashedly and wholeheartedly recommend it.[4]

One of our later experiences deserves special mention. For many years, the U.S. National Science Foundation maintained a 133-foot ship, the Alpha-Helix, especially fitted up for oceanographic and related scientific research. In mid-1975 the Foundation informed me that the proposed program for the ship for 1976 would put it on the Amazon, if enough good projects could be found. Was our group interested in taking advantage of its facilities as a floating base for fieldwork? We were, and began to plan. In January of 1976 came the moment of crisis without which the normal process of planning for fieldwork would be incomplete. Those responsible for scheduling the vessel had apparently made insufficient allowance for Brazilian participation and sensibilities as they plotted its Amazonian schedule, and the Brazilians had dug in their heels with respect to the necessary concurrences. Because of my extensive (by then) Brazilian experience (plus the fact that our own proposed project included strong Brazilian collaboration), I found myself leading a three-man "diplomatic" delegation to Manaus in January. The Brazilians were polite but firm (the amenities were on their side), I made the necessary recommendations to the Foundation, and all was soon well.

Our allotted time on the Alpha-Helix was 14 July–27 August 1976. We became a team of 13—9 Americans and 4 Brazilians. Our older son James joined the group; I had promised our two sons that when they could carry a full load they could join one of these expeditions, and Jim was now a year out of medical school. We all assembled in Manaus, from which the Brazilian Air Force flew us (and 63 pieces of equipment and baggage) to Tabatinga, just at the border with Colombia. From there we proceeded to nearby Leticia, Colombia's principal access to the Upper Amazon (technically the Solimões River), where the Alpha-Helix was tied up.

We pressed the Alpha-Helix's facilities to their maximum. Priscilla and I shared a cabin just off the bridge, twin to the captain's cabin. The rest of the group, and the equipment, fitted wherever they could, and, while we were in port, two of the group stayed ashore to provide a little more breathing room for the others. We divided into teams, flying with the Mission Aviation Fellowship north to the Baniwa and the Yanomama, south to the Panoa and Kanamari. Then, in phase II, the ship dropped down the river in stages, stopping at the riverside villages of the Ticuna, until we ran out of Ticuna villages. The arrangement was not only convenient and efficient, but because of the facilities of the ship, we could carry out some studies not previously feasible. Thus, in addition to the usual collection of genealogies, blood samples, and physical measurements, we did special studies on how well they handled a sugar load (the well-known glucose tolerance test) and the severity of their parasite burden, and prepared white blood cells for special studies back home. This was fieldwork in the style to which I would have liked to become accustomed some ten years earlier.

This is perhaps the appropriate place to discuss the selection of field colleagues for such undertakings as these. I have had several unfortunate experiences in the field with staff who were "pillars of society" and "paragons of excellence" in their professional setting, but rather overwhelmed in the field. There is a class of physicians who have for so long been functioning on the basis of constant social reinforcement—going back to premedical days—that they do not do well in a setting where not only is this reinforcement withdrawn but so are their other social landmarks. My early experiences led me to look carefully into the origins and hobbies of potential associates—how near to the farm were they, and how did they feel about hunting, fishing, and backpacking.

I personally have no use for hunting but a strong commitment to the pursuit of trout. Success is measured not so much by the size of the catch as the beauty of the stream on which one finds oneself. There is no more pleasing sight than the gentle blanket of mist that hovers over a good trout stream when you approach it at dawn. Trout fishing is best done alone. The rocks are always slippery. Any stream of moderate size has its holes; going over one's waders is a standard hazard. It's a good idea to tell someone where you're going, but it's not uncommon to change your mind and fish in an area where should mishap occur, it's a while before you're found. Several times after a fall I've momentarily wondered how I would get back to home or camp, as the case might be, but one does get back and then returns to the stream as soon as possible. Somewhat lonely experiences of this nature are good preparation for the jungle. Fortunately, by the latter half of this work I had a superb field team.

The fieldwork also drew heavily on temporary alliances with the sturdy, fundamentalist missionaries who are active among these tribes. As should be clear by now, I do not share their religious convictions. Even so, our common concern for the Indians provided a bond that let us work together effectively. On a number of occasions, one of the most valuable members of our team would be a young, second-generation missionary, born while his parents were out on leave, reared as a child with the Indians, and spending his school vacations back at the mission post with the Indians. Such people were thoroughly at home in the jungle, able to converse fluently with members of the tribe among whom they were reared. They had a bit of sustaining psychological edge on the scientist members of our team: they were absolutely convinced that in case of a serious mishap, they went straight off to a beautiful reward, to be joined in due time by their loved ones; we were much less sure of our ultimate destination.

The Tropical Rain Forest and Its "Dangers"

I will not attempt to describe the tropical rain forest in which so much of our work was conducted. Undisturbed, it is truly awe-inspiring, leaving one grasping for the appropriate descriptive metaphors. But, I must agree with Richards in the statement, "Tropical vegetation has a fatal tendency to produce rhetorical exuberance in those who describe it."[5] Any thoughts I had of attempting to capture the essence of the forest before turning to the results of our detailed studies of its people vanished when I stumbled on "Tropical Nature" by A. Forsyth and K. Miyata;[6] their fact-filled free verse has caught so perfectly (and so much better than I could) my overall response to it that, rather than offering a feeble paraphrase, I urge you to read the original.

Our lifestyle brought us very close to the forest. When we were traveling, we slept beneath a tarpaulin suspended over a frame of saplings and tied at its corners to trees, but open on all sides. On "location," whenever possible, we simply occupied a vacant section of the *shabano*. The forest itself is really very safe, but there are some precautions one learns to take.[7] There are tree-living "fire ants" (*Ponera* spp.) that literally throw themselves on you, mandibles first, should you jar their tree. Along rivers, one encounters several kinds of ground-dwelling wasps

144

with a sting that incapacitates for 24 hours (hence the local name, *"veinte y cuatro"*), but, sensitive to vibration, they buzz warningly when their nest is approached. In sandy areas, one encounters the larval form of a flea (*Tunga penetrans*) that rapidly burrows beneath the skin, where it expands into a pea-shaped nodule, often accompanied by local infection. We sometimes acquired a few when swimming at sites used by the Indians, and part of the ritual once back in Ann Arbor, when they had matured to a proper size, was digging them out of each other's feet. The rivers contain the freshwater stingray (*Potamotrygon* spp.). To avoid stepping on one when helping a boat across a sand bar, one carefully shuffles his feet along the sandy stream bottom. Finally, vipers of various species (*Bothrops*) are relatively common, and almost every village contains one or two Indians with badly scarred and poorly functional limbs, the result of an encounter which can on occasion be fatal. Considering the time the Indian spends in the forest, these encounters are really uncommon, and we, of course, had the added protection of shoes.

The most formidable denizen of the jungle is the jaguar (*Panthera orrca*). Since we traveled in groups, we were never in any danger, but one night I thought my time had come. We had made camp at the Catholic Mission at Platanal, highest mission on the Orinoco, where an addition to the mission was under construction. The framework for the addition was up, crowned by a tin roof. Space had been cleared around the addition and the excess felled timber had been consigned to several fires. We were invited to sling our hammocks under this new roof. In the early hours, I stumbled out of my hammock to answer the call of nature. As I walked off drowsily to a respectable distance, suddenly, in the darkness just outside the beam of my flashlight, were twin glowing orbs of red, just the right distance apart to be eyes. I froze, heart pounding, hair on end. Three seconds later came the realization that I was looking at the two remaining glowing embers in a dying fire.

With respect to the diseases to which we might be exposed, there were four primary causes of concern. The most immediate was *P. falciparum* malaria, a type relatively resistant to chemotherapy (and, incidentally, a major impediment to the exploitation of the rain forest in this area). We started a regime of prophylactic anti-malarials three weeks before going into the field, and slept under netting. Two other hazards emerged from our early studies of serum antibodies and stools. We found that serological evidence of infectious hepatitis was widespread among the Indians, and at least 50% were carriers of *Entamoeba histolytica*. Later we will discuss why the latter infection may not be attended by disease among Indians to the same extent as among members of agricultural or industrial societies. At any rate, since both these

145

organisms spread predominantly through contaminated water, we were careful about our water supply. Finally, yellow fever is endemic to much of the tropical rain forest; we were all vaccinated.

In principle, the Indians themselves must be classed as one of the potential dangers of the forest. The tribes among whom we did most of our work were not reservation Indians, but people who still regarded themselves as free people and sovereign. We were in their territory, on their terms if we wished to proceed. We were of course determined to observe their customs, to give no offense. Even so, there were a few little surprises. Coming around a bend in a trail one day, I looked up to see a young Indian some 30 meters away, bow drawn and arrow cocked and unmistakably aimed straight at my chest. I was virtually sure it was his idea of a practical joke, and walked smilingly on, but, just as in our society, there must be a few sick minds in any Indian tribe. Then there was the time, in a remote village, when the Yanomama who was traveling with Nap and myself overheard the young bucks discussing a plan to overpower us and steal our trade goods that night. At dusk Nap casually blasted the tips off a tree branch overhanging the *shabano* where we were sleeping, and we retired with the shotgun leaning against his hammock—to a quiet night.

Finally, an unsought distinction is being the only practicing population geneticist "assaulted" by an apparently irate Indian chief. We had finished our work in a Makiritare village and were packing to leave, myself sitting on a bench next to Charlie Brewer,[8] he cleaning and washing his dental casts in a bucket of water, myself rearranging the contents of one of our footlockers. Suddenly the village chief appeared, followed by about half the village adults. Apparently angry, he began a loud tirade. One of the Indians who spoke a little Spanish gave Charlie a running translation. "The chief says you have been stingy; you haven't given us enough in return for our blood samples." We had in fact paid off exactly as agreed. But the chief knew we had a supply of trade goods left, for the next village on our schedule. I knew it was his role to press for the maximum possible, and through Charlie and our interpreter I stated firmly, "We have done as we agreed!" The tirade continued. Charlie whispered, "This is serious!" Suddenly, the chief, a powerful man, stepped forward, grasped us by our adjacent shoulders, and shoved. Charlie's rear slid neatly into his pail of water; I sprawled. I was convinced he was bluffing, and, getting to my feet, said again, "The chief knows we have been fair." At this his apparent anger suddenly evaporated—the bluff was not working—he laughed and said, "Big Indian joke." I laughed and said, "Come to my country, and I will play a big American joke on you!" Charlie translated this into proper Spanish, his Indian friend put it in Makiritare, all

146

the Indians laughed, the tension was gone, and in the end they pitched in to help us load our bongos to travel.

Paradoxically, the greatest source of mental stress on these expeditions was the very factor that made our undertaking possible: airplanes. Airstrips have usually been in short supply where we have worked, and weather services nonexistent. Planes to get specimens or personnel out (or in) often were held up, and the prospect of losing a week's collection because of difficulties with transportation is guaranteed to wake one up several times during the night. How often have I envied the members of expeditions whose "collectibles"—plants, insects, artifacts—could be preserved indefinitely, permitting a much less driven schedule. Each time—no matter how often we had done it—that we got a shipment out in good shape, it was an occasion for a small celebration albeit, since I ran a "dry" camp, nonalcoholic. The alcoholic celebration came later, when we returned to Ann Arbor, and learned the specimens had not only gone out in good shape but arrived in good shape. In all, we only lost one shipment.

The reader should not be misled by this effort to introduce a little realism into this description of our fieldwork. The fact is that, day or night, I felt substantially safer in the rain forest than in many cities of the United States. We lost no days to significant illness during any of our fieldwork. The only "epidemic" to afflict our little group occurred during that expedition on which we brought in two cases of a canned orange-flavored drink, our touch of luxury for that trip. IVIC, our Caracas base, is at a comfortably cool elevation; the Orinoco basin at Esmeralda, where the Venezuelan Air Force deposited us, is just plain hot, especially if you are working at full speed to transfer supplies from plane to bongo, in an attempt to get upriver before nightfall. We grossly overindulged on orange drink, which, it developed, was heavily citrated. The bongo made several unscheduled stops going up river, and it was a long night, as our saline purge took effect.

Young Yanomama male awaiting his turn to have blood drawn (1969).

9

Health and Disease in Recently Contacted Amerindians

Modern medicine, in the great citadels of medical science, does reasonably well at measuring disease, but not that well at measuring health, which is as much a frame of mind as a set of physical attributes. How presumptuous, then, to hope to evaluate health and disease among unacculturated people under the circumstances just described. What we tried to develop were a series of revealing "health probes" which were acceptable to the Indians, who had little or no insight into our strange ways. These probes could be successfully pursued under field conditions (backed, of course, by the appropriate laboratory facilities). In what follows I will present a half dozen vignettes, snippets from a much larger collection of data, which collectively will illumine some impressive differences between the dietary and disease patterns of Indians and ourselves. In the final chapters I will have much to say about the care and feeding of the gene pool—this and the following chapters are meant to contribute perspective on how far modern man has departed from the circumstances under which he evolved and some of the possible genetic consequences.

A word has to be written about the Indians' receptivity of these studies. The pokings and proddings that accompanied our physical examinations, our interest in their blood, urine, and stools, must all have seemed very strange to the Indians. Granted that it was a "cash" transaction based on trade goods, I nevertheless feel that our subjects were much more accepting of us than we would be in a comparable situation (could we envision it) back in the United States. Indian culture, in short, impressed me as more tolerant in many respects than our own. This was certainly the experience of the early settlers. It was only after repeated abuses that the stereotype of the hostile Amerindian emerged.

Appearance

We have performed reasonably complete physical examinations on over 300 Shavante and over 500 Yanomama Indians, and looked at a lot more Indians of various tribes in the course of our studies in the 90 Indian villages in which we have now worked.[1] The males, in general, present a picture of exuberant vitality, an impression confirmed by their dancing and chanting frequently extending through most of the night. (We sometimes wondered when they slept, as we tried to meet our own needs.) The women, by contrast, often appear prematurely aged. Signs of trauma—to limbs, scalp, and eyes—were common to both sexes. Much of this was of human origin—in the men, scars from arrows and club fights; in the women, burn scars from the application of a glowing ember by an irate husband. In the examinations of the Shavante males, we were initially puzzled by a universally present callus on the right shoulder. Sudden insight came the day we witnessed our first *buruti* race. This is a relay race, involving two teams, the "baton" a section of palm tree trunk balanced on the right shoulder. After the race, we found one of the "batons" to weigh 75 kilograms and the other 85! All the adult males participate!

When I write of a picture of exuberant vitality, it must be remembered we are seeing only a snapshot in time. Furthermore, some of those who were ill might not present themselves for physical examinations. The true picture of health and disease can only be derived from a longitudinal study. While we have been in direct contact with Indian villages, there have been only a half dozen deaths, of both young and old. The records which some of the missionaries have kept are of value in deriving a more rounded picture; a little later we will present our current estimate of the Yanomama life tables. I have the impression that death when it visits moves quickly in these societies, with a speed not entirely explained by the sanitized prolongation in and out of hospital of a fatal illness in our own culture. The candle of life, burning brightly one minute, is snuffed out the next.

Blood Pressure

The most notable findings of the general physical examinations of the Yanomama were that, in contrast to the findings in industrialized populations, pulses were relatively slow (60–70/minute) and blood pressure leveled off at about 105/65 in the late teens, rather than continuing to rise during subsequent decades as it does in industrialized communities. The Shavante were similar, but with slightly higher blood pressures throughout adult life (about 120/70); this may relate in part to their more robust body build. True, there were not many senior citizens to be examined, but, as will become apparent, we believe hypertension, so common in the civilized world, is very uncommon.

Vision

Visual acuity, as measured by the Snellen Eye Chart for illiterates, was superb. Among the adult males, 62% tested at 20/15 (reading correctly at 20 feet what in our culture is normally read at 15 feet), and 23% tested 20/10.[2] This is an astonishing performance even when allowance is made for the fact that the Snellen test is less demanding than the corresponding test for literates (letters, numbers). When a lesser visual acuity was encountered in an Indian, it could usually be attributed to specific disease of the cornea, lens, or retina of the eye. The women did less well, in large measure since they often did not seem to understand or enter into the test.

Colorblindness (all types) was much less common than in most Caucasian populations, being some 2% in the approximately 1000 males we tested in this respect, in contrast to the 8% in most Caucasian groups. Almost half of the colorblind men we encountered were clustered in two adjacent Yanomama villages, some related as brothers, leading us to suspect a recent mutation in an ancestor of the group. Thus, our studies confirmed a number of other reports of the keen vision and low frequency of colorblindness in tribal people. (There are exceptions to this generalization, and the tests for defective color vision we used in the field are less demanding than those often used in surveys of civilized populations, but the generality stands.)

Caries and Malocclusions

We were impressed during our first round of fieldwork among the Shavante by the near absence of caries and the generally superb dental occlusion, a finding to be confirmed in later fieldwork. This led us to an interest in the epidemiology of dental caries, and the incorporation of a dentist into the field team on several occasions. The paucity of caries was all the more impressive because of the high frequency of gingivitis and receding gum lines and the poor dental hygiene (no toothbrushes, of course), and the marked erosion of the teeth due to the coarseness of the diet and its frequent contamination with soil.

This relative freedom from caries disappeared rapidly upon contact with missions. Among the Yanomama we defined a most unusual situation. The lowest point on the Orinoco at which the Yanomama are found is where the Orinoco is joined by the Ocamo River. At this confluence there is a Salesian mission, established in the mid-50s. Here dental caries were already rampant among children. In the course of our early fieldwork, our Venezuelan dentist-associate, Charles Brewer, had observed that as we ascended the Ocamo, caries were much less frequent at each successive Indian village with which we made contact. This led us, in 1971, to incorporate a dental epidemiologist from the National Institute of Dental Research, Charles Donnelly, into the team.

At that time, there was great interest in the role of a particular species of bacteria, *Streptococcus mutans,* in the etiology of caries. We wanted to test the hypothesis that the mission had inadvertently brought this bacterium to the Indians, with the resulting increase in caries. It was not an easy hypothesis to test, since *Streptococcus mutans* is a rather fragile bacterium, and we had to arrange special down-river carriers and plane flights to get our mouth swabs back to the appropriate laboratory in the National Institute of Dental Research as soon as possible. Our observations included a village at the Mission, a group of five seldom-contacted villages some two-days travel up the Ocamo river, clustered around a stretch of rapids through which we could not haul our heavy bongos, and two villages at intermediate locations between the Mission and the rapids.

Back in the laboratory, our specimens analyzed, our hypothesis rapidly collapsed: *Streptococcus mutans* was equally frequent in all villages.[3] The only possible alternate explanation of the finding (thus far) stems from the fact that the Mission was importing for staff use small amounts of refined flour and sugar, some of which undoubtedly found its way into the mouths of the Indians living in the village near the Mission. Sugar cane was also occasionally available at the Mission. Indians from the "intermediate" villages often visited the Mission, but those from the truly up-river villages did not. We postulate that these refined substances, lodging beneath the dental plaque so common on Indian teeth, disturbed a delicate balance. Given how very little refined food the Indians in the "intermediate" villages seemed to have access to, this balance would have to be delicate indeed.

Nutrition

Although some of the young women were pleasingly plump, we saw no obesity among relatively unacculturated Indians. In Africa, I had witnessed the result of chronic protein malnutrition among children, a disease termed "kwashiorkor." I never saw this among young Indian children. As we became more familiar with the food patterns of the Indian, we understood why. The Yanomama, for instance, will eat almost any form of animal protein. The larger animals—tapir, monkey, capybara, and armadillo—are favorite protein sources, but birds, lizards, snakes, frogs, and very small fish (the latter poisoned in streams) are quite acceptable. Honey is a greatly prized delicacy, but when a bee tree is found, the grubs are also eaten. In season, a type of caterpillar which clusters together in large numbers in communal webs is consumed in quantity, as are the beetle larvae which can be dug out of the trunks of decaying trees. The Yanomama also cultivate a palm tree, *Guilielma gasipaes,* whose fruits are a relatively good source of protein, and collect other jungle fruits and nuts. All of the Indians among whom we have worked are now deriving the bulk of their calories from the cultivation of plantain and/or manioc and/or maize. I can believe that prior to the introduction of these cultivars, there were periodic food shortages, but, given such an eclectic diet, doubt that *chronic* protein deficiency, so common in the Third World, was often a problem for early human populations.[4]

We have conducted a number of studies of the constituents of the blood serum of the Shavante and Yanomama. One of the striking findings was the low cholesterol values. At the time (1964–1966), the average cholesterol values we observed in the Shavante were the lowest on record.[5] Two other findings of note were the high–normal levels of serum albumin and the elevated levels of gamma globulin. The serum albumin levels confirm the apparent good nutritional status of the Indians. These observations are comparable to those by others on relatively undisturbed hunter-gatherers practicing limited agriculture. The gamma globulin fraction of the serum proteins contains the antibodies, and these high levels undoubtedly reflect the pressure of the many viruses and parasites to which the Indians are exposed.[6] As we will be discussing later, infant mortality is probably lower among unacculturated Amerindians than among many agricultural populations at the turn of the twentieth century. We attribute this lower mortality to the pattern of child spacing and prolonged breast feeding with, following birth, passive transfer to the infant of a relative immunity to many local diseases, due to the high gamma-globulin content of the mother's milk.

Among the reservation-dwelling Indians of the United States, diabetes mellitus is rampant.[7] For example, some 50% of the much studied Pima Indians of Arizona above the age of 35 meet the criteria for diabetes mellitus. This had led to speculation about a special propensity of the Amerindian to develop diabetes. However, if obesity is defined as a weight over 125% of what would be considered normal for height, then 43% of Pima males over 35 are obese, as are 79% of Pima females. Given the well-known correlation of obesity with diabetes, an alternative possible explanation of the high frequency of diabetes was that it followed the obesity. The special facilities of the Alpha Helix let us collect the appropriate samples to test how well unacculturated rain forest Indians metabolized glucose. We found that the glucose tolerance of a set of young male Indians, Marubo and Yanomama, whose average age was mid-twenties, was essentially normal and significantly better than that of a comparable sample of the Pima.[8] We have no data on the frequency of diabetes mellitus in the older Marubo and Yanomama. On the basis of this limited observation, however, we surmise that diabetes mellitus is rare among unacculturated Indians.

Some Differences in Mineral Metabolism

Given the unusual findings regarding blood pressure among Amerindians, and given the current interest in the relation between salt intake and hypertension, we decided to look into the salt balance of the Yanomama. We know that prior to the advent of the missionaries, the Indians did not add salt as such to their diet, and that there were still many villages without access to salt. In studies in which William Oliver played a lead role, we found that in these latter villages, the amount of salt in 24-hour urine samples from a group of male Indians was only 1% as much as in the urine of members of the expedition (whose values were very similar to those of the average U.S. adult). Even that average 1% value in the Indians was quite influenced by a few "outliers" with higher values, whom we suspected of surreptitiously helping the expedition's cook dispose of any excess food. Without these outliers in the series, the average value would be well below that 1%. Since the Yanomama were obviously in salt balance, this implied an amazingly low intake and highly efficient retention. Salt retention is normally regulated primarily by two hormones, namely, renin, formed in the kidney, and aldosterone, a product of the adrenal cortex. We found that the Yanomama urinary aldosterone excretion per 24 hours averaged 10 times higher than the U.S. normal, while plasma renin activity was twice as high.[9] In the U.S., when aldosterone values like that are encountered they usually indicate the presence of an adrenal tumor or some obstruction to the blood supply of the kidney. We believe this low salt intake plays an important role in the absence of hypertension that we documented.

This result with male subjects immediately set us to wondering how Yanomama women met the salt demands of pregnancy and lactation. A successful pregnancy requires the retention of some 30 grams of salt, and each liter of human milk contains some 0.3 grams of salt. On a later expedition, we were able to obtain urine and blood samples from a relatively small sample of Yanomama women, most of whom were lactating and four of whom were pregnant. The aldosterone and renin values were higher than any recorded in the medical literature. The human ability to conserve salt when necessary is fantastic. This is really not surprising, when we recall that our omnivorous primate ancestors (and presumably early man) of course added no salt to their diet and relied much, much more on plant than animal food. The high salt intake now so widespread throughout the world virtually shuts down a hormonal activity that was until recently vital to our survival.

A second aspect of mineral metabolism to attract our attention was the subject of trace metals. One of our colleagues in the School of Public Health, Bertram Dinman, interested in the increased exposure to a variety of trace metals alleged to accompany industrialization, persuaded us to take special collecting equipment into the jungle on the expedition of 1971. Back in Ann Arbor, the blood and urine samples were analyzed for their copper, lead, cadmium, and mercury content. We expected, of course, to establish low, low, baselines, against which to contrast the higher values resulting from the exposures of civilization. Lead and cadmium levels were indeed markedly lower than in a suitably matched population in Ann Arbor; copper was about the same. The surprise came in the mercury levels. Serum mercury levels were significantly higher in the Yanomama than in Ann Arbor residents, the finding largely due to the males of certain villages. We have no really satisfactory explanation of the finding, which we are convinced is valid, evidence that "natural" exposures to mercury long preceded the Industrial Revolution.[10]

Finally, some interesting findings regarding iodine metabolism should be mentioned. Goiter due to iodine deficiency has long been a problem to the populations of the interior of South America, including some of the Amerindians. In 1958 Marcel Roche of the Venezuelan Institute for Scientific Investigations, in an enterprising and difficult study, found among a very isolated sample of Yanomama and Makiritare Indians to whom a small dose of radioactive iodine had been administered, that on average 71% of the dose was concentrated in the thyroid within 24 hours. The "normal" standard, based on studies of civilized populations, is 30% to 40% uptake in 24 hours, and such high concentrating abilities as he had observed are usually seen in persons with goiter, with or without signs of hyperthyroidism. Only one of these Indians had a goiter, and none showed clinical signs of an overactive thyroid. The drinking water was found to be quite low in iodine, at the low level at which goiter commonly occurs in more civilized populations. In repeat studies in 1962 and 1968, Roche and colleagues found that the uptake had dropped to 60% in two different villages in contact with missions, but in a third, still very isolated village, the uptake was as high as before.[11]

We ourselves had one experience with goiter in an Indian village which left me impressed with how delicate the natural balance may be. We have performed physical examinations in three Shavante villages. In two of these, in keeping with the findings among the Yanomama, there was no thyroid enlargement, but at a third village, 25 among 209 Indians had definitely enlarged thyroids. All the enlarged thyroids were found among the 111 persons whose estimated age was greater than 14 years. The enlarged thyroids were all smooth and soft-to-rubbery to palpation, suggesting recent enlargement. Thus we had come to the village not long after some mysterious goitrogen had been introduced—but our limited epidemiological inquiries produced no clues.

Although the fact that in an experimental setting some foods were goitrogenic has been recognized since the beginning of this century, more recently evidence has been accumulating for a variety of other poorly understood "natural" goitrogens, some in the drinking water of towns and cities. We must assume that these goitrogens were originally absent among the Yanomama, and the high iodine uptake of their thyroids represents the norm for man in his early days. These poorly understood goitrogens had just been introduced to the village mentioned above. There is very little evidence for the occurrence of goiter in pre-Colombian times, the few illustrations in pottery figurines serving to document that sporadic cases were known but by no means establishing a high frequency of the condition. The effects of the goitrogens can be offset by increased iodine uptake. Thus, the goiter once so prevalent in the Great Lakes Region of the United States and other parts of the world has largely disappeared with the introduction of iodized salt. These studies, like those on salt metabolism, illustrate how very efficient the human body can be if the need arises.

Parasitic, Bacterial, and Viral Diseases

A standard approach to a quick look-in on the diseases to which a population is exposed is a "serological profile." With the remoteness of the Shavante and Yanomama, we truly didn't know what to expect.

Serum samples obtained early in the course of our studies were most instructive. Let me just mention some of the high points. There was abundant evidence of exposure to two groups of viruses of great medical importance, namely the enteroviruses, of which the poliomyelitis virus is a well known example, and the arboviruses, of which the yellow fever and encephalitis viruses are the best recognized. About 75% of the Yanomama appeared from serological tests to be chronic carriers of the hepatitis B virus, an unusually high percentage. A high proportion of those tested possessed antibodies against various types of *Salmonella,* a common bacterial cause of acute dysentery, and also against the well-known streptococcii.

We were at first puzzled to find antibodies against the diphtheria bacterium, when we (and others) could obtain no history of the characteristic and dramatic inflammation of the throat that characterizes clinical diphtheria. However, in our later studies of skin infections (tropical pyoderma) in the Yanomama, Dale Lawrence and his laboratory colleagues found the diphtheria organism to be growing in some of these skin infections, and we believe this to be the source of the stimulus to produce antibodies. It is tempting to speculate that such skin infections early in life resulted in the elaboration of antibodies protective against infections with the occasional more virulent strains of diphtheria later in life.[12]

We thus saw evidences of frequent past infections with the kinds of bacteria and viruses which either have animal reservoirs or remain active in people for long periods of time without apparent disease but during which the infected person is contagious. On the other hand, in the Yanomama, there was no evidence of past infection with the common epidemic diseases: measles, mumps, pertussis, and, of course, smallpox.

Finally, most of those tested showed antibodies against malaria, and examination of blood films identified all three of the common causes of malaria, *Plasmodium vivax, P. malariae,* and *P. falciparum.* The latter, which we discussed in connection with sickle-cell disease, was probably introduced into the New World with the early African slaves. However, although difficult to document with the desired accuracy, *P. falciparum* malaria may not have reached many Yanomama villages until relatively recently, so that our vital statistics (next chapter) are to some extent premalaria and certainly premeasles (see below). We encountered one village in which the uniformity of severe infection suggested this might be the very first contact of the village with *P. falciparum* malaria.

A second time-honored and standard approach to evaluating the disease pressures on a population is stool examination, supplemented by an effort to culture some of the kinds of bacteria found in the stools.

On two occasions we collected stool samples from a goodly number of persons in two Yanomama villages. One of course does not bring the entire sample back to the laboratory, but only a small portion, in a preservative. Processing these samples in the field, I was greatly struck by the highly fibrous but *relatively* odorless character of Indian stools. In this connection, I note that I grew up medically at a time when medical students did the routine blood, urine, and stool examinations on all patients. This led to a certain undesired expertise on the attributes of stool samples. Indian stools were different.

The examination of the stools for parasites was carried out by personnel of the U.S. Center for Disease Control in Atlanta, Georgia.[13] They were a parasitologist's delight. Eleven different parasites (or their eggs) were observed, with the number of different parasites per individual ranging from 0 to 9. Since only a single stool was examined, the data represent a minimal estimate of infection rates. Roundworm, hookworm, and pinworm were the most common parasites—not unexpectedly—but the finding that 50% of the Indians harbored *Entamoeba histolytica,* the cause of amoebic dysentery, was not anticipated. I find myself wondering whether on this high fiber, rapid-intestinal-transit-time diet, histolytic amoebae play a different and usually more benign role than attributed to them in textbooks of tropical medicine. As I write this, the advantages of a high-fiber diet are being rediscovered (and extolled) here at home as an antidote to the relative fecal stasis of modern man, this latter thought to play a role in the high incidence of bowel cancer. It is true that high-fiber diets lead to rapid intestinal transit times; I would have to guess that among the Yanomama the *E. histolytica* were scoured out with an efficiency inconsistent with establishing pathogenic colonies.

One can make a rough guess as to how heavy the body burden of a particular parasite is by the frequency of the parasite or its eggs in the stool. It appeared that although parasitization was widespread, the individual infections were not especially heavy. This was particularly true for roundworm and hookworm. In some parts of the world, roundworm populations build up in the intestines of children to such concentrations that they interfere with nutrition, and hookworms increase to the point where they may cause anemia. Both these infections are contracted by contact with feces-contaminated soil containing the eggs or larvae of the parasites. In time, in long-settled agricultural villages, where privies are not well designed, cared for, or utilized, soil contamination may be staggering. We explained our findings in the Indians with the thesis that although the intimacy of village life led to a sharing of parasites, the Indians' custom of moving the village every three or four years and of stooling wherever in the forest the need overtook him, together resulted in a low concentration of infective eggs or larvae in any given spot, in consequence of which the total body burden of parasites was held down.

We became acquainted with one of the most ubiquitous of Yanomama parasites quite serendipitously in the course of our study of Yanomama chromosomes. The preparations were made in Venezuela but the slides read in Ann Arbor. There was a great to-do in the lab one day when an experienced technician announced firmly that she had found a worm in one of her preparations. Among the incredulous crowding around was a Korean physician learning cytogenetic techniques, Kyoo Wan Choi, who immediately made the diagnosis that this was the immature form of a filaria parasite. These immature forms (microfilaria) are commonly found in the blood, whereas the mature forms live in the tissues. This little dandy had the misfortune to be suspended in the same blood sample obtained for the chromosome studies.

The laboratory then found many more of these microfilaria, and in due time we teamed up with two parasitologists, P. C. Beaver and T. C. Orihel, for their identification.[14] They proved to be of two types, *Dipetalonema perstans* and *Mansonella ozzardi,* both known to be endemic in South America. Nunes de Mello, one of the Brazilian participants in the fieldwork along the Amazon aboard the Alpha Helix, described in the preceding chapter, was a parasitologist. He demonstrated, again and again, the presence of dozens of these microfilaria in small snippets of skin obtained from apparently healthy persons. Even the adult forms of this parasite did not appear to cause serious disease. This is the hallmark of a successful parasite—it gets along well with its host.

One of the best-known bacterial inhabitants of the human gut is the hardy *Escherichia coli,* so extensively studied by the molecular geneticist. Although the organism is usually of no medical significance, occasional strains cause diarrhea. To identify these strains, as is necessary in understanding the source of an epidemic of diarrheal disease, antisera that will type all of the varieties of *E. coli* have been developed. In 1967, there were 147 antisera in the "standard" battery of typing antisera (the 0 series); these 147 antisera had been developed during the study of outbreaks of diarrhea all over the civilized world. That year we decided we should find out how the *E. coli* of the Yanomama compared with those described throughout the world. To this end we enlisted the help of W. C. Eveland, a bacteriologist in the University of Michigan School of Public Health with vast experience in these matters.

To make a long story short, we returned from the field with 72 stool samples, from which Eveland isolated 432 lines of *E. coli*. To characterize them, he had available all 147 of the then-described typing sera. These would have typed 99%-plus of the lines isolated from the populations of the world's great capitals. To our surprise, less than half of our isolates (204) were typable; these fell into 54 strains of which 8 had been associated with diarrhea in outbreaks in various parts of the world. But the scientifically exciting finding was so many untypable isolated. Eveland selected at random 13 of these untypable lines, from which he then developed typing antisera. These antisera all proved to be different, and with these additional antisera he could type another 50 isolates—leaving 178 isolates still untyped![15]

This was as far as we chose to take the problem. Obviously the *E. coli* of the Yanomama were a very strange lot. That this was not a Venezuelan peculiarity was shown by the fact that 169 *E. coli* isolates obtained from stool samples submitted for parasitological study in Caracas could all be typed with the standard battery of sera. There is a vast literature on how the bacterial inhabitants of one's gut influence in various ways one's functioning and well-being. I can say only that the internal milieu of the Yanomama, on the basis of this one probe, is as different from that of industrialized populations as are so many of the other factors I have been describing in the external milieu.[16]

So what did we learn from these studies? The foregoing (and other findings) suggest a population under heavy pressure from many diseases, and yet this was not the impression from our physical examinations. These Indians appeared as healthy as any U.S. population. Three specific points deserve comment. First, from the evolutionary standpoint, our observations on these Amerindians help define a norm for humans that has obtained for millions of years, departures from which may create new evolutionary pressures even as old pressures diminish.

Second, the great majority of the diseases to which these populations were exposed were endemic rather than epidemic. Epidemic diseases, like measles and smallpox, sometimes called "herd diseases," are usually diseases of short duration with no known reservoir in nonhumans, requiring for their survival large population concentrations in which new susceptibles are constantly appearing to take the place of those in whom infection with the disease has resulted in immunity. The concentrated populations of industrialized nations require immunization programs or readily available treatment to keep these diseases in check; when these immunization programs break down, as continues to happen, the results can be disastrous for the population. The small and isolated Indian populations do not present the conditions that support epidemic disease.

Endemic diseases, on the other hand, are those that can persist in small populations, either because like the parasitic diseases they are chronic and so transmissible over an extended period of time, or because the responsible organism has an animal reservoir from which from time to time it escapes into humans. This may have an important implication for the role of the bacterial and viral diseases as agents of natural selection. To the extent man was shaped by the pressure of contagious disease, it was in the early years of human evolution by endemic rather than epidemic diseases. A newborn comes in contact with these agents of endemic disease immediately following birth. However, since a mother transfers antibodies to these diseases in her milk, especially during the first week of life, a newborn is protected by a passive immunity to these diseases even as it builds up its own antibodies. There is thus a relatively smooth transition from passive to active immunity, or to a "steady" state, with less of a toll from fulminating disease than we now associate with many of these infections in a more civilized setting.[17]

The third major medical impression concerned nutritional balance. By virtue of a very catholic attitude toward what was edible, especially with respect to sources of protein, the Indians we have studied appeared to be very adequately nourished. Later in this book, I will speak to the scourge which chronic malnutrition has become throughout contemporary society. If we consider how the human condition has evolved since tribal days, it is difficult to take pride in a world some 20% of whose inhabitants are malnourished.

A Measles Epidemic

As noted in the preceding section, in the early years of our fieldwork our group studied rather extensively the serum antibody profiles of various Yanomama groups. One of the salient observations was that with the exception of a few members of one village, the Yanomama had not been exposed to measles. This was a striking testimony to their isolation but it also posed a major potential threat, since the tragic impact of measles on what is termed a "virgin soil" population has been well documented. In 1968 we learned, just before our departure for the field, that measles had been introduced to the Brazilian Yanomama. Accordingly, for the fieldwork of 1968 in Venezuela we brought with us 2000 doses of measles vaccine, with the intent that, although we ourselves would administer it to any of the very isolated villages we reached, in general we would place it in the hands of the missionaries, who would immunize the Indians with whom they were in contact. While this would not protect all the Yanomama, it would set up a "fire break" at the most obvious portals of entry of the disease.

This rather academic exercise abruptly took on a new significance when measles was introduced to the Venezuelan portion of the tribe coincident with our arrival in the field. The Orinoco, at about the junction of its upper and middle stretches, is joined by the Casiquiare River, which connects the Orinoco with the Rio Negro, a tributary of the Amazon. (I cannot vouch for the entrenched belief that the Casiquiare flows one direction part of the year, then reversed itself the remainder of the year.) This river, which in effect makes an island of northeastern South America, has been since earliest times a well-traveled inland waterway. Shortly before we had entered the field on this particular expedition, two young Brazilians had made the inland passage via the Casiquiare and then turned up-river, looking for employment at the Salesian mission at the juncture of the Ocamo with the Orinoco, already mentioned as the lowest mission on the river in contact with the Yanomama. Within a few days of arrival, one of them was prostrated by a high fever, not immediately, in its preeruptive stages, recognized as measles. He had numerous contacts with Indians at the mission.

When his disease, as well as the resulting disease in Indian contacts, was diagnosed as measles, what we had projected as an academic exercise in immunization suddenly became an urgent need to attempt to develop a wall of immunized Indians in the villages around the mission. This task was not simplified by the fact that if an Indian traveler from Ocamo, perhaps incubating the disease, brought word of the epidemic to a neighboring village, its inhabitants, terrified by the prospects of an epidemic, abandoned their *shabano* for a temporary camp in the forest. Much of our carefully designed protocol for that expedition was quickly scrapped as we dashed from village to village, organizing the missionaries, ourselves doing our share of immunizations but also treatment when we reached villages to which measles had preceded us. We always carried a gross, almost ridiculous excess of antibiotics—now we needed everything we had, and radioed for more.[18]

In the middle of our efforts to control that epidemic, there occurred one of those incidents that remind one how small is the margin between tragedy and comic relief. We had traveled much of the day to reach a remote mission, where measles had preceded us. Arriving at dusk, we found dozens of sick Indians, and together with the missionaries worked until midnight caring for the ill. Then it was time for some sleep, but the only place we could hang our hammocks was in a open-sided lean-to where the missionaries serviced their outboard motors. We turned in at midnight, I, for one, exhausted. In the jungle, you generally sleep in your clothes, but allow yourself the luxury of removing your boots, or whatever your footwear is. These I always carefully parked beneath my hammock, directly below my shoulders, within my mosquito netting, with my flashlight in my left boot. That night I was awakened from layer upon layer of deep sleep by a sensation of being jostled, accompanied by strange, snuffling noises. I reached for my flashlight, turned it in the general direction of the noise, and there was a monstrous beast, which appeared to be munching on my precious mosquito netting. It was a tapir. I did what any red-blooded American would do who was semi-conscious: hit him over the snout with all my strength with my flashlight. This was a mistake; it didn't seem to bother the tapir, but the flashlight never functioned again, and, in the jungle, a flashlight is very valuable property. In the dark, in a carefully modulated voice that probably carried two miles, I informed my companions we were being attacked. I finally got the attention of the jungle-born, second-generation missionary who was sleeping two hammocks away, our guide in this leg of our work. His response was uncontrollable laughter. Finally, thoroughly ventilated, he informed me the beast was the post pet, but nobody had thought to tell us about him.

That experience with measles, in conjunction with some follow-up studies the next year, has had a profound impact on how I think about the susceptibility of unacculturated human populations to the epidemic diseases of civilization. Estimates of the size of the native populations of the Americas at the time of first European contact are notoriously controversial. If we accept Denevan's[19] figure for America north of Mexico of 4,400,000, then the 1910 U.S. census figure of 210,000, plus allowance for Canadian Indians, reveals about 95% decimation, most of it during the eighteenth and nineteenth centuries, more from disease rather than warfare. It is a medical dogma that the isolated tribal populations of the world, who when first contacted some 500 years ago proved so susceptible to the epidemic diseases of civilization—measles, whooping cough, smallpox, tuberculosis, and syphilis—have a special inborn susceptibility to these diseases. This belief, even in recent years, has salved society's conscience as these populations have continued to exhibit higher death rates from these diseases than long-civilized populations. As a result of our experience I challenge this view as overly simplistic. In this connection, I point out that rarely if ever before has a medical team like ours been in a position to record an unfolding epidemic such as this one.

When, prior to the advent of an effective vaccine, a measles epidemic swept through a civilized population, only those not exposed during the last epidemic became ill. These were usually children, with immune parents to care for them. By contrast, when an epidemic hits a "virgin soil" population, everyone goes down at the same time. The febrile phase of measles, before the eruption appears, is prostrating for adults as well as children, as many a young military recruit learned in boot camp prior to the advent of vaccine. A group of Indians, all but a few simultaneously ill with measles, is paralyzed. Mothers and their children are both ill. The impact is especially severe on infants still at the breast (the normal state of affairs until age three). Not only are they highly febrile but, as the mother's milk supply fails because she too has measles, they are also deprived of liquids. There are no pantries in the Indian village; the larder must be replenished daily. In short, with virtually everyone ill simultaneously, there is no one to administer even minimal care or nutrition. Convinced they have been overtaken by malevolent spirits, the standard Indian response is to retire to a hammock to die; the jackknife position assumed in the hammock invites the collection of secretions in the base of the lungs, followed by bronchopneumonia.

The response to any virus, such as measles, can be analyzed in two stages. The first is the immediate, acute response, with its complications—for measles, principally bronchopneumonia and middle-ear infections. The second is the immunologic response, characterized by the appearance of the antibodies that confer a lasting immunity. Our impression was that the Indian was just about as sick in the primary phase as your standard Caucasian—no more, no less. We saw no complicating middle-ear infections. I believe the reason for this was the relative absence in Indians of enlarged tonsils partially blocking the Eustachian tubes and thus setting the stage for infection. On the other hand, we saw a great deal of bronchopneumonia. Although I believe some of this was precipitated by the Indians' postural response to the disease mentioned earlier, I would not exclude some greater innate susceptibility.

Unfortunately, "experiments of nature" always lack the control of laboratory experiments. Our observations are complicated by the fact that a wave of severe upper-respiratory infection had swept through these same villages one or two months previously. This may have set the stage for the pneumonias. In this connection, there is quite a bit in the writings of explorers and health officials about the greater susceptibility of the Indian to the ill effects of the common cold, a phenomenon we ourselves have witnessed. But here too the story is complicated. Like other epidemic diseases, the common cold cannot maintain a permanent foothold in the isolated villages of a tribe. Relatively uncontacted groups do not have the history of repeated respiratory infections characterizing civilized populations. I suspect that by the time we are adults, we have acquired a partial residual immunity to the common cold, an immunity lacking in the Indian.

With respect to the secondary response to measles, our records are clear. A year after the epidemic, we found both the vaccinated and those who had been ill with the disease to have developed protective antibody titers just as high as in Caucasians—even though this may have been the first experience of this tribe with measles.

This is not the kind of situation to generate accurate mortality figures. From data supplied by the missionaries our best estimate of the case fatality rate in Brazilian and Venezuelan villages that were not vaccinated before exposure to the disease, but in which the missionaries could organize limited medical care, was about 9%. This is of course high by the standards of the civilized world, but far below the 30–40% estimated to have occurred in the contacts of earlier centuries.

We are not the first to feel that it is what we term the secondary aspects of such an epidemic that is responsible for so much of the mortality. As long ago as 1877, Squire, describing the collapse of village life during an epidemic of measles in Fiji, wrote: "Excessive mortality resulted from terror at the mysterious seizure, and the want of commonest aids during illness; there were none to offer drink during the fever, nor food on its subsidence. Thousands were carried off for want of nourishment and care as well as by dysentery and congestion of the lungs. We need invoke no special susceptibility of race or peculiarity of constitution to explain the great mortality."[20] More recently (1968), Noel Nutels, a Brazilian physician who devoted most of his life to the medical problems of Brazil's Indians, reporting on a measles epidemic in the Xingu National Park, wrote: "Of the 654 patients, 114 died. Among those who received medical care, the death rate was 9.6%; among those who could not be treated in time, it reached 28.6%"[21]

A Cytogenetic Surprise

What I now consider potentially the most exciting discovery to come out of the "Indian Program" was completely unplanned. Cytogenetic studies in many nations had by the 1960s demonstrated that when certain white blood cells (the lymphocytes) of presumably normal individuals were recovered from the body and briefly cultured, one or two cells per 100 were characterized by serious chromosomal damage, such as a single chromosomal break, or breaks in two different chromosomes, sometimes followed by an exchange between the involved chromosomes, the exchange resulting either in a translocation or a chromosome with two centromeres and a fragment with no centromere. Since the centromere controls chromosomal behavior at the time of cell division, chromosomes with no centromere are lost, whereas chromosomes with two centromeres may be pulled apart if the centromeres move in opposite directions. Some part of this damage is presumed to result from exposure to noxious agents present in the environment—radiation, herbicides, pesticides, occupational chemical exposures, and the like. As one aspect of the fieldwork among the Yanomama of 1969, we decided to attempt cytogenetic studies, the expectation being that, with the tribe's remoteness from industrialization, we would encounter less visible damage than was the case for more civilized populations.

It was no easy trick to establish a temporary cytogenetic lab at the Venezuelan Institute for Scientific Investigations and get our samples (from two quite remote villages) back to the laboratory in time for good culture results, but all went well. The slides were analyzed in Ann Arbor. To our amazement, not only did we observe more "simple" chromosomal damage in the Yanomama (about 4.0% of cells), but, in addition, out of 4875 cells examined, there were 21 (0.4%) that exhibited chromosomal damage much more extreme than Arthur Bloom, my cytogenetic collaborator, had ever seen before—and Art had come to the Department of Human Genetics after a three-year sojourn in Hiroshima, where he had conducted cytogenetic studies on the survivors of the atomic bombings. Part A in the figure on page 167 illustrates the chromosomes of a normal cell as it approaches cell division, and part B one of the abnormal Yanomama cells.

We of course published the observations.[22] The opinions of our peers were of two sorts. One view—not very complimentary—held that the cells were artifactual, somehow resulting from the unusual conditions surrounding the study. The other view—to which we inclined—held that, since it was known that viral infections of cells maintained in tissue culture resulted in chromosomal damage, we were observing the sequelae of some jungle fever. It was imperative that we attempt to repeat the findings. In 1970, we sampled two more Yanomama villages and, in 1971, resampled one of the two original villages. This time, although we saw a few of these cells (3 in 15,649, or 0.02%), the frequency was so much less than previously that it threw the original findings in doubt.[23] Some artifact arising from the unusual circumstances surrounding our study seemed the most likely explanation.

There the matter rested until 1984 when D. P. Fox and his colleagues reported that 6 of 153 divers working in the North Sea oil field exhibited such cells. The following year Janet Tawn and her associates reported the same type of cell in 2 of 12 young persons who had recently been employed in the British nuclear industry (but not yet exposed to any radiation). Tawn was in a position to make sequential observations on

The appearance of normal chromosomes in a cell approaching division (A) and a rogue cell at the same stage from an American Indian (B), a Japanese (C), and a Ukrainian (D). Note the numerous small chromosomal fragments in the rogue cells. In addition, many of the rogue-cell chromosomes have multiple constrictions of the type that indicate the presence of a centromere (some examples are indicated by arrowheads). Chromosomes with multiple centromeres often fragment at the time of cell division.

these two young men; the cells had disappeared 50 days later, vindicating our own follow-up results.

The next chapter in my own involvement with this problem unfolded about the same time as the developments described in the preceding paragraph. In the course of my annual visit to Japan in 1984, A. A. Awa, in charge of the cytogenetic studies in Hiroshima and Nagasaki, informed me that over the years he had occasionally seen these cells in the control (unirradiated) Japanese population. One of them is shown in part C of the figure. Together we undertook a more extended analysis of the phenomenon. The cells were rare: 24 in 102,170 cells screened (0.02%). Two significant epidemiological observations emerged. Firstly (as had been true for the young Englishmen), the cells were not randomly distributed among individuals: if an individual was found to have one such cell, there was an above-average probability of encountering additional such cells. Secondly, there was no association between the presence of such cells in an individual and age, sex, or season of the year, the three standard epidemiological entries into such a problem. It was at this juncture that I decided to call them "rogue" cells, in the classical biological sense of a marked deviant from the typical observation.[24]

Then, in 1990, Awa's laboratory received for blood samples from three Ukrainian villages cytogenetic studies, as one aspect of a large cooperative study on the possible aftereffects of the Chernobyl disaster conducted by the International Atomic Energy Agency in Vienna, four years after the event. I happened to be in Japan when the laboratory personnel began to analyze the slides: there was understandably great excitement when rogue cells were detected, one of which is shown in part D of the figure. At that stage of the analysis, all of the slides were coded as to origin, so that we could not check the relation of the findings to exposure to the (relatively small amount of) radiation in the regions receiving fallout from the Chernobyl disaster. A year later, on the occasion of my next visit, the cytogenetic analysis completed, we broke the code. To our surprise (and joy), all the rogue cells were found in the members of a very small control village, so that the issue was not obscured by the question of a radiation effect.[25]

So now we know that in people all over the world there is periodically a shower of highly abnormal lymphocytes. We don't know whether these showers are a once-in-a-lifetime experience or are repeated at regular or irregular intervals. I estimate that at the height of such a shower, the frequency of these abnormal cells in a person exhibiting the phenomenon is at least 1 in 100 lymphocytes. Since it can readily be calculated that in the average normal adult about 12 trillion lymphocytes are in the blood stream at any one time, this implies that at the height of a shower, there might be some 120 billion of these abnormal cells in circulation!

The great majority of these cells are so badly damaged that they could not possibly undergo a successful cell division. This point is obvious from the figure. The many small fragments in these rogue cells would be lost at the time of cell division, and the chromosomes with two or more constrictions, marking the site of the kinetochore responsible for chromosome movement, will be torn apart if, at cell division, the two kinetochores move toward opposite poles of the dividing cell. Thus, the two daughter cells from the division of a rogue cell would usually have very incomplete gene complements. However, a statistical analysis of all of the rogue cells we have observed to date reveals that the amount of damage per cell follows a bell-shaped curve. At the lower (left) end of the bell, there are cells that could survive a cell division. Also, there must be occult damage that we did not visualize with the cytogenetic techniques employed, and some of these cells with hidden damage should be able to divide and be the point of departure for an abnormal cell line. If only 1 in 1000 of the abnormal cells appearing in one of these waves has the ability to reproduce itself, at the height of a wave there would be 120,000,000 of these slightly damaged lymphocytes in circulation.

The potential significance of this finding must be considered in light of the fact that chromosomal rearrangements are frequently encountered in malignant tumors, and the genetic damage accompanying the rearrangements is thought to play a role in the etiology of these conditions. We do not yet know in what other cells of the body this phenomenon occurs. Lymphocytes are an unusually favorable kind of cell in which to detect this phenomenon because these cells can be readily induced to divide in cell culture and their chromosomes visualized at the first cell division. It would be more difficult to detect these rogue cells in other cell types. There is, however, no reason to believe the phenomenon does not occur in other cell types. Thus, one of these rearrangements in a somatic cell might be the point of departure for a tumor. The prevailing view has been that the chromosomal rearrangements so often encountered in cancer cells are the result of the chance accidents that are bound to occur in structures as complex as chromosomes. Our findings suggest, by contrast, that there is some underlying process whose activation may result in much of this damage. This suggestion of course leaves room for other causes of chromosomal damage, such as radiation or chemical action. The implications if these cells were to occur in the germ line are especially intriguing. There is a "surprisingly" high frequency of chromosomal abnormalities of a variety of types in aborted human fetuses and newborn infants (see Chapter 12). This phenomenon may make a significant contribution to these abnormalities. Finally, one of the features of evolution in both the plant and animal world is chromosomal rearrangement, and again we must wonder if this phenomenon drives some portion of these rearrangements. (I realize that the foregoing constitutes something of a flight of fancy, but, if I am wrong, hard facts will bring me back to earth as our study of this phenomenon continues.)

At the moment, I entertain two working hypotheses of the origin of these cells. The first stems from the fact that there is an important class of viruses, the so-called retroviruses, that regularly insert in chromosomes, including those of humans, as one aspect of their life cycle. Moreover, material that is similar to but not identical with these retroviruses is also scattered through human chromosomes. Under certain circumstances, when the virus is termed a retrotransposon or jumping gene, they have been shown in experimental organisms to move about in chromosomes, sometimes carrying bits of the host's genetic material with them. As briefly mentioned in Chapter 1, certain of these retroviral-like elements are well known as a cause of mutation and chromosomal rearrangement in both somatic and germ-line cells in organisms as diverse as corn and *Drosophila,* and there is no reason to believe humans are exempt from their action.[26] These abnormal cells might be the result of the activation, for whatever reason, of quiescent retroviral-like elements in chromosomes.

The second hypothesis is that the damage is the direct result of a viral infection. For instance, the infection of human cells growing in culture with the SV40 virus (which normally infects monkeys rather than humans) results in chromosomal damage very similar to that encountered in rogue cells. There are several close relatives of the SV40 virus that commonly infect humans, as judged by the presence of antibodies in human serum, but the biological significance of the infection has been poorly understood. To explore the possibility that one of these, known as JCV, might play a role in this phenomenon, Awa and I recently sent Eugene Major, a specialist in these viruses, samples of blood sera from persons exhibiting rogue cells and suitable comparison samples. Antibodies to JCV were much stronger in the samples from persons with rogue cells, a sign of recent infection. The association is so strong that it is difficult to escape the conclusion that this common virus, in ways yet to be determined, is responsible for the rogue cells. We are currently busy following up on this clue. No matter what the final explanation of this finding is, these cells are a humbling reminder of how much is yet to be learned about factors influencing our genetic material.

More Ethical Considerations

Earlier I have written of the ethical considerations of the Japanese study. Similar considerations attend the Amerindian studies. The incredible brutality and callousness of the western world toward the people with whom it made first contact during the five centuries following the discovery of the New World and Oceania—not to mention the exploitation of Africa—makes sickening reading, even in the context of those harsher times. As we examined the Indians and collected our samples, all this the basis for learned papers that would ultimately contribute to our professional reputations, were we only the latest of the exploiters, now for scientific reasons? Students have on several occasions raised this point when I have lectured on these studies.

We took great pains to introduce no disease. We treated the sick as we traveled. At the end of each period in the field, we submitted detailed reports and recommendations to the appropriate authorities of Brazil and Venezuela and wrote general accounts of our findings.[27] In 1968, I arranged a Symposium—subsequently published—for the Pan American Health Organization, entitled "Biomedical Challenges presented by the American Indian,"[28] at which a variety of health issues were discussed. I have no illusions about how effective any of this was in the long-range sense. On the other hand, further and less benign contacts than ours are absolutely inevitable, as each year efforts to exploit natural resources drive more and more people into the interior of South America. Great and final disruptions of the remaining tribal populations are imminent. Did we ameliorate the situation, even if by ever so little, and simultaneously collect data of some scientific value?

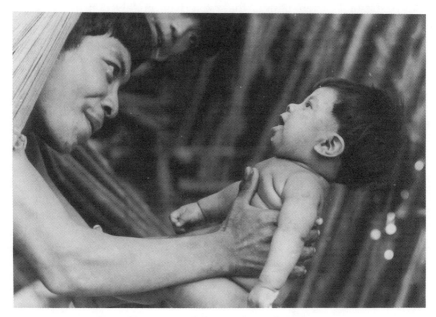

Yanomama father and son (1968).

10

Tribal Demography

A description of the demography and breeding structure of a human population cannot compete for reader interest with a description of the sexual behavior, warfare, or quaint customs of the same group. But these data are of paramount interest to the population geneticist, setting as they do the constraints within which the evolutionary process functions, and they will be important to us later when we consider how far our species has departed from the selective forces that shaped us. In the early 1960s, I could find in the anthropological literature precious little demographic data "hard" enough to incorporate into genetic thinking. To be sure, cultural anthropologists had devoted a major effort to describing the kinship and prescriptive marriage patterns of a wide variety of tribal groups, and had collected extensive genealogies. On the other hand, at that time there really was very little detailed information from these populations on birth and death schedules, the departure from ideal marriage patterns when the culturally prescribed mating was impossible, the extent to which the presumed ("legal") father was not the true father, or the amount of inbreeding. Krzywicki, in 1934, had made a valiant effort to pull together what was known,[1] but the data were simply inadequate for the needs of modern genetics. The shortcomings in the data have been in part due to the lack of written records, but also because some of this information is of relatively little interest to the culture itself, or even, as in the mention of deceased persons, subject to certain taboos within the culture. In the last chapter we discussed disease as an agent of natural selection; the other major potential agent for genetic change is differential fertility among those surviving to maturity, but this has genetic implications only if related to inherited attributes. So in this chapter we first discuss our efforts to define "breeding structure," then see what some of its consequences might be.

The Yanomama Age Pyramid

In retrospect, I must confess that when I decided to devote a major effort to attempting to develop an appropriate demographic data base, I could not realize in how unorthodox a fashion we would have to proceed. It is impossible to obtain a reproductive history from a Yanomama woman; the best one can hope for is a count of living offspring. Accordingly, to establish pregnancy rates, I have been forced to palpate the abdomens of all the women of several villages, to determine how many had uteruses at or above the umbilicus, and also to collect urine samples from these same women, on which, back in Ann Arbor, the usual pregnancy tests were performed. Part of our program has involved very extensive determinations of blood and enzyme types on as many members of each village we visited as possible. While these were collected primarily in the interests of quantifying village genetic differences, as we shall be discussing later, the data can also be used in the context of "paternity tests." Some missionaries have attempted to maintain birth records for villages with which they were in contact, and have generously made this information available to us. The Yanomama have a great reluctance to mention close relatives who have died recently—lest their omnipresent spirits hear their names spoken and are displeased—but there's no taboo on mentioning deceased who are nonrelatives or only distantly related. To construct even relatively limited pedigrees, Chagnon had to work through informants not closely related to the parties of interest, and then check, check, check.

One of the simplest but time-honored ways to characterize a population is by its "age pyramid." Even in this respect we encountered difficulties with the Yanomama. Since there is no calendar or written records, all ages are estimated. With children, ages can be estimated rather accurately, but such estimates become progressively more difficult as individuals become older, especially for women. In estimating the ages of adults, it helps the process to place people in their family relationship. For instance, a woman who has a granddaughter entering puberty will rarely be less than 50 years old; one can refine that estimate by considering all her grandchildren.

The figure on page 175 presents our estimate of the age pyramid for the Yanomama, based on as complete a census as possible of 29 Yanomama villages, involving some 2622 persons. This is a small sample

AGE

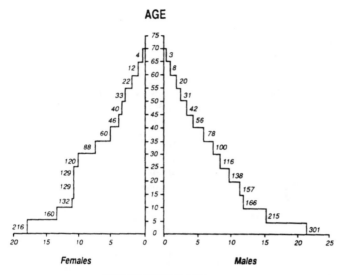

Females   *Males*

PERCENT OF POPULATION

The Yanomama age pyramid in the 1960s.

AGE

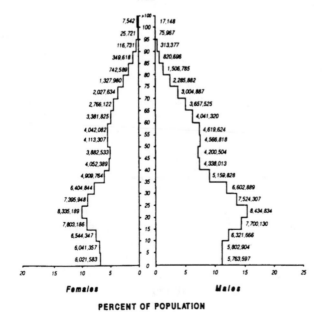

Females   *Males*

PERCENT OF POPULATION

The U.S. age pyramid in 1980.

175

by the standards of the demographers working with national populations but one of the largest censuses ever taken of people at this cultural
level. In this census, Nap Chagnon, the anthropologist, played the leading role, but each of us who examined a Yanomama also estimated the
age, and we assigned final age only after considering all our estimates.
The pyramid is presented separately for males and females. The one for
males exhibits the step-wise regularity characteristic of such pyramids,
but that for females shows a vertical "plunge" in the 10–30 age group
that we really don't understand. Even so, there are two striking features.
The first is the imbalance in the sex ratio among the young, which on the
basis of the observations of the missionaries we attribute primarily to
female infanticide. From this imbalance, plus the anecdotal accounts of
the missionaries, we estimate that in addition to the 5% of the newborn
of both sexes who are killed because of real or fancied congenital defect,
insufficient lapsed time since birth of last child, or at the husband's
instigation because of suspicions regarding paternity, an additional 20–
25% of female infants are killed, simply because of their femaleness.
Infanticide is allegedly accomplished by the attendant at birth (often the
woman's mother), who places a stout stick across the child's throat as it
lays on the ground following delivery and then stands on the ends of the
stick for the requisite time. The second notable point is the broad base
to the pyramid, accompanied by a paucity of persons aged 60 or greater
(only 1.0%). To assist in the appreciation of how dramatic the "demographic transition" has been, compare the age pyramid by sex for the
U.S. in 1980, with 15.9% aged 60 or more. Average estimated age among
the Yanomama was 20 years; for the U.S. in 1980 it was 33.7.[2] The
"bulge" in the age pyramid for the U.S. is the result of the post-WWII
baby boom and is a transient phenomenon. When allowance is made for
this, what is notable in comparing the two age pyramids is the narrowness of the base and the greater height of the U.S. curve.

The Reproductive Patterns of Yanomama Women

All Yanomama women, unless very severely handicapped, marry. From
the approaches mentioned earlier in the chapter plus the missionaries'
records, we estimate that the average Yanomama woman completes a
pregnancy every three to four years, with (sexes combined) approximately 85% of liveborn children permitted to live. Inability to conceive
seems very rare among Amerindians; virtually all women contribute to
the next generation. Since such birth rates are well below the natural
fertility of women, we must ask how they come about. Marriage (and

sexual intercourse) occur at an early age. To what extent "adolescent sterility" usually protects against early conception in this culture is unclear. We have no good data on age at first pregnancy. I have already mentioned the fact that Yanomama children are breast fed for some three years. The nursing is truly demand-style; a Yanomama mother will nurse her baby almost every hour during the day and frequently during the night. (If the mother is not available and the baby cries, someone else will frequently nurse the child.) There is good evidence that nipple stimulation produces surges of a hormone (prolactin) which is very effective in suppressing ovulation. The more frequent the surges, the better the suppression of ovulation. Since lactation seems a way of life for Yanomama women following the birth of their first child, Yanomama women should profit maximally from this natural control of fertility. On the other hand, ovulation eventually breaks through this suppressive mechanism.

But there are also conscious measures to control the entry of new life into the community. A Yanomama woman leads a very active life. By the time the next child arrives, the last should be quite mobile. Furthermore, foods suitable for infants are in short supply in the rain forest; there is literally no substitute for mother's milk. The response is a rather elaborate system of population control. Firstly, there are intercourse taboos of approximately one year's duration following the birth of a child. How well they are observed is difficult to say. Secondly, if a women does become pregnant "too soon," abortion may be induced by rupturing the fetal membranes through various types of trauma to the abdomen. Since this is not even a semi-effective method of inducing abortion until the uterus is above the pelvic brim, most abortions occur relatively late in pregnancy. Finally, should a woman deliver a child "too soon," infanticide may be practiced.

There is, then, a deep commitment to reconciling the rate of entry of new life into the population with the lifestyle of the culture. Such a commitment, however arrived at, is characteristic of many tribal peoples. Please note that I relate the commitment to lifestyle rather than to resources. With the advent of agriculture and the great religions, that commitment was lost. To some extent that loss of commitment was initially offset by the higher infant death rates that accompanied the "agricultural revolution," but with modern sanitation and medicine, that particular biological control of population increase is now gone, with the tragic results of overpopulation with which so much of the world is now confronted.

The social pressure on women for the child spacing just described seems to be great. Even so, given that reproduction begins in the teens, this pregnancy spacing would permit on average about eight live births to a woman who reaches menopause. The age pyramid makes it clear that many women do not reach menopause; we estimate that only 35% of females who reach age 15 will live to age 50. The mean number of live births to women reaching age 15 is about 6.

Abortion at an advanced stage of gestation and infanticide are to many a horrifying and totally repugnant thought. But viewed in the context of the culture, I must wonder who under these circumstances would be exhibiting the greater compassion, the woman who quickly dispatches a life not yet begun, that another child's life not be imperiled, or a mother who, bearing a new child "too soon," either condemns its predecessor to malnutrition or creates an increased probability of death for both children if the milk supply is divided. I find myself at one with a greatly respected anthropologist, Raymond Firth, who 50 years ago, before the problem of population control was so glaringly evident, in discussing the disturbance in population balance in the Pacific island of Tikopia when Christianity was substituted for the ancient mores, wrote:[3]

> It might be thought that the so-called sanctity of human life is not an end in itself, but the means to an end, to the preservation of society. And just as in a civilized community in time of war, civil disturbance or action against crime, life is taken to preserve life, so in Tikopia infants just born might be allowed to have their faces turned down and to be debarred from the world they merely glimpsed, in order that the economic equilibrium might be preserved, and the society maintain its balanced existence. (p. 376)

The Reproductive Patterns of Men

The problems in determining the reproductive patterns of the women are insignificant compared with collecting accurate data on male patterns. There must be a gene of universal distribution that leads men to exaggerate their sexual prowess. Nevertheless, it is clear that one consideration dominates male reproductive patterns: whereas the females strive for a certain culturally imposed reproductive uniformity, male conduct is designed to maximize male reproductive differentials. This is primarily because of the institution of headmanship. Every village of 50 to 150 Indians has its acknowledged headman—sometimes two. Competition for this position begins early; a headman usually makes it by age 30. Just as every American boy is alleged to dream of becoming president, so every young Indian male is presumed to dream of becoming a headman (or, alternatively, a shaman). In fact, all American boys don't dream of becoming president, and all Indian boys probably don't dream of becoming headman, but the competition is brisk nonetheless. The significance in our context is that in this nonmaterial culture, the most tangible reward of leadership is multiple wives. A headman usually has two or three. In this woman-poor culture, if the headman has multiple wives, someone goes without. To be sure, there is the equivalent of divorce, and most men have been married at some time in their lives. Nevertheless, the investigations of Chagnon lead to the estimate that whereas the headmen over estimated age 34 can claim an average of 8.6 living children, the nonheadmen of the same age claim an average of 4.2. The most striking example of headman fertility I have encountered was in the very first (Shavante) Indian village I ever visited, where the headman, Apewĕ, old by Indian standards, had 5 wives and 23 surviving children at the time of our study.

Because of the possibility of what we would term marital infidelity, and because it is only hospitable for a headman to make his wives available to a visiting headman, to understand gene flow in this culture we had to estimate how many children had parents other than as represented to us. Our genetic typings were of great value in this regard. They indicated that about 9% of children did not belong in the biological sibship to which they had been assigned. I believe the true frequency may be lower than 9%, since, because a small gift was involved when we obtained a blood sample, an Indian couple may have temporarily claimed a sister's child if she was absent from the village at the time of study. Additionally, given the frequency of divorce, a woman might be confused as to which of two men was responsible for her impregnation. Furthermore, our blood samples were not always in the best of condition when they reached the laboratory; occasional typing errors could result in false exclusions. We conclude that, in general, family relationships were as represented.

Since differential fertility is potentially so major an avenue of natural selection, we felt we must pursue its role in genetic change further. Given the absence of records, we resorted to a computer simulation of a Yanomama population.[4] Jean MacCluer and Francis Li, on the basis of data supplied by Chagnon and myself, in effect moved a complex of 4 Yanomama villages into a computer, where for a simulated 400 years the individuals lived, married, reproduced, and died on schedules determined by our actual field data. The program even allowed for headmanship and all the prerogatives thereof. The output of the program was thought provoking. In the population entered into the computer, there were 86 males who were between one month and nine years of age. The output of the computer simulation program was such that we could determine how many grandchildren each had. The results were striking. Fifty-seven of the Indian males in that cohort of 86, because of early death in the first or second generation, left no grandchildren. Among those with grandchildren, most had between 1 and 5, but one individual had 33! While this is only a simulated result, in general, where we could check, the computer output was a reasonable facsimile of the real Yanomama world. This simulation demonstrates the kind of Russian roulette to which a newly arisen mutation is subjected in these cultures: it would be quickly snuffed out about half of the time, but should it occur in the "right" individual, it is equally quickly very securely established.[5]

The Yanomama Life Table

From all of the foregoing information, Ken Weiss and I tried to develop "life tables" for the Yanomama.[6] The table that best agreed with the data is shown on page 181. The sexes are shown separately because their expectations are so different. These tables, plus the other data, let us draw the following conclusions: Infant and early childhood mortality rates from *natural* causes are about 25%. We have already commented on mortality from infanticide. The starting point for these tables was pregnancy rates, determined as described earlier, rather than observed birth rates. The estimate that 27% of males and 57% of females die during the first year thus includes deaths from induced abortions and stillbirths, as well as from infanticide and natural causes. The figures are inflated if we have overestimated the amount of infanticide for either sex. Natural mortality plus infanticide reduce survivorship to age 15 to about 50%. In general, during childhood, female mortality rates from natural causes are somewhat lower than male. This fact, plus the loss of young men in warfare, results in the sexes becoming numerically equal by the end of the third decade, despite the early excess female infanticide. The life expectancy at birth is about the same for males and females, approximately 20 years. For those surviving to age 15, there is an expectation of 25–30 more years of life. Between 45–50% of the population is under age 15, with only 7–10% over age 50. The average age at which women reproduce is 27 years, and this is of course also the approximate interval between generations.

The figure on page 183 presents the combined survival curves for males and females in graphic form and, by way of contrast, the survival

The Yanomama life tables for males and females, after Neel and Weiss (1975).

The most satisfactory male life table for the Yanomama

Age	Q(X)	P(X)	l(X)	L(X)	T(X)	E(X)	C(X)
0	0.267	0.733	100	83	2147	21.5	4.6
1	0.160	0.840	73	262	2065	28.2	14.3
5	0.110	0.890	62	291	1802	29.3	15.3
10	0.088	0.912	55	262	1512	27.6	13.2
15	0.148	0.852	50	231	1250	25.0	11.2
20	0.152	0.848	43	197	1018	23.9	9.1
25	0.156	0.844	36	166	822	22.7	7.4
30	0.160	0.840	30	140	655	21.5	6.0
35	0.164	0.836	26	117	515	20.1	4.8
40	0.168	0.832	21	98	398	18.5	3.8
45	0.173	0.827	18	81	300	16.8	3.0
50	0.177	0.823	15	67	218	14.8	2.4
55	0.213	0.786	12	54	151	12.5	1.9
60	0.282	0.718	10	41	97	10.2	1.3
65	0.362	0.638	7	28	56	8.2	0.9
70	0.470	0.530	4	17	28	6.4	0.5
75	0.612	0.388	2	8	11	4.9	0.2
80+	1.000	0.000	1	3	3	3.8	0.1

The most satisfactory female life table for the Yanomama

Age	Q(X)	P(X)	l(X)	L(X)	T(X)	E(X)	F(X)	C(X)
0	0.430	0.570	100	52	1982	19.8	0.0	3.2
1	0.118	0.882	57	210	1920	33.7	0.0	12.7
5	0.066	0.934	50	243	1710	34.0	0.0	14.2
10	0.042	0.958	47	230	1467	31.2	0.0	12.8
15	0.130	0.870	45	210	1237	27.5	0.147	11.2
20	0.134	0.866	39	182	1027	26.3	0.132	9.3
25	0.137	0.863	34	158	845	24.9	0.179	7.8
30	0.140	0.860	29	136	687	23.5	0.181	6.4
35	0.143	0.857	25.2	117	551	21.9	0.101	5.3
40	0.147	0.853	21.6	100	435	20.2	0.031	4.3
45	0.150	0.850	18.4	85	335	18.2	0.031	3.5
50	0.154	0.846	15.6	72	250	16.0	0.0	2.9
55	0.188	0.812	13.2	60	178	13.4	0.0	2.3
60	0.251	0.749	10.8	47	118	11.0	0.0	1.7
65	0.327	0.673	8.1	34	71	8.8	0.0	1.2
70	0.432	0.658	5.4	21	37	6.9	0.0	0.7
75	0.570	0.430	3.1	11	16	5.2	0.0	0.4
80+	1.000	0.000	1.3	5	5	3.8	0.0	0.2

The Yanomama life tables for males and females, after Neel and Weiss (1975). Age is age at the beginning of each age class. Q(X) is the chance that those reaching the age class will die before reaching the next age class. P(X) is the chance of surviving the age class. l(X) is the number of survivors left at the beginning of the age class out of every 100 born, and L(X) is the number of person years lived in the age class per 100 born. T(X) is the total number of person-years left to be lived from the beginning of the age class until all who are alive then have died. E(X) is the life expectancy of those surviving to the beginning of the given age class. F(X) (for women) is the annual chance that a woman in the age class will bear a daughter. Finally, C(X) is the proportion of the total population which is in the given age class.

curves for India at the turn of the century and for Japan in the mid-1960s. Infant mortality in India from natural causes was about twice that in Yanomama. On the basis of these data and those of the preceding chapter, as well as contacts with the rural populations of South America and Africa, and a consideration of the literature, we have argued that the health of tribal populations deteriorated with the advent of civilization, a thesis recently generalized by Cohen[7] through a consideration of prehistoric remains. At present, at least in the developed nations, health—as measured by longevity—has more than recovered. It has, however, a different character than previously. It is also, as we will discuss in Chapter 17, precariously balanced. The epidemic "herd" diseases that so contributed to the decline in health with the advent of civilization (and whose control then resulted in increased population densities) are for the moment no threat to health, but the appearance of AIDS suggests that this control cannot be taken for granted. Japan illustrates the survival curve of a fully developed country, with particular reference to the excess of senior (nonreproducing) citizens.

A word of caution concerning the interpretation of the Yanomama life table is necessary. There is a discrepancy between the picture I have painted of generally healthy Indians and the early acquisition of antibodies against the endemic diseases of the locality, and the amount of midlife mortality reflected in the life tables. During the course of our fieldwork, Chagnon[8] made a special effort to elicit cause of death for the recently deceased. Traumatic death (warfare, snake bite, and accidents) accounted for the demise of about 35% of adult males and 10% of females. The bulk of the remaining deaths were vaguely attributed to witchcraft or epidemics, a term employed when multiple members of the village are simultaneously ill and die. As noted earlier, we believe that *P. falciparum* malaria is only now spreading into this region, and the heavy toll of this disease on adults as well as children when it reaches previously uninfected areas is well known. This may be the primary agent in the present midlife mortality of women mentioned earlier. Thus I would argue that prior to the recent advent of malaria, the Yanomama survival curve compared even more favorably with the curve for India, especially with respect to mid-life, than was the case at the time of our studies.

From all of these data we can make some deductions about the population balance, although because of the probably recent advent of malaria we must be a little cautious. The growth rate of the Yanomama should be between 0.5% and 1.0% annually. The net reproduction rate for women is estimated at 1.25, that is, each female born will on average replace herself 1.25 times. Those familiar with the principal of com-

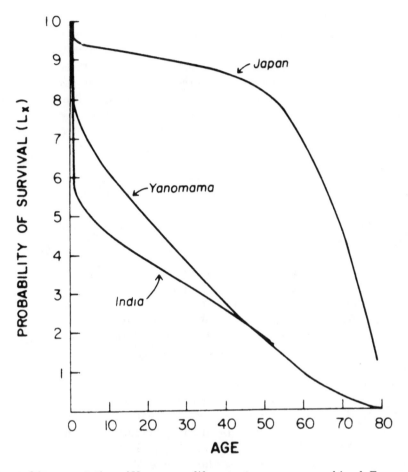

A graphic presentation of Yanomama life expectancy, sexes combined. For purposes of comparison, the life expectancy curves for India at the turn of the century and Japan in the 1960s are also presented. Note the much higher frequency of early death in agricultural India and the deferral of death until a relatively advanced age in Japan, with the Yanomama in a somewhat intermediate position.

pound interest will recognize that these figures make the Yanomama a relatively rapidly expanding group. Early human populations did not over any considerable period of time expand that rapidly. The estimates of the total population of the world at the end of what geologists term the Pleistocene Period (i.e., just before the advent of agriculture some 10,000 years ago) ranges between 3 million and 10 million persons.[9] Were it possible to specify the magic moment when our anthropoid ancestors crossed the threshold to Australopithecus some five million years ago, it seems doubtful that more than a few thousand individuals were involved. Granting all the various sources of error in both of these estimates, the *average* annual rate of increase in the human population for those five million years preceding the advent of agriculture would be something like 0.001–0.003% per year. This is a very slow rate of increase. It implies, for example, that a band of 100 should increase by two or three persons in an entire generation.

Does this discrepancy between the Yanomama rate of population increase in the short run and the slow *average* rate of increase of early human populations invalidate the use of the Yanomama as a general model? I think not. No early population could possibly regulate its numbers generation after generation to the average annual rate of increase of 0.002% suggested by the estimates quoted earlier. There must have been good times, when groups within the population were increasing relatively rapidly, and bad times, when group numbers plummeted. We happen to have come upon the Yanomama in relatively good times. There is reason to suspect that their staple crop, the plantain, is a post-Columbian introduction. Until three or four generations ago, they probably relied on the small stone axes found all through their territory to gird trees in the fashion essential to slash-and-burn agriculture. These axes could not possibly have been very efficient. but for some generations now, even the most remote villages have had access by trade channels to pieces of machete blades—much more efficient for girding trees. Finally, most of the tribes in contact with the Yanomama, especially those to the south, have been badly decimated by their contacts with neo-Venezuelans and Brazilians; the Yanomama have been free to expand into contiguous areas without some of the bloodshed usually attendant on such expansion.

Inbreeding

With our long-established interests in inbreeding, it was inevitable that we should try to establish how frequent it was amongst Amerindians. The Yanomama recognize male-descent lineages; a man should marry outside his lineage. A highly preferred form of marriage is for men of two lineages to exchange younger sisters as brides. In the following generation, the female offspring of such an exchange must marry outside the lineage. A second highly preferred form of arranged marriage, which meets this requirement, is for a man to marry the daughter of his father's sister, or of his mother's brother (technically, a prescriptive bilateral cross-cousin marriage system). Thus, the preferred marriage involves certain types of first cousins. When such a marriage is not possible, a man (or a woman) will try to marry within the village, which of course contains many of the man's more remote kin. This marriage system, if observed, should result in a high level of inbreeding. Unfortunately, one can only be certain of the fact of a first-cousin marriage if the identity of all four grandparents is known. Despite Chagnon's best effort, he could only establish the identity of the 4 grandparents in 37 of the 124 marriages represented in the 4 villages where he knew the genealogies best. Thirteen of these 34 marriages involved first cousins. This was a high frequency, but was it representative? Again, we resorted to computer simulation, to try to determine how rapidly inbreeding would build up under these circumstances. The answer was, quite rapidly, by our contemporary standards. The key was the small geographical extent of the marital quest and the differential fertility we have just discussed. For instance, the "grandchildren" of the more prolific headmen would all be first cousins, and they would be concentrated in several adjacent villages.

We believe that the level of inbreeding that we encountered in the Yanomama is not a recent development, but one that goes far back in time. Accordingly, aided by the computer program, we could ask the question, if this pattern of inbreeding was in place when the Indian entered the Americas, just how inbred had these populations become by now? Our best estimate was that the *average* marriage in an Indian village represented a level of inbreeding at least five times as large as the inbreeding in a first-cousin marriage. This is, in fact, greater than the inbreeding in a brother–sister union. This conclusion was so surprising that we have gone back to reexamine it from every possible vantage point, and from every possible vantage point it seems to hold. It rests on the fact that if one could construct a pedigree extending back 40–60 generations, one would find that any two marital partners in a village are connected by literally hundreds of loops of relationship. All of these loops together, when added up, cumulate to a degree of inbreeding greater than that which would be achieved by a brother–sister union when the parents of the brother and sister were unrelated.[10]

Chance and the Loss of Favorable Mutations

Evolution is the replacement of "satisfactory" genes with genes that enable their possessor to perform even better in the competition to survive and reproduce. The observations on the population structure of the Yanomama led us to appreciate what a major role chance could play in whether any particular gene survived or was lost. Earlier in this chapter, I referred to a computer simulation of four Yanomama villages. Mutations (some 280 in number) were introduced into the 10–19-year-old members of the founder population for this simulation. In this specific situation, if the gene was neutral in its survival value, only 8.6% were still in the population after 16 generations, the others lost in the genetic transmission process or by the early death of their possessor.[11] Next we attempted to study the role of positive selection in retaining genes with favorable effects. We assigned the 280 mutants a survival value of 10%. As expected, this increased the percent still present in the population after 16 generations, but only to 18.6%. Even with a selective value of 10%—which in evolutionary theory is a large selective value indeed—81.4% of all such "good" mutations would be lost! This was thought provoking; we will return to the implication for human evolution later.

Genetic Implications of Changing Demographic Patterns

The advent of civilization and the emergence of national populations has resulted in major departures from the demographic patterns we documented among the Amerindians. Let us assume that the Amerindian patterns are in their broad outline a reasonable approximation to the breeding structure of early man. (We shall speak to some possible differences in what follows.) How, then, has the population structure under which we evolved altered with civilization.[12]

Firstly, although there are of course still fertility differences between individuals in Western culture, these are narrowing by comparison with Amerindian patterns, and the basis for such differentials as persist are dramatically altered. The most striking factor in the demographic transition is the loss of a polygamous headman for each relatively small population unit. Headmen—not just among the Yanomama but probably in all tribal cultures—emerge by a combination of attributes. They are well versed in the tribal history and lore, and, since Amerindian cultures operate largely by consensus, must be superior and persuasive speakers. They must acquit themselves well in battle, and be skillful hunters. The intimacy of life in an Indian village is such that there can be none of the discrepancies between public image and private conduct with which political leaders in the United States and elsewhere so regularly surprise us, nor can there be a delay of 20 or 30 years in recognizing the consequences of a hideous misjudgment on the part of a leader. Everything anyone in such a village has ever done is known to all the other members of the village. Dummies don't become headmen. It is difficult not to perceive headmen as a superior group, some of which superiority should have a genetic basis. *They have twice as many children as the average Indian.* Given the population structure we have just reviewed, an average village of 100 persons might have only 15 men of an age appropriate to assume headmanship. The fact that the one headman enjoys such a reproductive advantage must be construed as a powerful agent for positive natural selection, now largely lost in our own culture, where fertility differentials are predominantly related to socioeconomic circumstances. Incidentally, the office of headman is in no sense hereditary. It is true that a son of a headman may have some advantages in the competition. For instance, he can manipulate the marriages of his relatively numerous sisters to form alliances favorable to his cause. Basically, however, when the existing headman begins to lose his grip on the situation, the jockeying to replace him is wide open.

Reproductive differentials among women are much less than among men in polygamous societies. Virtually all Yanomama women marry, and, as noted, sterility is rare. Given the emphasis on child spacing, such differences as do exist in number of children reared to maturity are apt to reflect premature death of the woman more than any other factor. (This is to be contrasted with the situation in the western world, where because of failure to marry, inability to conceive, or a conscious decision not to reproduce, at least 10% of women are childless.) I am led to the inescapable conclusion that natural selection in these cultures operated far less through the women than through the men, a situation now in transition as recently, in many nations, women begin to individualize their own reproductive patterns according to their convictions.

A second obvious genetic departure of most of the civilized world from tribal societies is the relaxation of inbreeding. A discussion of the consequences of such relaxation rapidly becomes complex, and we will consider only the simplest case, involving diseases due to completely recessive genes with quite deleterious effects, incompatible with reproduction. The frequency of these genes is the result of a balance between their entry into the population in consequence of mutation and their elimination as a consequence of the deleterious effects of homozygosity. Long isolated, endogamous tribal populations should be in a balance between these two forces. When inbreeding is relaxed, as is now particularly the case for Christian communities, homozygosity for genes of this type decreases, and there should be a decrease in the diseases associated with these genes. This, however, is only temporary. Mutation pressure continues, and the gene frequency will very slowly build up, until finally the frequency of homozygotes will again come into balance with mutation pressure. However, the relative frequency of heterozygotes in the population is now greater than before. Should this population ever revert to high levels of inbreeding, it would, so to speak, "pay the bill," i.e., the gene frequency would have risen above the frequency consistent with the new level of inbreeding, and there would now temporarily be more of whatever disease is associated with the genes in question than would be the case had inbreeding continued at the original levels. Furthermore, there is evidence from experimental genetics that the heterozygotes for these recessive genes are sometimes themselves slightly disadvantaged, so that a relative increase in the frequency of the heterozygous carriers of a deleterious recessive gene is not to the advantage of the population.

A third point of difference between many tribal cultures prior to contact with civilization and much of the civilized world today is the loss in many parts of the latter of a commitment to regulate the entry of new life into the culture to a rate consistent with the functioning of the culture. The superficial result is the tragic level of overpopulation that is resulting in an ever-increasing vandalization of our planet with no real prospect of resource renewal. This overpopulation also has a rather subtle effect that is not well appreciated. In resource-poor countries, this overpopulation is accompanied by increased infant and childhood mortality. At first glance this would appear to increase natural selection as only the fittest will survive. But, in fact, the total population is victimized. Since in these countries malnutrition is widespread and even the survivors are malnourished, the genetic potential of individuals often will not be realized, and the entire malnourished population is disadvantaged in the competition with other populations. I have never been able to understand how it can be construed by some religious sects to be the will of a loving God that a woman whose contraceptive efforts have failed should be denied access to a properly managed, early abortion, given not only the socioeconomic facts just mentioned but the tragic abuse and neglect suffered by so many unwanted children.

I've spoken several times of those rare moments when one feels briefly suspended in a timeless void with no reference points, from which one emerges with all the emotional catharsis induced by a first-class performance of Verdi's *Requiem*. I will introduce only a single additional example. On one of our expeditions in which the collection of data on pregnancy rates was a particular focus, we'd made camp well up a remote tributary to the Orinoco, across the river from a Yanomama village. I had slung my hammock on a bluff beside the river. Slumping into it that night, looking off across the river with an unobstructed view of the incredible richness of the tropical stars, the stars and I were suddenly one. Man is forever wondering how he fits into the intricate web of life; these are the moments when he is part of it, free of debate between the committees of the mind. I wondered at the time, and at rare moments thereafter, if this was evidence of some dangerous instability that might ultimately prevail. It's the kind of experience you don't share with your "hard science" friends. Then I read Loren Eisely's "Immense Journey,"[13] in which he describes his own similar experiences so beautifully and equates them—on a lesser scale, to be sure—with the visions and insights sought by man from time immemorial in the pursuit of his understanding of the universe, the experience often encouraged by prayer and fasting in great solitude. I now presume these epiphanic experiences are an almost universal human attribute, their occurrence greatly facilitated by some reprieve of the senses from the usual mind-numbing onslaught of daily trivia.

Children playing in the center court of their *shabano*, a circular, lean-to village.

11

Genetic Differences Between Villages:
Stepping Stones to Evolution

A s stated in the Preface, a primary objective of this book is to consider the current human genetic condition, through eyes focused by past experiences. Now it is time to introduce still another set of the facts that will contribute to the ferment of such a discussion, and like some of the preceding facts, there will be a few surprises. This chapter will describe what is for me one of the most exciting insights to come out of the Amerindian fieldwork. I will draw heavily on it later. But somehow, of our various findings, the significance of this particular one has been the most difficult to project to those who are not professional geneticists.

Village Microdifferentiation

Thus far, for good reason, I have avoided defining the word "tribe." We all think we know what the word means, but it is used very loosely. I will now define a tribe as a group of people enjoying a more-or-less common culture and language. The members of a tribe aggregate in bands or villages. As brought out in the last chapter, marriage is, if possible, usually contracted within the band or village, but may involve a member of a neighboring group. The more remote from each other any two villages of the tribe, the less the likelihood of an exchange of marital partners. On the other hand, a gene in one corner of a tribal distribution may in principle, in the course of generations of marital exchanges between villages, reach a rather remote village in another corner of the tribe. Later in this chapter we will consider the evidence that the diffusion of genes encounters distinct difficulties at tribal

191

boundaries. Since within a large tribe there are dialectical differences and regional cultural peculiarities, there is frequently the question of when to recognize subtribes. Likewise, since adjacent tribes may have shared a common region many generations ago, still speak closely related languages, and have similar cultures, it can be difficult to set precise boundaries to a tribe. Thus the term tribe when closely scrutinized can never have the precision of legally defined entities such as town, city, or nation.

A common stereotypic belief concerning tribes involves a chief reigning over large areas. This has in general been true only in tribes with an advanced civilization, such as the Mayas, Aztecs, or Incas of the Americas, or some of the African tribes at the time of first European contacts. If we consider the total sweep of tribal history, paramount chiefs must have been very much the exception. The functional population unit in early human evolution was the band (sometimes called horde), and later villages loosely organized into tribes. Although there were temporary and shifting alliances between villages, these were basically competing units, each with its own headman.

Most studies of the genetic composition of Amerindian (and other) tribes have been restricted to a few villages, often after these have, as a result of contacts with non-Indians, come upon hard times and several villages have been forced to fuse into one to maintain critical mass. The Yanomama were not in that situation. We had an unusual opportunity to study what might be termed the internal structure of a vigorous tribe. Our chief approach involved genetic typings. There are by now hundreds of "genetic systems" in which, in addition to the normal protein gene product, one or more genetically variant forms of the product have been recognized. Perhaps the systems best known to most readers are the ABO and Rh blood types. With respect to the ABO types, human red blood cells may be typed as O, A, B, or AB. With respect to Rh, blood cells may be positive or negative, and there are many types of positivity. Where two or more of the variants that may occur at a genetic locus are relatively common, we speak of a genetic polymorphism; the sickle-cell gene is an example of such polymorphism. In general we speak of a genetic locus as exhibiting polymorphism when, in addition to the normal gene, there are one or more variants each occurring in 2% or more of the population. All these variants, of course, ultimately trace to variation in the DNA, and recently the study of polymorphisms has moved to the DNA level, but the necessary DNA techniques were not readily available at the time of our studies.

The ABO system is of little value in the study of South American Indians (except to indicate admixture), since in the absence of admixture they are all type O. But the Indians presented at least four different Rh types and, in addition, varied with respect to five other red blood cell genetic systems for which we could type. We also routinely subjected some 25 proteins of the Indian blood samples to electrophoretic examinations as described for hemoglobin in Chapter 3, and four of these proteins exhibited genetic polymorphisms. The figure on this page shows the range of variation when we compared 48 Yanomama villages with respect to 16 allele frequencies in these 9 systems. For instance, the

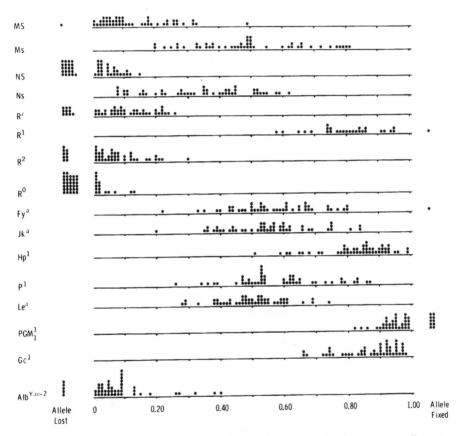

The variation in the frequency of 16 different genes in 48 Yanomama villages. The first 15 genes are found all over the world, but the last, an albumin variant, is found only in the Yanomama. Note that five of these genes have actually been lost in certain villages. This is graphic evidence for the degree of genetic differentiation between Yanomama villages.

village frequency of the Ms allele (a blood-type allele of the MN system) varied from 20% of the genes in this system in one village up to 80% in another; the frequency of the JK^a allele, a member of another blood-group system, showed an equal range.

Geneticists, when comparing two villages (or other population groups), have techniques for combining the gene frequencies for numerous systems into a single index figure, which can then be compared with other such figures. Using allele-frequency differences between villages such as those shown in the figure as the basis for the comparison, R. Ward and I derived the average genetic difference (or "distance") between a set of Yanomama and also between a set of Makiritare villages,[1] as well as the corresponding differences between the Amerindian tribes we had studied, and then compared these distances with those between representatives of 15 major ethnic groups,[2] using the same markers throughout. The astonishing fact is that, as we geneticists measure difference, the average difference between Yanomama villages was 90% as great as the difference between the other Amerindian tribes, and tribes in turn were on average 50% as different as the average difference between representatives of 15 major ethnic groups. We refer to this intratribal (i.e., village) variation as microdifferentiation. Our studies showed that this microdifferentiation with respect to marker genes was also observed for anthropometric, dermatoglyphic, and linguistic traits.[3] It is far greater than we had suspected at the outset of our studies. When we come to discuss human evolution, microdifferentiation will be the cornerstone of my thinking.

"Private" Tribal Genes

The typing sera used to define blood groups would rarely recognize a new blood type should it occur in a tribe. By contrast, the technique of electrophoresis will demonstrate the presence of any variant form of a protein if this protein has a different electrical charge from the normal. Altogether, employing electrophoresis, we have examined representatives of 13 tribes for, on average, electrophoretic variation with respect to 25 different proteins. Here again there was a major surprise. As mentioned in the preceding section, we did find that some of these proteins were characterized by genetic polymorphisms shared by all the tribes. In addition, we encountered a number of very rare inherited variants of various enzymes, similar to those we had been encountering in population surveys in Ann Arbor. The surprise was the extent to which we also encountered examples of what we have termed "private polymorphisms." We now have eight examples of this phenomenon. Private polymorphisms are alleles—by our various criteria, unique alleles—that in a single tribe, or perhaps two closely related tribes, occur in 2% or more of the population. They owe their occurrence to a mutation in a remote tribal ancestor, a mutation that persisted and spread throughout the tribe. In no case do we know whether this mutation confers some advantage, or is essentially neutral in its effects by comparison with normal, or even confers a slight disadvantage. Some of these private polymorphisms are present in 10–20% of all the members of the tribe.[4]

An example may help clarify the finding. One of the proteins we examined by electrophoresis was serum albumin. By and large, unlike the findings just described with respect to the blood groups, we encountered relatively little genetic variation in albumin. The Yanomama were an exception. Here there was a distinct type whose frequency from village to village was highly variable, with as many as 60% carriers in a few villages in the northwest part of the tribe, as contrasted with 0% carriers in some villages in the southwest (see the figure on page 193). We did not encounter this particular variant in any other tribe.

The observation of these "private polymorphisms" lends itself to a rather startling calculation. The foregoing figures indicate that when we examine the product of any particular gene from several hundred members of an Amerindian tribe, there is one chance in 40 we will encounter one of these private polymorphisms. The tribal individuality at the DNA level should be even greater. Now, we know only very approximately how many gene products each of us elaborates, but a common conjecture is that each of us has some 50,000 genetic loci yielding protein products. Should this be the case, and if our findings are representative, then in an average Amerindian tribe there should be $50,000 \times 1/40$, or approximately 1250 of these private polymorphisms. This says that not only is there this marked microdifferentiation of villages with respect to shared genes, but any given tribe (and the villages of which it is comprised) also differs absolutely in many genetic respects from its neighbors.

The occurrence of these private polymorphisms is informative in a second way. The fact that the private polymorphisms almost always stop at tribal boundaries suggests very little genetic exchange between neighboring tribes. The linguistic and cultural differences between tribes appear in the past to have discouraged tribal intermarriage rather strongly. I emphasize this because in many of the accounts of the North and South American Indian written by explorers and anthropologists, you encounter descriptions of tribal fusions or intertribal marriages. It would appear these were relatively uncommon before the tribal disruptions occasioned by European contacts. Earlier I discussed how marriages were contracted within villages if possible but with members of neighboring villages if necessary. Now we have good biological evidence that only rarely indeed was that necessity allowed to extend across tribal lines.

The Heterozygosity of Amerindian Tribes

One way to measure the amount of genetic variability present in a group is by the average heterozygosity at a series of genetic loci. This is called the heterozygosity index. Recall that if there are two alternative forms of a gene (A and a) that may occur at a genetic locus, and the frequency of A is p and of a, q, then the three possible genotypes (AA, Aa, and aa) occur in the proportions p^2, $2pq$, and q^2. The term $2pq$ represents the frequency of heterozygotes, and the heterozygosity index is simply $2pq$ averaged across all the genetic loci studied. A slight modification of the formula accommodates the situation where there are three or four or more alternative forms of a particular gene. The index is usually converted to percent by multiplying the value obtained by 100. When several different populations have been studied with respect to variation at the same set of loci, the heterozygosity index is a convenient way to summarize and compare the amount of genetic variation present in these populations.

For the cosmopolitan and quite admixed Caucasian population of London, the heterozygosity index based on 23 proteins examined for electrophoretic variants was 7.8%. In the cities of Hiroshima and Nagasaki, the index for the same proteins was 7.7%. Finally, for these same proteins in the Indian tribes we studied, the average index was 5.4%.[5] Thus by this measure the amount of genetic variability within these Amerindian tribes was 70% as much as in very cosmopolitan populations. Electrophoresis measures only a fraction of the genetic variation present in these populations, but the ratios of these three indices to one another should be a valid measure of the relative amounts of genetic variation present in the three populations. Given the many tribes that contributed their genes to the urban gene pools of London, Hiroshima, and Nagasaki, it is not surprising that these indices are higher than in an isolated Indian tribe. Rather, what is impressive is that by this yardstick, the amount of variability in these long-isolated and inbred tribes is so relatively similar to that present in very admixed populations. We return to this matter shortly.

The Causes of Microdifferentiation

We have seen that there is marked village and tribal genetic microdifferentiation for both common and private polymorphisms. How does this come about? We have been able to identify two factors that together can account for most of this microdifferentiation. The first is the manner in which a village originates. Characteristically a new village comes into being because tensions within an existing village, which has been slowly expanding in numbers over several decades, reach the point where the village splits up. This may occur because two strong men, one a headman, one a challenger, are vying for leadership, or alternatively, because at the "retirement" of a chief, there are two strong challengers. In any event, one loses out, one wins. The loser, gathering together his followers, will often leave the village to begin a new one. These "followers" will, in general, be his wives and children, his brothers with their wives and children, their parents, plus various other persons who feel more comfortable with the new group.

If a small population were to split up at random, it would be rare for the two villages to be identical genetically, simply because of the laws of chance. The process I have just described, however, superimposes on the laws of chance a highly structured pattern of splitting. We have been fortunate in our fieldwork among the Yanomama to come upon two pairs of sister villages, each pair of which had risen within the past decade from the kind of split I have just described. This means that in each case, by pooling our genetic findings from the two sister villages, we can for the genetic markers for which we have typed more or less reconstitute the genetic composition of the original village before the split and then measure, by the same kind of statistical functions I have just been describing, just how much the two sister villages differ. In both instances, the difference was striking, well beyond what we would expect from any hypothetical "random" split.[6]

The second factor in the microdifferentiation is the differential fertility we have already discussed. Within these small villages, some men—usually the headmen—will be much more prolific than others. In particular, should the headman of the new village and his brothers be the descendants of a headman of the previous generation—and the headman would have more brothers than usual if his father were a headman—then any unusual genetic attribute of the previous headman would be strongly represented in his sons and their children. Thus the "lineage effect" will tend to produce village differences at the time of a split, but the village that is splitting will already differ genetically from other villages because of the differential fertility that characterizes the dominant figures in the village.

Some Implications of these Findings for Human Evolution

We are all fascinated by the question of our origin, whether it be perceived as the result of the chance concentration of very simple chemicals in a small pool of water some four billion years ago or, somewhat more recently, through divine creation. With reference to the appearance of our species, within the past 60 years there has emerged a much better fix on the time at which the line of descent common to the chimpanzee and ourselves diverged from the other primate lines. Prior to 1930, the prevailing view was that our ancestral line had split off from the other primates early, perhaps in the Oligocene Epoch (now dated 37–26 million years ago). Within the past 30 years, a large increase in finds of early hominid fossils, in situations where they can be accurately dated, has drastically altered this viewpoint. It now appears that human evolution has been relatively rapid. If is, of course, quite arbitrary to designate the precise time at which our ancestors had diverged sufficiently from the other primates that the term hominid became appropriate, but an estimate currently in favor is 5–7 million years before the present (BP). Although the details remain in dispute, there is reasonable consensus that, within this time frame, our primate ancestors first evolved into *Australopithecus africanus*, then *Homo erectus*, next *H. habilis*, and finally *H. sapiens*.[7] At an average generation time of 20 years, this is three "speciations" in some 250,000 to 350,000 generations. It is important to bring generation time into consideration because only at meiosis and fertilization do new gene combinations come into existence. This recognition of the relative speed of human evolution must be regarded as one of the major developments in biological thought of this century.

The possible implications of the Amerindian studies for human evolution are perhaps best illustrated by a return to the evolutionary genetics of my student days. At that time our thinking was dominated by what I might term "large-population genetics." Breeding groups were usually represented as consisting of at least several thousand individuals. When a population is this large, there is a great deal of inertia in the gene pool, which implies that the frequencies of specific genes would not vary greatly from generation to generation. If a favorable mutation were introduced into a population such as this, it might by chance be lost in the early generations, but, if it survived and achieved even a low frequency, then selection took over, and it proceeded to increase up to some optimum frequency, but very slowly.

What we have documented in the studies of the Yanomama is a population with much more genetic flux than we could have anticipated. We, of course, are not the first to recognize that human tribal populations are usually subdivided into small, semi-isolated breeding units. But, by a fortunate selection of a tribe to study, we have been able to show the genetic consequences of this population structure. Each time a new village comes into being, it represents a combination of genes (packaged into individuals), the exact likes of which has probably never existed before. A major cause of this microdifferentiation is the kinship system, leading to nonrandom village fissionings. Our species is unique in the way the kinship system determines population aggregation. We are therefore led to suspect that tribally organized human populations show more microdifferentiation than most other animal populations, but the studies on animal populations that would demonstrate this point are difficult because of the problem of defining the breeding unit in an undisturbed animal population, if such an animal population can be found. I repeat again, the village is the unit of genetic competition. If a village prevailed over other villages, the genes of that village would increase disproportionately. However, at the time of the next split, this favorable gene combination would be disrupted. But even so, the daughter villages should retain some of the favorable gene combinations of the parent village as they began their competitive existence.

What I am suggesting is that the tribal gene pool of early man, however defined, was much more drastically and regularly reshuffled into genetically diverse, competing units than has previously been thought. Early human populations, if the Yanomama are any guide, were so structured that there was continuous change in the composition of the ultimate (the village) gene pool—even at the risk of disrupting at times "good genetic combinations," or losing promising genes. Geneticists like to picture evolution as leading to a situation, termed genetic equilibrium, in which each gene is represented in a frequency determined by its selective value, and genes occurring at different (interacting) loci are in optimal proportion with respect to each other. Our studies make it clear that because of the constant reshuffling of genes at the level of the ultimate functional unit, the band or village, equilibrium was seldom if ever realized, and if it was realized, not for long.[8]

From time to time, as human populations spread over the earth, a band or village, or several related villages, already quite different from other bands of the tribe, must have become detached from the tribe to which it belonged and become the nidus for a new tribe. From what we have observed, the new tribe could be quite different from the parent tribe at the moment of origin.[9] Some of these offshoots should possess better combinations of genes than others. Since, ultimately, tribes as well as villages are competing, we again see a process that ensures that the competition will involve very different gene sets. Imagine, for instance, how different the initial situation would be if, at the time the nidus of a new tribe came into being, each village of the old tribe, by some process, contributed genes to the new tribe. Then the initial tribal differences would be very much smaller.

To What Extent Can We Generalize from the Yanomama?

Again and again I have warned against taking the Yanomama as an exact model for early human societies.[10] On the other hand, we know of no better approximation. Although the Australian aborigines and the South African bushmen when first contacted relied less on agriculture than the Yanomama, and so should reflect the hunting-and-gathering stage of human evolution much better than the Yanomama, their cultures were much more quickly disturbed than that of the Yanomama at the time of our studies, rendering it more difficult to generalize from these groups than from the Yanomama. The various New Guinea tribes, on the other hand, have achieved a population density that was certainly not typical of early man, and which now greatly constrains tribal movements. Whichever of these societies anthropologists and geneticists study, the extrapolations to the earlier human condition must be done with caution. Nevertheless, it is a reasonable postulate that in the remote past the world's tribal populations were constantly sending out exploratory pseudopods and trying out new gene combinations. Now, with massive population amalgamations under way, the current population structure resembles a large, increasingly homogeneous, quivering blob of jelly, which, though it may shake a bit, is unlikely to spawn a detached and unusual offshoot that will persist long enough to establish an identity.

The Wright Model

Beginning in 1931, Sewall Wright, a commanding figure in the genesis of modern population genetics, attempted to envision the circumstances most favorable to rapid evolution.[11] His model entailed the subdivision of the population into local units, in whose individual allele combinations and collective gene frequencies the stochastic process, because the subpopulations were relatively small, played a major role. Competition between individuals within these demes (read band or village for humans) was a way of life, as a result of which the better-adapted individuals left a disproportionate number of progeny. In turn, demes also competed, as a consequence of which the deme with the higher proportion of adapted individuals prospered disproportionately, displacing the others. An important aspect of the Wright model was that no single group would represent *the* perfect response to the selective pressures; there were multiple "adaptive peaks." The environment, using the term broadly, was constantly shifting. An evolutionary danger was that a population representing an adaptive peak in one environment would be maladapted as the environment changed. However, because of the internal dynamics of the population, aggregates of new gene combinations were constantly being created, and if one of these were better adapted to the new conditions than the previous best adapted, it would then take over, at least temporarily. I suggest that of all the plant and animal species whose population structure has ever been carefully studied, Amerindian tribes most closely respond to the Wright model. This correspondence does not in any sense "prove" the correctness of the Wright model, but, for those who accept the essential correctness of this model (as I do), it seems more than coincidence that the human species, which evolved relatively rapidly, appears to have had the population structure Wright postulated as conducive to rapid evolution.

Punctuated Equilibrium

In recent years, led primarily by paleontologists S. J. Gould and N. Eldredge, there has been a good deal of discussion of the concept of "punctuated equilibrium" (as contrasted with "phyletic gradualism") in evolution.[12] The basic argument is that the fossil record shows that many species remained morphologically static for millions of years, then *suddenly* evolved into new species. Gould and his adherents have applied this concept to human evolution, arguing from what, to me, is a very inadequate, even if rapidly expanding, human fossil record that between evolutionary surges there were static periods as long as one million years in the human line of evolution. As to the biological basis for the alleged bursts of evolution, the fossil record is of course of no value, and the advocates of punctuated equilibrium have been conspicuously vague about the process, keeping the argument on the phenomenological level. This of course is a luxury denied to the geneticist, who cannot think of evolution apart from questions of mechanism.

In large measure based on our own studies, I have suggested that we now have an adequate basis for rapid human evolution in the population structure and mutation rates of the hominid line, without invoking mysterious bursts of evolution.[13] A critical component in the process may be the human kinship system, the guiding factor in the large differences in the genetic composition of current human tribal bands or villages. I believe that the kinship system emerged rather early in human evolution, and, in fact, elements of it are to be seen in some nonhuman primate societies. A key development in its emergence may have been the point at which, in contrast to other primate females, the human female became essentially continuously sexually receptive. In a highly provocative essay, P. J. Wilson identifies the continuous sexual receptivity of the human female as a biologically adaptive development to the need for paternal–maternal bonding created by the prolonged infancy of humans.[14] Cultural anthropologists increasingly identify the evolution of the role of "father" as a critical event in human socialization, the first step toward a kinship system. A chimpanzee male, for instance, has no

Yanomama mother and children, standing in area recently cleared for planting by the slash-and-burn technique.

way to identify his offspring, given the fact that a chimpanzee female, beginning with the α-male, copulates successively with most of the adult males in the band at the time of oestrus, but will not accept a male at other times.

The kinship system plus the role of chance ensured that human bands or groups of allied bands—the basic units of human competition—differed remarkably from one another. Kinship may create a sense of group coherence, which at the same time intensifies the competition with nonkin groups. The "kinship effect" may be stronger in human evolution than in the evolution of other animals and would, in a technical sense, represent an extension of the Wright model as applied to human evolution. Earlier, I suggested that in the course of human evolution new tribes probably arose from old because a band became so detached from the parent tribe that it became the basis for a new tribe. Some of these tribes survived and some did not. Visualize this process repeated thousands of times, with the offshoot village each time being as nonrepresentative of the total tribe as any single Yanomama village would be of the whole Yanomama tribe. Given that natural selection favored the band–tribe with the best complex of genes, then we have the basis for what can be termed rapid step-wise evolution, although each step would be relatively small. When a population is expanding, as when *Homo sapiens* moved out of Africa, the opportunities for step-wise evolution are especially prominent.

Earlier, I commented on the fact that, as measured by electrophoretic studies, an Amerindian tribe contained 70% as much genetic variation as a cosmopolitan population. The electrophoretic approach of course reveals only a fraction of the total genetic variation present but does provide a convenient yardstick. The important point in the context of the present argument is that this variation is sufficient for successive population fissionings along lineal lines to result in a rate of genetic divergence that, viewed through the veil of an imperfect fossil record, could appear to be a mysterious evolutionary burst.

Concurrently, the mutational process, as I will soon argue, was feeding a greater amount of "new" variation into the population than we had realized until recently. The new mutations would make a small contribution to any single generation, but we are talking about 250,000 to 350,000 generations. Natural selection should conserve the favorable mutations, but genetic theory suggests that, by chance, unfavorable mutations can also persist for many generations. Thus, if we could reconstruct the succession of bands that connect any early hominid group or band to the emergence of *Homo sapiens,* we would find that what with the vagaries of tribal origins and mutation, evolution has proceeded in erratic steps, sometimes backward but more usually forward (however that be judged). This viewpoint allows for more small discontinuities in the process than the more classic view of a slow, unidirectional change, but falls far short of punctuated equilibrium.

In Chapter 9, I described the cytogenetically highly damaged "rogue" cells first encountered as a population phenomenon among the Yanomama but now also detected in Caucasians, Japanese, and Ukrainians. They may be the wild card in the evolutionary deck, the least damaged of these cells, when and if they occur in the germ line, representing a class of evolutionary experiments of a different type from the spontaneous mutations we will discuss in the following chapter. The systematic study of so rare a phenomenon as these cells represent is really quite difficult, but, in the present context, it is important to emphasize that although these cells do occur in individuals in bursts, their detection throughout the world's populations suggest that the process resulting in their appearance is ongoing, i.e., they are not occurring in populations in very rare bursts such as might contribute to punctuated evolution.

In closing out this discussion of "punctuated equilibrium," I should say that the "equilibrium" aspect of the concept gives me no less difficulty than the "punctuated." Considering human social structure and our own demonstration of microdifferentiation and private polymorphisms, I find it highly improbable that in the past the human evolutionary process could stand still for long. It is far more likely that the process was continuous, if somewhat jerky, but that, given the still relatively poor human fossil records, the portion of the (highly subdivided) hominid range in which the changes connecting one species to another were occurring has not yet yielded the key skeletal remains.[15]

The time of arrival of the first Amerindians in the New World remains a topic for lively discussion and, sometimes, acrimonious debate.[16] But no matter whether it was 30,000 or 15,000 years ago, there is agreement that the Amerindians very quickly expanded throughout a vast region, their populations repeatedly fissioning. The genetic distances and the private polymorphisms described earlier are an attempt to document the rate of human divergence. We should not, however, in our enthusiasm for these new statistics, overlook the evidence supplied by morphological features. There are repeated reports of villages of blue-eyed, fair-skinned Indians, such as the Mandan Indians of the Dakotas, leading to lingering conjecture concerning the descendants of lost, pre-Colombian Caucasian explorers. I have pictorially documented a Yanomama village where most of the eyes were blue-brown (hazel), the skin and hair decidedly lighter than usual, and the epicanthic fold less

A Yanomama teenager whose lower lip, corners of mouth, nasal septum, and ear lobes have been pierced with a sharp thorn, to create receptacles for whatever seems most appropriate.

pronounced. This village almost merits the term quasi-Caucasian. In some villages in the northern aspects of their distribution, the Yanomama are very small, warranting the term pygmoid. For instance, in one village the men averaged only 147.7 cm in height and the women only 137.5 cm. Likewise, I have encountered a village of the Kanamari whose people, if transplanted to Japan, would elicit no comment. If this degree of microdifferentiation could develop during the Amerindians' relatively brief history in the New World, as a result of processes well understood, why indeed postulate mysterious burst of evolution during the long period of human prehistory?

I would not like to seem to imply that because the population structure of the Yanomama corresponds to the Wright model we now understand all aspects of the human evolutionary process. The Wright model supplies a framework, but there remain many perplexing details, as is the case for the evolution of all other species as well. In Chapter 10 we considered the high rate of loss of mutations from a population. Even with a selective value of 10% above the average, mutant loss, while half that of a mutant with neutral selective value, was still very high. Under these circumstances it is very difficult for favorable mutations to get established. We have just considered how dramatically the genetic composition of villages is regularly reassorted, leading to the loss of favorable gene combinations. Later we will consider the new insights into the complexity of the gene pool and the potential for genetic mishap inherent in the convoluted structure of DNA. All of this is necessary for developing the appropriate humility when, in the final chapters, I attempt to develop a worldwide view of the human genetic condition and suggest what, in light of these unknowns and uncertainties, constitutes an appropriate genetic program for the future.

I will come back to the Indians when we discuss human mutation rates, but this is pretty much the end of the Amerindian presentation. The experience of gathering these data has been simultaneously intensely gratifying and intensely frustrating. Gratifying because I believe we did significantly extend our understanding of the biological basis of subdivided tribal societies while at the same time meeting the ethical obligations to such studies, and frustrating because of how much we didn't learn and now, with the rapid disturbance of these remaining societies, may never learn. If all goes well, we can pick away at the secrets of our DNA for generations to come, but by contrast, ours is the last generation to be able to work with relatively undisturbed tribal populations. How many times has the realization of that fact, and the intellectual agonizing over whether I was really making the "right" observations, kept me thrashing in my jungle hammock, until finally I would get up in disgust, to gaze at those wondrous tropical heavens until peace came.

The picture that emerges from the studies described in this and the preceding two chapters is of a biomedical world profoundly different from that in which modern civilizations are currently functioning. That there were differences came as no surprise, but the extent of these differences did. Currently, either of two viewpoints of the implications of these differences can be argued. One is that humans are almost boundlessly adaptable, and the adjustment to modern civilization entails no significant biological price. The other, to which I subscribe, is that man is indeed highly adaptable, but recently civilization as it has developed has so altered the milieu for survival and reproduction that new selective pressures (or lack of pressures) are now at work, pressures that students of human biology are just beginning to understand. Society must address the possible implications of these developments more forthrightly. It does not follow from the fact that a certain development becomes possible that society's interests are served by pursuing it. As argued in those essays of 35 years ago (pp. 31–33), isn't it time that humankind, guided by some of these newer genetic insights and confronted by a culture motivated predominantly by expediency rather than reason, begins more consciously to match the culture to the animal? I will return to these issues in the final chapters.

Since I was last among them, the Brazilian Yanomama have fallen on truly difficult times. In 1973, the Perimetral Norte road cut through Yanomama land in southwest Roraima Territory, the first step in rendering this area much more accessible. The subsequent discovery of gold and cassiterite in traditional Yanomama territory about 1980 has culminated in an influx of very poor and equally desperate miners into the area, with not only the widespread introduction of disease but violence as well. Estimates of the number of miners streaming into the area run as high as 40,000 persons. It is estimated that as many as 2000 of the approximately 9000 Yanomama in Brazil have already died of disease in consequence of these contacts. In addition, the current techniques of gold mining have resulted in widespread pollution of streams with mercury. In the summer of 1993, Brazilian gold miners are alleged to have massacred some 50 Yanomama. Events as they are unfolding are unfortunately not so different from those in the American "wild west" of 150 years ago, where, with government remote (or even when present), the mistreatment of the Indian constituted one of the most sordid chapters in U.S. history. Without a moral position, we can only appeal to the Brazilian government to enforce a more enlightened policy than we of the U.S. can claim as our heritage. Public outcry over the plight of the Yanomama has been enormous and, perhaps in partial response to this, the Brazilian government, after considering several quite inadequate plans, formally established in 1991 a large Yanomama "reserve" of some 94,000 sq. km, from which miners are being expelled, and the Venezuelan government the same year set aside an approximately equal-sized, contiguous area. How well these governmental actions can be implemented remains to be seen.[17] Acculturation is inevitable but now, perhaps, it can proceed at a more appropriate pace.

On the Flowers of Orchids and the Wings of Butterflies

As a boy, I was an inveterate collector of insects, especially butterflies. When I put a magnifying glass on the wing of a butterfly, I wondered at the delicacy of the shadings and transitions of color in, for example, the wing of a tiger swallowtail. Given (as I later learned) the *probable* rather coarse resolving power of the insect eye for detail, why the evolution of such delicacy? It really shouldn't matter to an insect seeking a mate, while the various predators would scarcely pause to appreciate the pattern before they swooped to the kill. This same type of question surfaced in a more demanding form when the jungle trips began: I became hooked on orchids. It was difficult to believe that such aesthetically pleasing flowers could emerge from such generally scrubby looking plants. It's true that most of these flowers had to attract insects for their pollination, but this is thought to be more by odor than by the beauty of the floral parts, which should be lost upon an insect. What is the possible utilitarian value of the delicately punctate magenta on the pinkish background of a Phalanopsis blossom, the fringed lip of some of the Epidendrum flowers, or the elaborate petals of *Oncidium kamerianum*? From admiring orchids in their natural setting, I progressed to bringing a few home, then more. The area of the corner of the living room in which they sat expanded. And in due time, Priscilla gave me an ultimatum. Build a greenhouse or else. So we did—a really little one, but 15 minutes in it at the end of a hard day in the academic trenches has remarkable curative powers.

In this fascination with orchids, I feel I have at least one bond with Charles Darwin. It would be very easy, reading many of the accounts of Darwin's life, to visualize him as one who after the voyage of the Beagle ended in 1836, spent the rest of his life as an armchair scientist, engaged in the great synthesis. He was, however, fascinated by the range of adaptive mechanisms exhibited by the flowers of orchids, publishing in 1862, only three years after "The Origin of Species," a book entitled, "The Various Contrivances by Which Orchids are Fertilized by Insects," a book that is a model of careful observation and, given the times, ingenious experimentation. Reading it again recently, I was impressed that while Darwin's observations still stand, there is no evidence that to be effective these contrivances require the frilled lips and the nuances of color, so aesthetically pleasing to an animal like man, who wasn't even around when they were well evolved.

It is in "The Various Contrivances by Which Orchids are Fertilized by Insects" that the usually mild Darwin permits himself a beautiful jibe at the Creationist controversy, raging then as it is now. After discussing the structure of the pollen masses in *Monachanthus viridi,* he writes:

> Thus every detail of structure which characterizes the male pollen masses is represented in the female plant in a useless condition. Such cases are familiar to every naturalist, but can never be observed without renewed interest. At a period not far distant, naturalists will hear with surprise, perhaps with derision, that grave and learned men formerly maintained that such useless organs were not remnants retained by inheritance, but were specifically created and arranged in their proper places like dishes on a table (this is the simile of a distinguished botanist) by an Omnipotent hand "to complete the scheme of nature."[18]

Periodically, flushed with new knowledge, I stop to ponder the "unnecessary" delicacy of butterfly wings and orchids. My designation of "unnecessary," of course, violates all scientific canons, assuming as it does that other animals see orchids as we do and that these patterns do not trigger some very basic response of their nervous systems. But there is another context in which orchids give pause. Evolutionists in general subscribe to the concept of "genetic coadaptation," that natural selection in time brings together a set of genes attuned to each other, working in precisely coordinated harmony in morphogenesis to produce a functional organism. But in hybridizing orchids, one can bring together gene sets which have been separated for perhaps 100,000 years, involving phenotypes as diverse as that of an elongate, many flowered Epidendrum and that of a pseudobulbous, several-flowered Cattlya; or that of a Eulophiella with very large pseudobulbs and a creeping rhizone with that of a slender-leaved, compactly pseudobulbous Cymbidium, and the result is a fully functional plant, different from both parents, but doing all the right things at the right time. This is my ultimate example of the still poorly understood homeostasis inherent in living material.

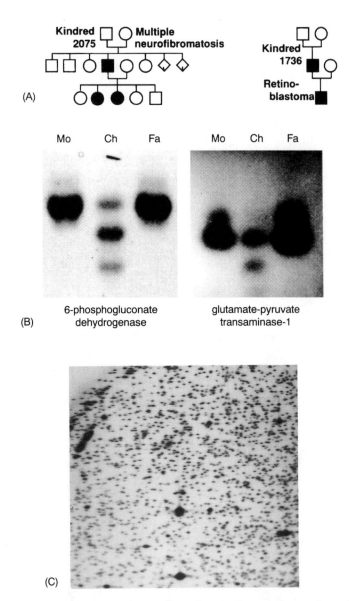

(A)

Kindred 2075 — Multiple neurofibromatosis

Kindred 1736

Retino-blastoma

Mo   Ch   Fa   Mo   Ch   Fa

(B)   6-phosphogluconate   glutamate-pyruvate
       dehydrogenase   transaminase-1

(C)

The changing faces of mutation-rate studies. A. Two pedigrees from the muta-tion-rate studies on retinoblastoma and neurofibromatosis described in this chapter, illustrating the *de novo* appearance in a family of a trait that is there-after dominantly inherited. B. Two examples, from the biochemical studies of the genetic effects of the atomic bomb, of mutation resulting in an enzyme pattern in a child not present in either parent. (The enzymes in question are indicated on the figure.) C. A two-dimensional display of the results of the enzymatic digestion of the DNA of a single person (courtesy of Dr. Asakawa). A fragment found in a child but not present in either parents is a potential indicator of mutation.

12

Spontaneous Mutation

As we have seen, the sole source of the variation in the genetic material upon which inherited characteristics and evolution depend is the mutational process. It is difficult to think of a more basic phenomenon in biology. My commitment to the study of mutation rates dates back to that sudden burst of mutation in my *Drosophila* stocks at Dartmouth, a commitment then confirmed with my participation in the studies in Japan. Those Japanese investigations, however, have been only part of my involvement with this subject. Now I shall discuss a series of studies carried out in Ann Arbor from 1949 down to the present on spontaneous mutation rates, and then try to place these in juxtaposition with some of the findings from Japan and South America.

Mutation and Human Disease

The human geneticist who studies mutation operates under constraints not shared by the experimentalist. Thus, whereas the experimentalist can develop special test strains of *Drosophila* and mice and then manipulate these subjects as he sees fit, the human geneticist has to take mutations on a catch-as-catch-can basis. More specifically, whereas the experimentalist can set up crosses to detect both recessive and dominant mutations, until very recently it was only feasible to study the latter type with any accuracy in human populations, and then, as we shall see, only for selected traits. There is, on the other hand, one great advantage in the study of mutation in humans. A number of diseases of medical importance are maintained in the population by mutation pressure. People with these diseases reproduce at a lower rate than usual, so that without mutation the diseases would die out. A dominant mutation manifests itself by the appearance in a normal family of an affected person who, upon reproduction, passes the trait on to half the children. The advantage in studying these mutations is that there is a medical network busy detecting these people—one has only to decide the best ways to tap into it.

My first effort at estimating a human mutation rate, in a collaboration during 1949–1950 with H. F. Falls, (mentioned earlier for the important role he played in the fledgling days of the Heredity Clinic), involved retinoblastoma.[1] This is a highly malignant tumor of the eye encountered primarily in children, the average age at diagnosis being two years. The medical literature at the time of our study showed that although most cases were sporadic, i.e., the first known occurrence in the family, there were instances in which a person who had survived the disease had produced similarly affected children. This observation suggested that the condition was due to a dominant mutation in the germ line; if a person with the mutation survived (because the affected eye or eyes were removed before the tumor spread), the mutation could be transmitted to his/her offspring. We wished to determine how often that mutation occurred. For this we needed a large population base, which we decided, somewhat arbitrarily, should be the children born in the State of Michigan between 1936 and 1945. Step 1 was to establish as complete a roster as possible of children thought to have the disease and born during this period. This required combing hospital records throughout the state, consulting the records of ophthalmologists and pathologists, examining the records of the State School for the Blind, and even a little newspaper publicity. The next step was to check these diagnoses out and obtain good family histories as well as examinations of the immediate relatives, almost always, for their convenience, in their homes. There went more weekends than I care to remember. Finally, then, when we had sorted our cases out, we had to consult the state records on births for that ten-year-period, so that we could calculate the exact frequency of this mutation per 100,000 children born.

Fieldwork of this type has its rewards but also its trials. Most families were delighted in the interest we took in the disease and were happy to cooperate. On the other hand, for some families the experience, culminating in the death of a child, had been so traumatic, and there was such resentment over the perceived inadequacies of medicine, that any discussion of the past was extremely difficult. Since the disease was no respecter of occupation or education, we saw a true cross-section of Michigan.

In the end, we determined that among the 1,054,985 children born in Michigan during this period, there were 52 who ultimately developed the condition. Of those, 49 were born to unaffected parents and were presumed due to mutation. This was 4.6 mutations per 100,000 children, or, since each child presumably had two genes at risk for this condition, 2.3 mutations per 100,000 genes each generation. This estimate—one of the early estimates for a human mutation rate—stood for about a decade, but then had to be revised substantially downward. Other subsequent investigators, with greater time depth to their records, could demonstrate that although, as we knew, there were many examples of a child who survived transmitting the disease, if the reproductive records of a large collection of children with retinoblastoma were analyzed, not as many who reproduced transmitted the condition as should be the case if it were always a dominant mutation in the germ line. This led to the recognition that the mutation in some of these cases involved only the *somatic cells* of the child, specifically, the cells of the retina of the eye. These mutations presumably arose in a single somatic cell during embryogenesis, a somatic cell whose descendants did not contribute to the germ line. These persons will not transmit the disease. Other mutations occurred *in the germ line* of one or the other parents, so that the mutation was present in all the cells of the child, but found preferential expression in the eye. These persons will transmit the disease. Our estimate— meant to be a germ-line estimate—had to be revised downward, by approximately one-half, to exclude the fraction of cases due to a somatic mutation.

On the basis of a hypothesis advanced by Knudson in 1971,[2] it develops that even the apparently dominantly inherited retinoblastomas have a more complicated basis than previously surmised. With respect to inherited retinoblastoma, it is postulated that the affected child inherits a mutated gene from a parent, but the gene finds expression only if there is a second (somatic) mutation in a retinal precursor cell, involving the sister (homologous) allele of the gene already mutated. This second mutation can be shown with current techniques to be any one of an array of events, including loss of part of the chromosome carrying the normal counterpart of the retinoblastoma gene. The noninherited type of retinoblastoma appears to be due to independent mutations of the two retinoblastoma genes in a single somatic cell, in this case, a retinoblast of the eye. Since both retinoblastoma genes must be mutated for the tumor to develop, the normal allele is technically a recessively acting, tumor-suppressor gene. Our data themselves have stood the test of time, but the precise interpretation has undergone successive modifications.

In collaboration with various colleagues (M. Shaw, J. H. Chandler, H. F. Falls, T. E. Reed, W. J. Schull, and F. Crowe), I was involved between 1950 and 1960 in the investigation of four more dominantly inherited traits that are maintained in the population by the pressure of mutation.[3] The first was multiple polyposis of the colon, a condition in which the large bowel is studded with thousands of polyps, in one or more of which malignancy develops by an average age of 40. The second was multiple neurofibromatosis, to which I return shortly. The third and fourth were Huntington chorea and aniridia, the former a degenerative disease of the brain which usually is diagnosed between ages 30 and 50, the other a congenital absence of the iris of the eye, which results in extreme light sensitivity. In each study the routine was similar to that mentioned earlier for retinoblastoma but modified as appropriate to the disease. None of these diseases brings any joy to the family, so that we were constantly treading on delicate ground. Given my concern with the possibility that increased exposure to radiation (and certain chemicals that might make their way into the environment) might augment mutation rates, these studies increased awareness of the impact of any such increase.

As mentioned, one of our efforts to estimate a human mutation rate was concerned with multiple neurofibromatosis. In 1951, Jack Schull and I decided that this syndrome, with a frequency of about 1 in 3000 persons, would be an appropriate condition for a mutation-rate study. Even today, although the gene involved has now been identified, neurofibromatosis remains a poorly understood disease, characterized by scattered, light brown areas of cutaneous pigmentation (café-au-lait spots) and nonmalignant tumors of the nerves, spinal cord, and brain. The café-au-lait spots appear shortly after birth but the tumors later, in the teens and twenties, and then gradually. Although they only rarely develop into a cancer, the tumors are disfiguring and can cause difficulties as they expand. Since the tumors, which are the hallmarks of the disease, are not present in children and since we needed an accurate estimate of the frequency of the condition, we had to develop good diagnostic criteria.

The first step was to recruit a young dermatologist, Frank Crowe, to work with us. Frank's background was unusual. The year preceding his projected entry into medical school, he had accepted employment with a construction company developing fortifications on a remote island in the Pacific, Wake Island. It wasn't much of a job, but the pay was good and would help finance his medical education. When the war with the Japanese broke out some months after his arrival, Wake, defended by a small military detail, was captured early on. The civilians on the construction project were considered to be military personnel, and treated as POWs. Frank survived four gruesome years in captivity and returned to complete medical school and specialize in dermatology. He was a person of remarkable equanimity; I was never certain whether this was what helped him survive his ordeal, or whether it was developed during the ordeal. Be this as it may, we settled into a very systematic appraisal of over 200 persons with the condition. Out of these studies emerged a new diagnostic criterion for the disease. We determined that in those cases where the diagnosis was certain because of the presence of the characteristic tumors, 80% of the people had six or more café-au-lait spots that were more than 1.5 cm across. This meant that in children who had not yet developed tumors, a presumptive diagnosis could usually be made simply by counting the number of the café-au-lait spots. To my knowledge, this is the only diagnosis in medicine to be made by a simple counting procedure.

After three years of work, we developed an estimate of the frequency of the condition and the proportion of sporadic cases, from which we could estimate a mutation rate. That rate, of between 14 and 16 events per 100,000 germ cells, remains the highest yet reported for any human condition. Although I now believe we may have erred a bit on the high side, it still seems likely that as dominant mutation rates go, this one is an outlier. The realization of the relatively high frequency of this poorly understood syndrome—greater than hemophilia, greater than Huntington chorea, greater than childhood muscular dystrophy—finally led in 1978 to the formation of a National Neurofibromatosis Foundation. Frank showed such an aptitude for clinical research in the course of this study that the Chairman of the Dermatology Department (and ourselves) made strenuous efforts to direct him into a career in academic medicine. He, for his part, had so dreamed during those years of captivity of returning to his native Boise, Idaho to practice medicine that nothing we could say had any effect and, back in Boise, he fulfilled that dream for the rest of his life.

Other people were simultaneously working on such estimates, for these and other diseases. To date, some 20 medical diseases have been studied with respect to the frequency with which they arise *de novo* each generation because of mutation. The average frequency with which, each generation, mutation results in the appearance in the population of persons afflicted with any specific one of these conditions is about 4/100,000 births.[4] With 20 such conditions, the total impact would be roughly 80/100,000 births, i.e., close to 1 in 1000 newborns is destined to develop one of these 20 chronic conditions because of spontaneous mutation. This of course is not the complete story of the impact of dominant mutation each generation, because by no means all of the possible studies have been carried out. The number of liveborn infants who at birth or later will be *significantly* handicapped by freshly arisen, dominant mutation is surely several times over the estimate just given.

For a number of reasons we have to be very cautious about thinking of these individual mutation rates as "typical." On the one hand, a geneticist can't study the rate with which mutation results in a medical syndrome unless there are enough mutations to make such a study feasible. One will not study the rate of occurrence of dominant mutations in the instance of a gene that never gives rise to a dominant mutation. This should bias the estimates upward. On the other hand, the same genes giving rise to dominant mutations should also be giving rise to some (recessive) mutations whose phenotypic effects are not immediately apparent; the study of the rate of appearance of these dominant traits did not tell us anything about the frequency of the hidden, recessive mutations of the same gene. Finally, a given disease could be due to mutation at two (or even more) different genetic loci, in which case the true rate per genetic locus involved in the trait is substantially less than the estimated rate. What these studies did constitute was a partial demonstration of the impact of spontaneous mutation on the medical burdens of a population. If the mutation rate of a population were to increase by 100% because of some type of mutagenic exposure, this medical burden of chronic disease would be doubled.

The Study of Mutation Goes Biochemical

In the 1960s, a better way to study mutation began to present itself. This was the use of electrophoresis to detect variant proteins. The great advantage of this biochemical approach for studies of mutation was that we could now choose to study for the rate at which it was altered by mutation, any protein that could be accurately examined electrophoretically. We could thus escape the bias inherent in the study of dominant mutations, created by the fact that if a gene did not give rise through mutation to a *clear-cut clinical entity,* one couldn't study its mutation rate. Furthermore, now we could study the previously hidden recessive mutations; most of the changes we could detect electrophoretically were not, in heterozygotes for the allele, accompanied by any gross departure from normality or by illness and so technically were recessive in inheritance. From the studies on experimental organisms, we knew recessive mutations to be more common that dominant. On the other hand, no single approach to the study of mutation rates will detect all possible types of mutations and, as already noted, electrophoresis detects only those changes in the genetic code that result in the substitution in a polypeptide of an amino acid that changes the charge of the protein molecule.

As will be described in Chapter 13, in 1975 we initiated in Japan a study searching for mutation resulting in electrophoretic variants in the survivors of the atomic bombings who had received significant amounts of radiation, as well as in suitable controls. The data on the controls would provide an estimate of the spontaneous mutation rate; the data on the exposed would test the hypothesis of an increase in this rate caused by the radiation exposure of the survivors. At about the same time we began a parallel study in Ann Arbor, limited to a search for spontaneous electrophoretic mutations in children born in the University's Women's Hospital. For this purpose, after obtaining parental consent, we collected samples from the placentas of the newborn infants plus venous samples from each child's father and mother. The child's sample was subjected to the same battery of determinations employed in Japan; in both studies, whenever a variant was encountered that was not one of the common genetic polymorphisms the parental samples were examined to determine whether the variant was inherited or had just arisen anew through mutation.

In 1986 we summarized all our data to date, from Japan and the U.S., on the rate with which spontaneous mutation resulted in electrophoretic variants.[5] We could also include in this summary the extensive comparable data from the Galton Laboratory in London, gathered by H. Harris, E. Robson, and D. W. Hopkinson,[6] and also the data from West Germany gathered by K. Altland and his colleagues.[7] Altogether, four mutations resulting in proteins with a changed electric charge had been detected in 1,226,097 tests. This corresponds to a mutation rate of 3.3 per 1 million gene tests each generation, *for this kind of mutation.*

As just noted, electrophoresis will not detect all the changes in a protein that may result from mutation. From knowledge of the genetic code and the amino acid composition of proteins (see later in this chapter), it may be calculated that for every mutation resulting in an amino acid substitution that changes the electrophoretic mobility of a protein, there is at least one additional substitution that is undetected because it does not alter protein mobility. In addition, some changes in the genetic code (the so-called synonymous changes) will not induce a change in amino acid composition, and others will result in an incomplete polypeptide; these will not be detected by electrophoresis. Finally, from studies in an experimental setting, it appears likely that corresponding to every mutation that can be detected by electrophoresis, there is *at least* one other characterized by some kind of rearrangement of the genetic material, in consequence of which no gene product (polypeptide) is formed at all. Thus the true adjusted mutation rate is—conservatively—of the order of at least 1 in each 100,000 genes tested per generation (rather than the 4 per 100,000 genes suggested by the study of dominantly inherited traits). I have to emphasize that this estimate is still not as exact as we would like. Only a very small fraction of the genome has been sampled for this estimate; the real figure could be half that given, or even twice, but we are certainly getting close to the true value. Incidentally, the possibility that these apparent mutations are not really that, but due to a discrepancy between legal and biological parentage, can be (and has been) rendered remote by the appropriate genetic studies.

As I set out this account of my life and times, I find myself realizing better than ever before how my various activities have interdigitated with and reinforced each other. Now it was time for the Amerindians to get into the mutation act. At the outset, in 1962, there had been a variety of justifications for the Indian studies, but the study of mutation was not one of them. In discussing the Indian studies, I have mentioned the extent to which we employed electrophoresis to detect genetic variants of some of the serum proteins and erythrocyte enzymes, with a view to characterizing the nature of the genetic differences between tribes. It must have been sometime in 1970, after reading a paper by the Japanese geneticists M. Kimura and T. Ohta,[8] that I realized that we could employ our data on electrophoretic variants among Amerindians to generate estimates of mutation rates, but now the approach would be indirect rather than the direct approach discussed thus far.

In the direct approach, as practiced with the Japanese and Ann Arbor populations, one examines children for the kind of electrophoretic variants that might result from mutation and, when one is found, checks the parents to see whether the variant is inherited or, if both parents are normal, whether it results from a mutation. The indirect approaches—there are basically three—are both more subtle and more subject to error. They are applicable only when one has data on electrophoretic variants from a highly isolated population for which one can estimate the number of individuals contributing children to successive generations, such as an Amerindian tribe. One makes the assumptions (1) that these variants have no effect on their bearer's survival, (2) that tribal numbers have been more or less constant for many generations, and (3) that the input of these variants from mutation equals their loss through chance failure to be passed on. These three assumptions add up to the postulate that, for these purposes, the tribe is in genetic equilibrium. This postulate cannot be strictly true for any tribe we have ever studied, nor, for that matter, for any tribe that ever existed. In applying formulae that assume equilibrium to results from these tribes, then, we are accepting an error the exact magnitude of which is difficult to specify. I have explored the effect upon the estimates of mutation rates of reasonable departures from our estimates of tribal size and from the assumptions of the method, and accept that our final indirect estimates (see below) may well be in error by a factor of two because of incorrect input to the calculations.

All three of these indirect approaches require an estimate of the average survival time in generations of a "neutral" mutation in these Indian populations, the term neutral implying that the mutation neither helped nor hindered its recipient. This we could get from the computer-simulation model of an Indian tribe I have described in Chapter 9.[9] The first (Kimura–Ohta) indirect approach needs in addition to the foregoing only an accurate estimate of the average number of these rare variants for each protein and tribe examined and an estimate of the number of adults in the tribe. The second approach, developed by Elizabeth Thompson (then of Oxford) while spending a summer with our group, needs this information but also an exact estimate of the frequency of each variant.[10] Finally, the third approach, developed in collaboration with Ranajit Chakraborty, requires in addition an estimate of tribal size.[11]

I will not burden the reader with the actual calculation. When all the data had been tidied up and the estimates were in, the first indirect approach yielded an estimate of 13 *electrophoretic* mutations per generation per 1 million genes tested; the second yielded an estimate of 7 electrophoretic mutations per generation per 1 million genes; and the third, an estimate of 11 mutations per 1 million genes.[12] The average of these estimates—10/1,000,000—is some three times higher than the results obtained on different populations by the direct approach, but because of the errors to be attached to the results of all the methods, we cannot be sure the estimates are really different. A conservative position would be that for these loci, in these populations, mutation results in an electrophoretically altered protein at a rate of at least 7 per million genes per generation, and the rate at which all types of mutations arise is of the order of 20 per million genes per generation. Our findings raise the possibility that these tribal populations actually have higher mutation rates than industrialized, civilized populations. This is not an unreasonable proposition. I have already indicated the many viral diseases to which these populations are subject; it is well known that certain virus infections tend to produce gross chromosomal damage, and whatever produces chromosomal damage may result in more subtle mutations. Furthermore, we are now aware of the occurrence of many "natural" mutagens in foodstuffs (a topic to which we return later), and the Indian certainly consumes a number of exotic and still inadequately analyzed items, and uses a variety of natural hallucinogens. On the other hand, the error in the two estimates of mutation rate (i.e., for industrialized and for tribal populations) is considerable, and they cannot be said to differ significantly. We may never be able to prove a difference in mutation rates between tribal and industrialized populations. At the very least, however, the data provide an important perspective when, later, we consider the problem of present mutagenic exposures in industrialized populations. The current data provide no reason to believe that mutation rates have generally increased with the advent of civilization, industrialization, and all the attendant new, potentially mutagenic exposures.

Chromosomes Mutate, Too

The mutations I have been discussing thus far are what the geneticist calls "point mutations." This is a term for highly localized genetic events, involving a very specific region of the chromosome concerned. Examined under the microscope, the chromosomes in which these events have occurred are not different from normal chromosomes (which fact does not exclude small chromosomal deletions). There is, on the other hand, a much grosser type of mutation, involving such large chunks of a chromosome that the event is visible with an ordinary light microscope. These are termed "chromosomal mutations," and have been well known in plants and animals since early in this century, but it was not until the late 1950s that cytological techniques permitting the detailed examination of human chromosomes began to become available. With these better techniques, a number of investigators set out to examine consecutive, unselected newborn infants for the occurrence of chromosomal abnormalities. The results were astonishing: about 6 per 1000 newborn infants carried a major chromosomal defect. The results were even more surprising when early spontaneous abortions were similarly surveyed. Approximately 50% exhibited gross chromosomal abnormality. Obviously, most of the developing fetuses with gross chromosomal defects died at an early stage of development. From these figures, it could be calculated that *at least* 6% of newly fertilized eggs carried a major chromosomal abnormality, with approximately 5.4% terminating as spontaneous abortions and 0.6% born alive.[13] I say "at least" because it has been difficult to recover and study early abortuses much before three to four weeks of gestation. Conceivably, the elimination of chromosomally abnormal children is even higher during the very early stages of development. These findings put a new light on the biological significance of abortion: in rejecting fetuses, the human uterus has not been acting capriciously but often with a strong biological purpose.

What Do these Findings Suggest About Gamete Mutation Rates?

All of the developments described in this chapter permit some very interesting if approximate calculations. The true number of functional genes in humans is unknown, but a figure commonly suggested is 50,000. If the average mutation rate per generation is (conservatively) 1 in 100,000, then the average sperm or egg cell carries 0.5 new mutations in a functional gene. The average (diploid) human carries one new mutation of this type. It must be emphasized how preliminary these estimates are. On the other hand, even if these estimates were too high by a factor of two, these calculations suggest that mutation is, so to speak, a fact of life.

This calculation can be carried one step further. To do so, we must turn to current concepts of the genetic material. The concept of the gene that obtained during my early years in genetics (as a proteinaceous bead) and the concept of a chromosome as a sequence of such beads joined by a nucleic acid string, was completely replaced in the 50s and 60s by one of the most dramatic developments in the history of science. Now a chromosome has been shown to be essentially one continuous, double-stranded, helically coiled super molecule of deoxyribonucleic acid (DNA), a molecule comprised of joined alternating molecules of a sugar (ribose) and a phosphate, with either a purine or pyrimidine molecule attached to each sugar (see figure on page 222). The purine moiety may be either adenine or guanine; the pyrimidine, either cytosine or thymine. This ribose–phosphate–purine (pyrimidine) complex is termed a nucleotide, which in the double strand of DNA pairs with other nucleotides according to specific rules. These paired elements are referred to as base pairs (bp), and a thousand of them in sequence as a kilobasepair (kbp or kb). Three such sugar/phosphate/purine or pyrimidine moieties in sequence function as a codon, specifying which amino acid occurs at a particular position in the string of amino acids that constitute a polypeptide. The three-dimensional structure of these intricate molecules is illustrated in the figure on p. 223. A gene consists of any sequence of these moieties that codes for a genetic message, plus the DNA necessary to regulate the gene's activity. Most genes are translated into polypeptides through a complex process involving the intermediation of another kind of nucleic acid, ribonucleic acid (RNA), but some genes fulfilling vital functions do so more directly, through DNA.

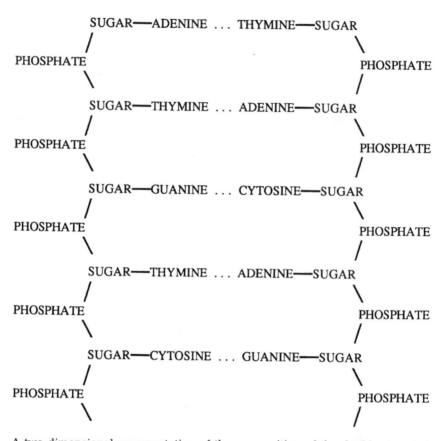

A two-dimensional representation of the composition of the double stranded DNA molecule. The two strands match up so that adenine (A) pairs with thymine (T) and guanine (G) with cytosine (C). There is a loose chemical bonding (hydrogen bonding) between adenine and thymine, and cytosine and guanine, and this bonding gives the double strand stability. The complementarity of the two strands is propagated at the time of cell division because when the double helix replicates, each of the two strands serves as a template for the other.

Mutation consists of a spontaneous or induced change in the DNA sequence, either in the form of a substitution of one purine/pyrimidine for another or more gross changes in the nucleic acid chain.[14]

We may now proceed to make one of those calculations on which geneticists thrive. Following the above, let us assume that in round numbers it should require between 1000 and 2000 nucleotides to code for the average protein; it is mutation in these nucleotides that we have been measuring by electrophoresis. If we assume that the nucleotide mutation rates measures thus far are representative of all the DNA, then the rate of mutation per nucleotide per generation should thus be the rate per gene divided by the number of nucleotides per gene; the resulting number is between 0.5 and 1.0×10^{-8}. Now, the total number of nucleotides in a human gamete has been estimated to be some 3×10^9. Only a fraction of these nucleotides are found in genes devoted to the elaboration of a protein product. It is further assumed that the mutation rate in this remaining DNA is comparable to that which we have measured in the DNA coding for the proteins of blood serum and erythrocytes, then the total mutation rate for nucleotide changes in a human gamete should be something like $3 \times 10^9 \times (0.5$ or $1.0) \times 10^{-8}$, or 15–30 per gamete. This is a conservative estimate, in the sense that the higher estimates of rate from the Amerindian studies do not enter into the calculation. Even so, this implies 30–60 new mutations involving nucleotide substitutions in a fertilized egg.

A three-dimensional representation of the DNA molecule, the famous double helix. This spiral structure arises from the internal tensions of the molecule. The hydrogen bonding that stabilizes the helix is represented by the letter "H."

There is reason to believe there may be an approximately equal number of mutations resulting in small deletions, duplications, or re-arrangements of the DNA. As for estimates of individual mutation rates, so these estimates of total rates carry the possibility of major errors, but the order of magnitude seems likely to be correct. Finally, there are the gross chromosomal mutations mentioned earlier. To those of you only now making an acquaintance with genetics, the total number of muta-tions these findings imply may seem very reasonable. To those of us who have spent the past half century in genetics, that number is heretical. In the past, the argument has been that since mutation is a random change in a finely balanced organism, most mutation must be harmful. That simple nucleotide substitutions *can* be harmful is well documented by the sickle cell and hemoglobin C mutations we discussed so extensively earlier. Some years ago a noted English geneticist, J. B. S. Haldane, argued that each deleterious mutation (and most mutations were pre-sumed deleterious) was eventually responsible for one "genetic death."[15] How can the human organism sustain this "load of mutations" (to borrow a phrase from Muller)?

At first blush an escape from this dilemma would seem to be pre-sented by the fact that no clear function has yet been determined for most of the DNA. Thus, if there were 50,000 active genes and each one was comprised of 1000 to 2000 nucleotides coding for protein plus some adjacent DNA necessary to the gene's functioning (say, 1000 nucleo-tides), then only some 3–5% of the total genetic material is essential. Unfortunately for that line of reasoning, there are now increasing ex-amples of mutations that have a profound effect on gene expression occurring several thousand bp from the DNA that actually codes for the amino acids in a protein. The regions in which these mutations occur must be considered part of the functional gene. I will address this phe-nomenon in Chapters 14 and 15. For now, it suffices to say that the amount of the DNA that can be regarded as of no functional significance is constantly being whittled down as the DNA is studied more closely.

In addition to this empirical basis for believing that much more of the DNA than that just necessary for coding proteins is functional, there are theoretical grounds for doubting that mutation in chromosomes can remain neutral in its effects for very many generations. There are now numerous examples of mutation that has resulted in a loss of DNA. While "in the beginning" some of these mutations could be harmless, after a while all the redundant DNA should be lost, so that any further loss is deleterious. A simple but imperfect analogy is of a very ornate clock that comes equipped with superfluous bells and whistles and even spares or backup systems for some of the critical pairs. It can continue running for some time in the presence of losses, but sooner or later they catch up with a vital part. In the long run these losses are met only if DNA *of the "right" sort* is being constantly generated through selection or as a result of mutations that replace this DNA just as rapidly as it is being lost. There is as yet no critical evidence on "gain" mutations, but should this point be correct, then our already high estimate of the total mutation rate must be increased (to include these gain mutations).

Two Distinct Types of Mutation Events

It is customary to envision the mutational events we have been discussing as resulting from the inherent instability of the complex genetic material, and especially from slip-ups when at mitosis and meiosis this material must replicate itself. There may, however, be an important second cause for spontaneous mutation in humans. Earlier, I recounted my experience with an epidemic of mutation in one of my *Drosophila* strains and mentioned briefly the various explanations of such phenomena that were subsequently developed. For some time, one of these explanations, mutation as a result of the activation of transposable genetic elements inserted in chromosomes, was thought to be a rare phenomenon, but now a wide variety of such elements has been identified in experimental organisms, with, in *Drosophila,* such picturesque names as "hobo," "gypsy," and "copia." M. M. Green, a long-time student of "jumping genes" in *Drosophila,* now estimates that as many as half of the spontaneous *Drosophila* mutations are associated with this phenomenon.[16] There is no reason to suppose that humans are immune to this mechanism of mutation, and indeed, the frequency of retroviral footprints in human DNA (see Chapter 15) certainly suggests that the physical basis for this phenomenon exists in humans.

Previously, I described the rogue cells we first encountered in Amerindians but which were subsequently also observed in normal Englishmen, Japanese, and Ukrainians. We have suggested that these might either be a manifestation of transposable (jumping) gene activity or of some acquired viral infection. Most of these cells are so abnormal that they could not complete a mitotic cell division, but a few of the least damaged may make it through the process. If this phenomenon also occurs in germ cells, it could be a cause of germ-line mutation in humans. As we will see when we consider the detailed structure of genes, it appears that a salient feature of evolution has been genetic rearrangements of pieces of "old" genes to build "new" genes. Such rearrangements constitute a kind of mutation, which I now believe may be driven

225

by the kinds of mechanisms that create rogue cells. This kind of mutation will be far more difficult to study than what we have measured in the past. The mutations characterized by the substitution of one nucleotide for another in the DNA strand are much more apt to result from a simple copy error at the time of DNA replication. Currently, what may ultimately be regarded as a third distinct type of mutation is emerging. This is the appearance within the exons of genes (and elsewhere in the genome) of multiple repeats (expansions) of specific two or three nucleotide sequences that normally are present in only a single sequence (or a very limited number of repetitions). Some seven inherited diseases (including Huntington chorea) have now been shown to be due to this phenomenon. With this demonstration, it is clear that mutation disturbs the genetic material in every possible manner.

It is an axiom of population genetics that if a species is to survive for long, the input from mutation in each generation—good, bad, or indifferent in its phenotypic effects—must be more or less balanced by a corresponding loss of the results of mutation in past generations. If this is not the case, the products of the mutational process will increasingly accumulate in the species' genome, and since, as we argued earlier, the logic of the situation demands that much mutation has deleterious effects, the species will deteriorate. We have just estimated that there may be some 60 to 120 new mutations in a fertilized egg. In Chapter 6, however, we saw that the results of studies of inbreeding effects suggest that the average Japanese—who cannot be much different from the species average in this respect—carries only two lethal equivalent genes finding expression between about the fifth month of gestation and maturity. (Two lethal equivalents can be achieved by two alleles at different loci that are incompatible with survival, or four alleles at four loci each of which when homozygous would result in a 50% reduction in survival, etc.) Even with allowance for the action of several more such genes before the fifth month of gestation or after maturity is attained, this seems a very small accumulation considering the input from mutation in each generation. One possible explanation stems from the fact that as many as 50% of all fertilized human eggs fail to survive past the fifth lunar month of pregnancy. Perhaps much of this loss is for genetic reasons, the chief mechanism for balancing the mutation–selection books. If this is not the case, then we simply do not understand at present how the critical balance between mutation and selection is maintained.

As will be apparent from the foregoing, the study of mutation is laborious and slow-moving, at a time when the new DNA technologies provide much quicker scientific return. If, however, one's basic interest is in how the DNA got that way, there are really only two factors to be reckoned with: mutation and selection. As I have already made clear, with the rapid dissolution of tribal cultures, it is now virtually impossible to mount significant studies of the selective forces that shaped us. If, on the other hand, mutation rates have not changed greatly with civilization, and this seems to be the case, then it is in the study of the kinds and frequency of mutations that we should be able to generate the most solid insights into the evolutionary process currently possible. This is what has kept me returning to the study of mutation despite the laboriousness of such studies.

This (high) total mutation rate provides a perspective against which to view the relative magnitude of the danger posed by current exposures to such mutagens as radiation and certain chemicals. A genetic generation ago, when mutation was envisioned as a much rarer event than the evidence now suggests, H. J. Muller,[17] envisioning the human species as delicately balanced as regards its ability to cope with (i.e., eliminate) deleterious mutations, could adopt an extreme viewpoint concerning the genetic risks of increased exposures to radiation. But if the species is already coping with substantial "spontaneous" mutation rates—to be sure, in ways not well understood—this alters the perspective on the risks of radiation. It is appropriate that we now turn to the final chapter on the study of the genetic effects of the atomic bomb.

The author (left) and W. J. Schull, with the buildings of the Radiation Effects
Research Foundation in the background (1991).

13

New Approaches to the Effects of the Atomic Bombs

In Chapter 5, we left the study of the genetic effects of the atomic bombs with the publication in 1956 of a monograph summarizing the data collected during the "morphological era." From 1955 until 1971, the genetics program at the Atomic Bomb Casualty Commission (ABCC) continued only as a study of the survival and sex ratio of the children born to persons exposed to the atomic bombs, and my scientific energies went largely into the Indian studies. It was also during this interval that we had carried out the studies on consanguinity effects in Hiroshima, Nagasaki, and Hirado described in Chapter 6. With each of these consanguinity studies we had taken advantage of our presence in Japan to update the analysis of sex-ratio and survival curves among the children being born to survivors, but these were really rather minor activities.[1,2]

A Cytogenetic Program

There was, however, during this time one very important genetic development at the ABCC. Beginning in the 1950s new techniques had made it possible to examine human chromosomes much more carefully and precisely than in the past, and with this development had come the amazing series of discoveries regarding the frequency of abnormalities of human chromosomes mentioned in the last chapter. It had been firmly established during the 1930s, from experiments on a wide variety of plants and animals, that radiation produced many types of visible chromosomal damage in the somatic and germinal cells of all the plant and animal species properly examined. Now with the new techniques it was possible to inspect the cells of survivors of the bombings for chromosomal damage, and also to examine the F_1 cohort we had been following for survival with respect to a new indicator of genetic damage transmitted by the parents, namely, certain types of chromosomal abnormalities.

A systematic cytogenetic study of a sample of the F_1 was initiated in March 1967 by Arthur Bloom and Akio Awa. This study used white blood cells harvested from a venous blood sample; for a variety of reasons, it was felt inappropriate to attempt to obtain the requisite blood samples before a child had reached the age of 13. Because of the age limitation, as well as the time elapsed since the bombings, the study would not yield good data on the frequency of those chromosomal abnormalities that result in life-threatening defects. On the other hand, there are two types of abnormalities for which the data should be adequate. One is disturbances in the number of sex chromosomes. With respect to sex chromosomes, the normal female has formula XX and the male XY. But a number of departures from this normal pattern were discovered very soon after the improved cytological techniques became available, such as individuals with a single X or three Xs, or XXY, or XXYY, etc. Abnormalities of this type are relatively common chromosomal mutations that are seen in increased numbers in the offspring of radiated fruit flies and mice; they therefore comprised a legitimate subject for the studies in Japan. The other type of chromosomal damage on which this study should yield good data are reciprocal translocations. These occur when the radiation breaks two different chromosomes, A and B, and then the free ends reattach improperly, the free end of A attaching to B, and vice versa. When no genetic material is lost in the process, the affected animal appears normal, although these abnormal chromosomes do cause certain difficulties at the time of gamete formation. It was a reasonable expectation from experimental studies that these reciprocal translocations would be increased in frequency among the children of exposed survivors. This search for chromosomal abnormalities was to continue for some 20 years.[3]

It was also during this period that T. Furusho and M. Otake made an extensive effort to correlate the physical development of Hiroshima school children with the radiation histories of their parents.[4]

A New Genetic Program is Born

In the chapters on Amerindians I have described how we employed protein variants to characterize different Indian villages and tribes. Beginning in 1964 with 11 proteins, the number of proteins in our test battery (mostly erythrocyte enzymes) had increased to 20 by 1970, as various investigators described reliable ways to search for electrophoretic variants of additional proteins, techniques that we then adapted for our program. With a half-dozen years of experience, we became quite confident of our ability to differentiate between the results of laboratory errors or protein degradation, on the one hand, and true genetic variants, on the other hand.

In the Amerindians, in addition to the genetic polymorphisms so useful in the study of village and tribal differences, we also encountered very rare variants, present in only one or two families. These rarer alleles had arisen only recently in the tribal history, as a result of mutation. In 1970 an idea whose time had come presented itself: why not mount a search in Japan for mutations altering the electrophoretic mobility of proteins, centered on these rare variants. This would entail screening a large number of the F_1 for electrophoretic variants; when a variant (other than a common polymorphism) was encountered, a family study would have to be performed, to determine whether the variant was inherited (the usual situation) or was an isolated occurrence in the family, the result of mutation in one parent or the other.

Electrophoresis would not detect all the mutations present in a protein, but only those in which an amino acid substitution had altered molecular charge but (for enzymes) had not interfered with the function of the protein. It can readily be calculated that only about one third to one half of all possible amino acid substitutions are detectable by electrophoresis. We would thus miss a high proportion of any mutations. There were no experimental data indicating how sensitive an indicator of radiation-induced genetic change these substitutions might be. Nevertheless, in my opinion the need for more—and more objective—data on the genetic effects of the bombs was such that we should pursue all potentially useful approaches. Even if we found no increase, it would be important to know what the bomb did *not* do.

It was clear from the outset that if such a study were to be undertaken, it had to be on a substantial scale. It was impossible, however, to decide precisely how large a study of this nature should be to have "significant" scientific impact since, in 1970, there were still no data on the baseline frequency of the event of interest, i.e., the rate at which spontaneous mutation results in electrophoretic variants of proteins. On the other hand, as we have seen, from studies of experimental organisms plus some limited work on human mutation rates, it could be estimated that each generation mutation occurred at any given genetic locus of the order of once in about 100,000 gametes; since each child results from the union of two gametes, about one in 50,000 children might carry a new mutation at any given locus. We did not know, however, exactly what fraction of these mutations would be detected by the electrophoretic approach.

It would have been impractical with the limited facilities to attempt to examine with these techniques all the children registered in the course of the previous genetic study. Furthermore, by now we knew that most of these children were born of parents who really had received little or even no radiation. We decided to examine as many children as possible of parents within 2000 meters of the hypocenter, now well accepted as the zone of significant radiation, matching them as the study progressed with a similarly sized panel of children whose parents were beyond this point. The first set of parents came to be known as the proximally exposed, the second set as the distally exposed.

Jack Schull was at that time still in Ann Arbor (he was to leave shortly, to direct a new unit in genetics at the Houston branch of the University of Texas). After a certain amount of statistical hand-waving, we decided that a worthy objective would be to examine for variants the protein products of some 500,000 genes from the children of each of the two groups. From the individualized radiation dose estimates, which by then had—finally—become available, and from what was known about the sensitivity of mouse genes to radiation, we could guesstimate that there might be of the order of twice as many mutations in the children of proximally as of distally exposed *if* human genes responded like mouse genes. This translated to not more than ten mutations among 500,000 tests on children of proximally exposed as compared to not more than five among 500,000 tests to the other children, some fraction of which would be detectable by electrophoresis. It was going to be a great deal of work for a small outcome. Again the need to know and the possibility that our forecast was wrong drove us on. By this time it was abundantly clear to ourselves and our critics that, in many respects, the study of mutation often required a stronger back than mind.

Selling the idea was a two-phase affair. The first phase involved the approval of the Atomic Energy Commission (since the undertaking would involve the Ann Arbor-based activity that they supported) and also the approval of the National Academy's Advisory Committee to the ABCC. Presentations were made to these groups in the spring of 1971, the necessary concurrences were forthcoming, and, in November, I took the concept to Japan for the second phase of gaining its acceptance.

Here the going quickly became more difficult. There were two basic reasons for the resistance to the new program. First, the facilities and the budget of the ABCC were rather fully committed; this new program would impinge on both. It is one thing to support a development in principle; it is quite another thing to support that development when it impacts on an ongoing program in a way that may restrict other research projections. Fortunately, the AEC was willing to provide add-on funding. But there was, from the standpoint of then-director George Darling a second special reason for delay. Up until this time, about 90% of ABCC's budget had been derived from the Atomic Energy Commission. While this was appropriate in the immediate post-war years when Japan was still so devastated, it seemed to the Commission increasingly inappropriate as the Japanese economy recovered. Negotiations with the Japanese for a drastic administrative reorganization were just being initiated. It quickly became clear that George was in no mood for any new development, especially one which might to the Japanese appear to increase the complexity of what they were being asked to support.

I chafed at the delay. Now that a new look at the bomb's genetic effects was possible, the sooner implemented, the better. The oldest "children" were already 25, and, with the great internal mobility of post-war Japan, leaving the two towns in increasing numbers. Their parents (whose availability was essential to a study of mutation) were aging and soon would be entering a time of higher mortality. After a certain amount of tenseness over the delay, George and I reached a compromise. I could organize a "pilot study," not involving blood samples from the children of survivors (which would require a new activity) but rather, samples from the adult survivors who were being examined on a regular schedule and from whom blood samples were routinely obtained for various clinical studies. We would have access to what was left of this blood sample after the "standard" determinations to which it was subjected. Such a pilot study would let us introduce to the ABCC, from the Ann Arbor laboratory, the necessary techniques and provide a first estimate of the frequency of the variants that would have to be examined as regards their possible mutational origin.

The pilot study required the rehabilitation into laboratory facilities of space used for other purposes, and recruiting and training personnel in the electrophoretic procedures we had been using at Michigan. Fortunately, our Japanese associates in past studies came to our support in the recruitment of Japanese personnel, and for month-long periods between 1972 and 1975, two of my associates in the Amerindian studies at Michigan, Robert Tanis and Robert Ferrell, both highly skilled in the techniques of biochemical genetics, rotated through the Japanese operation, setting up the lab and training staff, while I busied myself developing the complicated records system appropriate to the study. The Pilot Study continued for three years. In addition to familiarizing us with the kind of genetic variation electrophoresis would detect, it revealed

that the rare variants that had to be investigated as candidates for mutation occurred with a frequency of 1 or 2 per 1000 electrophoretic determinations.

At length the contemplated reorganization was effected, and the
Atomic Bomb Casualty Commission was replaced by the Radiation Effects Research Foundation, equally funded by Japan and the U.S., with
a Board of ten Directors, half American, half Japanese, and another
Board of ten Scientific Counselors, also half American, half Japanese.
All procedures were formalized to a much greater degree than previously; my protocol had first to be approved by the Scientific Counselors, then by the Directors. As I waited for the first meeting of the
Counselors, in April, 1975, I could not help but reflect back on the days
when, so to speak, I had simply been turned loose in Japan to develop
a genetics program. Be that as it may, the Counselors, and then the
Directors, did approve the protocol, and I was again involved with a
major study in Japan. Since, however, the new (RERF) organization,
slimmed down a bit from earlier, still employed some 600 persons, the
proposed genetics program was only one among many, rather than the
dominant program of the early years.

The Biochemical Genetics Study (BGS)

The protocol for the BGS was straightforward. By 1975, when the study
began, we had, through the other genetic studies, identified a potential
cohort of some 27,000 children born to proximally exposed parents, a
cohort that we could match with an equal number of children born to the
distally exposed. By no means all of these children would be available
for study, however; from the experience of the Cytogenetics Program
described earlier, we knew that about 40% of these children had either
left town or would refuse to participate in the study. Furthermore, since
in the event of an unusual finding we must examine both parents to
determine whether the variant was inherited or was a newly arisen
mutation, we decided to concentrate our studies on children both of
whose parents could be shown to be alive and still in the city. This, and
the convention of not attempting to obtain a blood sample from children
less than 13 years of age, would result in further sample attrition, which
we could only estimate as we got into the study.

The plan, then, was simply to examine blood samples from as many
of the children of the proximally exposed as possible, plus samples from
matched controls. In the beginning we examined the samples for variants of 1) four blood-serum proteins, 2) the major and minor components

of adult hemoglobin, and 3) 18 enzymes found in the red blood cell, a total of 24 proteins. As the study progressed, four more red cell enzymes were added to this original battery. The study required a staff of some four professionals and a dozen technicians, plus the clerical back-up to handle the voluminous records and the field staff to contact (and persuade) our subjects. A device that proved very helpful in the recruiting of professional staff was the rotation of personnel through the Ann Arbor laboratory for periods of a year, for additional training and studies. As the project evolved, Chiyoko Satoh played an outstanding role, with strong assists from N. Ueda, K. Goriki, M. Fujita, N. Takahashi, T. Kageoka, R. Hazama, and J. Asakawa.

In 1979, we introduced another biochemical technique to the search for mutations, namely, an attempt to identify genetic variants of enzymes characterized by the absence of any enzyme activity, employing in this search newly developed equipment that permitted the search to proceed relatively rapidly. Harvey Mohrenweiser, who by then had joined us at Michigan, had worked for three years to develop a battery of a dozen selected enzymes that best lent themselves to this type of search. In *Drosophila* and the mouse, a high proportion of the mutations produced by radiation had proved to be lethal in the homozygous condition (i.e., incompatible with survival). It was presumed that this finding was due to gene inactivation, accomplished in a number of different ways. In the Japan study, we proposed to screen for children who exhibited only 50% of the normal activity of a particular enzyme. This finding would presumably result from the inactivation of one of the two alleles at the gene locus in question. When such a child was encountered, we would then examine its parents. A mutation would manifest itself as a child with 50% activity, both of whose parents had 100% activity. Our battery of enzymes had been carefully chosen so that the distinction between 50% and 100% activity could be reliably determined.

During the eight years of these biochemical studies, usually accompanied by Priscilla, I would visit Japan annually for six to eight weeks. Our travels together were much easier now with the children grown: Frances a social worker licensed for counseling, Jim a physician, and Al well on his way to the same goal. Each visit required an exhaustive review of the program, checking for all the possible sources of error and bias that could cause difficulty in the final analysis. By 1984 we had pretty well exhausted the two study groups, one group, as noted, born to parents who were within the zone of radiation at the time of the bombings, the other a matched set born to parents who had received no radiation from the bombs. In the first group, there were three mutations in a total of 667,404 tests for electrophoretic mutations and one mutation among 60,529 tests for enzyme-activity mutation. In the second group, there were three mutations in a total of 466,881 tests for electrophoretic mutations, and zero in 61,741 tests for activity mutations. (The results of the electrophoretic studies on the controls were introduced in the last chapter, in the calculation of spontaneous-mutation rates.) When we combined the results of the electrophoretic and activity studies, the mutation rates were essentially the same in the children of exposed and of controls, namely, 0.6×10^{-5}/locus/generation. We had put another brick in place for the case we were building, but as with the previous studies, it had been hard won.[5]

New Estimates of Survivors' Exposures

In 1980 there came a surprising development. Serious questions were raised concerning the accuracy of the estimations of the survivors' radiation doses with which the RERF was working. Since 1965, the ABCC–RERF had been employing an algorithm for computing dose based on the presumed best thinking of the physics community, the Tentative 1965 Dosage, Revised (T65DR). The challenge to this algorithm came on diverse fronts. At the one extreme was a reevaluation, in light of atomic bomb test-shot results, of the power of the bombs detonated over Hiroshima and Nagasaki and of the spectrum of radiation they released, a legitimate reason for reevaluation. There was a major revision downward in the estimate of the neutron release of the Hiroshima bomb. At the other extreme, however, came the belated realization that the greater humidity of the air over Hiroshima and Nagasaki than over the desert of the U.S. Southwest, where the test shots that followed WWII were largely performed, would in the Japanese experience result in increased absorption of radiation before it reached the ground. There was also a substantial reassessment (upward) of how much radiation the superficial tissues of the body would absorb before the radiation reached target organs such as the gonads. These miscalculations, like the failure of the designers of the bombs to anticipate large numbers of radiated survivors, have gone far to persuade me that physicists share in the fallibility with which I am so familiar among biologists. At any rate, after a great flurry of activity, a new system of assigning doses came on line in 1986, termed Dosage System, 1986 (DS86).

DS86 treated the radiation released by the atomic bombs as principally gamma in character, but with a small neutron component, the latter larger in Hiroshima than in Nagasaki. In the early days of the A-bomb studies, exposure was expressed as body-surface doses, in *r* (roentgen). By this point in time, it had become important to work with estimated absorbed doses to individual organs, such as the testis or ovary. By recent international convention, the use of *r* units is superseded by a system in which gamma organ exposures are expressed in gray (Gy), one Gy corresponding to one joule of imparted energy per kilogram of tissue, and we will follow the new terminology in our calculations. One Gy is under many circumstances approximately equivalent to 100 of the previously employed *r* units. Neutron radiation, because of the greater density of its ionization track, has a higher genetic effectiveness than gamma. To combine gamma and neutron exposures into a single dose figure, one must estimate the relative biological efficiency for genetic effects per unit ionization for the neutron component (as contrasted to the gamma component); based on experimental data, we employed, for these low neutron exposures, a factor of 20. Formerly, the dose of a mixed gamma–neutron radiation exposure was expressed in "roentgen equivalent man" units (*rem*), calculated as gamma dose in *r* plus neutron dose in *r* × 20. Now, by international convention, that type of exposure is expressed in sieverts (Sv), one Sv equivalent equal to 100 of the previously employed *rem*. One Sv of ionizing radiation from an atomic bomb should closely approximate in effect 1 Gy of gamma radiation, the type used in the mouse experiments to be reviewed later.[6] With this new dosage schedule, the combined gonadal dose of the average set of parents, one or both within 2000 meters of the hypocenter at the time of bombing (the proximally exposed), is 0.44 Sv equivalent; 31,159 children born to such parents have been registered and are available for various analyses.

Bringing the "Old" Studies up to Date

The availability of new estimates of gonad doses meant that all the previous analyses relating outcomes to exposure had to be redone. The expectation was not that any of the earlier findings would actually be reversed, but rather that the quantitative relationships to radiation would be altered. For this purpose, the data were reorganized and extended. The early, clinical data were combined so that the unit of analysis became an Untoward Pregnancy Outcome (UPO), defined as a child who was stillborn and/or congenitally deformed and/or died within the first two weeks of life, and the data reanalyzed in collaboration with M. Otake and Jack Schull. The data on survival in these children after the first two weeks, plus the survival among children liveborn subsequent to 1953, were brought up to date through 1985, and analyzed by Y. Yoshimoto, H. Kato, Jack, and myself.[7] Because of the growing evidence that some cancers of childhood—most notably retinoblastoma and Wilms' tumor, but possibly others—were related to mutation in one or the other parents, a special analysis of malignant tumors in the offspring was conducted by Yoshimoto, Schull, Kato, and myself.[8] The data on chromosome abnormalities collected by Awa and his colleagues and the data on protein abnormalities were also reanalyzed according to the new doses. By this time Awa and his colleagues had examined some 16,298 children for chromosomal abnormalities, the subjects evenly divided between the children of the proximally exposed and the (distally exposed) controls. This is surely the largest cytogenetic study of this type ever carried out.

The only data not reanalyzed were those on physical development (at birth, nine months, and school age) and on the sex ratio. Our own data on child's measurements at birth and at nine months of age had revealed no difference between the children of exposed and of controls, nor had the data of Furusho and Otake shown any difference among school children in relation to parental exposure history.[4] The sex ratio data were similarly negative.[1] Unfortunately, for technical reasons the analysis of these data could not yield the kind of output necessary to the approach to combining all the data that was beginning to shape up in our minds, so that nothing new would be gained by a reanalysis.

A New Perspective on the Data

Over the decade of the 1980s, Jack and I began to develop a new perspective on these data. In the early years, as noted, we had felt the conservative and responsible course of action was to limit our conclusions to what the data could exclude in the way of effects, with no speculations as to what else the data might imply. But as the data accumulated and none of it suggested a statistically significant effect of radiation, even at doses that were being accepted as sufficient to double the mutation rate in the mouse, a line of thought began to take shape. Perhaps this total lack of statistically significant findings was suggestive that the doubling dose of radiation for humans was actually considerably higher than had been projected from the mouse data, i.e., perhaps humans were more resistant to the genetic effects of radiation than was being assumed on the basis of extrapolation from experiments with mice.

A difficulty in developing this thesis was the scientific tradition that one does not draw conclusions concerning effects unless one has a statistically significant different between the "treated" and the controls.

238

But then I realized that this approach applied to hypothesis testing. We were not in Japan to test the hypothesis that radiation induced genetic damage. There was an enormous literature to that effect, on a wide variety of organisms. True, the magnitude of the effect might vary, according to the organism and how the radiation was administered, but there could be no doubt that the exposures of Hiroshima and Nagasaki had produced some added mutations. Our challenge could be restated: take all our data at face value, as the most appropriate data for calculating the genetic effects on humans of exposures to ionizing radiation to be available in this century, and see where a proper analysis led us. That "proper analysis," as it turned out, was to become rather involved.

Calculating a Genetic Doubling Dose

The impact of mutation on a population can be presented in many ways. As we have seen in the preceding chapter, mutation is a fact of normal life, occurring both as a result of the spontaneous instability of the complex gene and (we believe) of its interaction with either stray radiation or the mutagenic chemicals always present to some degree in our drinking water and food, or generated endogenously in the course of normal metabolism. Even viral infections may alter DNA. Gradually, as recounted in the last chapter, an appreciation of how mutation impacts each generation of a human population has been built up. This suggested that the most appropriate perspective from which to view radiation effects was as a doubling dose, i.e., that amount of acute radiation that would result in the same impact of mutation as results from spontaneous mutation each generation.

At this point, something must be said about the nature of the indicators with which we were working as the basis for calculating a doubling dose. An increase in "untoward pregnancy outcomes" or an impaired survival, would be the result of genetic effects that were "dominant" in the classical sense, appearing in the first generation, effects to which legislators and regulators could relate. These would generally be the result of "large" genetic lesions. Chromosomal studies of the type carried out, and the biochemical studies, were intended to assay more subtle damage, much of it technically "recessive" in nature, whose impact might be distributed over many generations. The battery of indicators we had developed should give reasonably good but not perfect coverage of the potential genetic effects of the bombs. The statements in this paragraph are illustrated graphically in the table on page 240.

The actual calculation was difficult. Each of the data sets had to be

The spectrum of genetic damage that should be detected by the past and proposed studies of the children of atomic bomb survivors. The first column lists the end points studied in the children of survivors; the headings of the remaining columns list the types of damage that the end points should reflect. A black dot indicates that a given indicator would reflect the type of damage listed in the header to the column.

Indicator Employed in Japanese Studies	Loss or Gain of Entire Chromosome	"Large" Chromosomal Duplications/ Deletions	Unbalanced Chromosomal Translocations	Balanced Chromosomal Translocations	"Point" Mutations	
					Small Deletions/ Duplications	Nucleotide Substitutions
"Untoward" Outcome of Pregnancy	•	•	•		•	•
Survival	•	•	•		•	•
Sex-ratio	•				•	•
Cytogentic Abnormality	•			•		
Protein Studies					•	•
Cancer in F_1					•	•
DNA					•	•

expressed in the form of a linear regression of indicator (UPO, early death, etc.) on estimated gonadal dose. We then had to estimate, from the world-wide genetic literature and our own findings, how much of the indicator value in the controls was the result of spontaneous mutation in the preceding generation of parents. Such an estimate was now possible because of the great advances in cytogenetics and clinical genetics since our treatment of the subject in 1956. The specific estimates of the contribution of parental mutation to the indicator varied widely, from 100% of the endpoints for sex-chromosome aneuploids to between 2–4% for childhood cancer. As noted, for two indicators (sex ratio and physical development) such an estimate was not possible. For a third indicator (balanced reciprocal chromosomal exchanges) there was only one mutational event in the children of exposed and one in the children of controls. There was thus no difference but there were simply too few events for a regression analysis.

We were thus left with five indicators: UPOs, mortality through an average life expectancy of 29 years, sex-chromosome abnormalities, cancer with onset before age 20, and protein mutations. The first two of these indicators exhibited a slightly positive (but not statistically significant) relationship to dose; the regression of the other three indicators was very, very close to zero. With particular reference to childhood cancer (because the subject will come up in Chapter 18), the data were subdivided into leukemia and related disorders, malignancies (such as retinoblastoma) with good evidence for origin from a germinal mutation, and malignancies with no evidence for a contribution from germ-line mutations. None of these three subcategories showed any relation to parental radiation history.

Since all these regression estimates were based on the same defined population of children, they could be combined. The total value of these indicators as a result of *spontaneous* mutation in the control parents was estimated to be between 0.63–0.84% for all births. The cumulative (summed) regression of all these indicators on dose amounted to 0.375% per Sv equivalent. The amount of acute radiation in Sv equivalents that would double the background rate, obtained by dividing the first by the second figures, was between 1.69–2.23 Sv equivalents. Since the data, which for technical reasons could not be factored into this estimate (physical development, sex ratio, and reciprocal chromosomal translocations), reveal an even weaker suggestion of an atomic bomb effect than those used in the estimate, we feel that the estimate is conservative.[9]

This estimate is for an "instantaneous" exposure to the radiation emitted by an atomic bomb. What society wants to know is the result of more everyday low-level, extended exposures: gonadal exposure during diagnostic X-ray procedures, exposure from nuclear installations or from industrial activities, etc. One of the salient findings of the experiments with mice conducted during the 1950s and 1960s had been that if a given dose of acute radiation was administered in many small bursts or chronically over several days or weeks, the yield in mutations was much less than if the dose were given all at once, the exact reduction depending on the endpoint under consideration and the dosage schedule. In the mouse studies, a "dose rate factor" of 3 had emerged as a conservative extrapolation from the effects of acute to chronic radiation.[10] However, the gonadal doses used in the mouse experiments were usually 10–20 times greater than the average doses received in Japan, although there were a few relatively high human exposures. Considering the estimated gonadal-dose distribution in Hiroshima and Nagasaki and the statistical treatment of Abrahamson and Wolff,[11] we felt a dose rate factor of 2 to be appropriate for our extrapolation. The estimated doubling dose for chronic radiation therefore became of the order of 4.0 Sv equivalents.[12] Both of these estimates rather startled us: they were substantially higher than conventional wisdom based on animal experimentation.

It is important to place these rather abstruse statements concerning doubling doses in a more familiar framework. In Chapter 4, note 6, I provided estimates of the amount of radiation received in the course of a number of procedures in diagnostic radiology. Since diagnostic radiology delivers repeated small doses, it can be considered from the genetic standpoint as chronic radiation. It would, on the basis of the data in note 6, require approximately 300 X-ray examinations of the lower spine or abdomen of a male in the reproductive years (and somewhat fewer for a female) to deliver a doubling dose of radiation to the gonads. Otherwise stated, again on the basis of the data in note 6, the average U.S. citizen annually receives from diagnostic radiological procedures radiation of significance to the next generation equivalent to 1/8000 of a doubling dose, if our calculation is correct.

A Reanalysis of the Experimental Data

The discovery that ionizing radiation induced mutations was made, using *Drosophila*, by H. J. Muller in 1927. The advent of atomic bombs and nuclear energy was an enormous stimulus to the experimental study of the genetic effects of radiation. Since one objective of these new experimental studies was to extrapolate to human risks, a paradigm more closely related to man than the fruit fly was obviously desirable. The laboratory mouse had been the obvious choice, and in the last 45 years a very major effort has gone into studies of the genetic effects of radiation on mice. The genetic sensitivity of mature mouse gametes to radiation was early on shown to be substantially greater than that of immature gametes; it is the latter (spermatogonia, early oocytes) that will be the chief targets of a human exposure to radiation, since mature gametes will be functional for only a few months after an exposure. Thus all comparisons of the mouse data with the findings in Hiroshima and Nagasaki involve the results of the irradiation of immature mouse germ cells.

On the basis of extrapolation from a subset of the mouse experiments, various national and international (United Nations) committees have during the past 30 years adopted the position that the doubling dose of acute radiation for the immature germ cells of humans was about 0.35–0.40 Gy. As noted in the preceding section, a salient observation in the course of the mouse experiments, by W. L. Russell and colleagues,[10] had been that at low dose rates the genetic effect of radiation was only one-third that at high doses, so that in mice the doubling dose for chronic radiation became about 1 Gy. Our estimate for humans was thus about four times higher than the "official" rate, no small revision.

I had, for obvious reasons, been following the mouse data closely over the years, and as the difference between our doubling dose estimate for humans and that commonly extrapolated from the mouse began to emerge, I was troubled. True, there could indeed be species differences, but it was equally possible that some of the assumptions we had been forced to make in developing a doubling dose from the human data were off. Accordingly, in 1990, Susan Lewis, who specializes in mouse genetics, and I undertook a systematic examination of all the accumulated data on murine radiation genetics.[13] These murine data were of two types. One set dealt with the response to radiation of what I will call "vital statistics"—congenital defect in the offspring of radiated mice, litter size, survival of these offspring to weaning, and the fertility of offspring reaching the age of reproduction. The other set dealt with the response of more specific indicators, such as mutations at specific genetic loci or mutations resulting in specific phenotypes (skeletal abnormalities, cataracts, and protein variants), or chromosomal abnormalities. Some of the latter data were comparable to the studies in Japan on protein mutations and cytogenetic abnormalities; the former data were more like our studies on UPOs and survival. In the early years of the mouse studies, the effort was about equally divided between the two kinds of studies, but as the "vital statistics" approach yielded very little in the way of effects, whereas some of the studies on specific genetic traits were yielding very clear results, conventional wisdom said the former was "too blunt an approach." More and more attention was then centered on the results of the more specific indicators, and especially on the very objective and clear-cut data resulting from a system developed by W. L. Russell for measuring mutation at seven specific genetic loci.

Susan and I found we really couldn't compare the mouse data on "vital statistics" with the human data on UPOs and survival, primarily because the mouse gives birth to litters in which, both before and after birth, there is strong competition for survival, as contrasted to the single-born and pampered status of the vast majority of human births. Genetic defects compatible with survival until birth in humans might, therefore, result in early fetal death and unnoticed loss in the mouse. In addition, the mouse is born in a developmental state which corresponds approximately to a human fetus 100 days old, which means that new-born mice are examined at a stage not available for examination in humans. We therefore decided we could not use mouse "vital statistics" in this comparison. With respect to the remaining mouse data involving specific genetic endpoints, in our analysis we simply averaged all the results from the eight studies of this type. Somewhat to our surprise, we found that when all the data available were analyzed (and all data sets given equal weight), including some results that had been looked at askance because they didn't reveal much of an effect, the average doubling dose for acute radiation was 1.35 Gy. Applying the dose conversion factor of 3 mentioned earlier, to extrapolate to the doubling dose of low-level chronic radiation, we obtained a value of 4.1 Gy, in unexpectedly good agreement with the conclusions from the human data.

It should be made clear there is no *a priori* reason to expect the doubling dose estimates for mice and humans to agree. We would have stood by our estimate of the human doubling dose even if the mouse estimate had remained at the same (lower) value after our reanalysis. The sources of error in both estimates are such that in truth they could still differ by a factor of two. On the other hand, given the uncertainties in the assumptions we were forced to make in the analysis of the human data, it is intellectually comforting to find the mouse estimate now weighing in so much closer to the human.

Summing Up the Radiation Story

So, after some 48 years of study, we conclude that the genetic risks from radiation are less than the previous working assumptions adopted by most national and international committees. I am happy to say that in the three years since this viewpoint was published, it has not been seriously challenged. Jack and I have recently collected 13 of the most significant of the papers published during the 45 years over which this study has extended, with added commentary, into a book.[14] We specifically recognize the possibility for error in our doubling dose calculations—but almost surely less error than in past extrapolations from mice to humans. In addition to the errors inherent in our assumptions and our statistics, there is one further possibility for error, thus far unmentioned. The dose rate factor reveals that the cell possesses quite efficient DNA repair mechanisms. In making our calculation concerning a doubling dose, however, we assume (as do others in the same situation) that there is no level of radiation that does not result in some unrepaired damage, i.e., that there is no threshold to radiation damage, and that damage is directly proportional to dose. This assumption is based on the fact that in various experimental organisms, radiation has produced germinal mutations at the lowest doses at which it has been tested (0.25 Gy). This is an assumption one is forced to make in such calculations. On the other hand, it is possible that the repair of genetic damage following a mutagenic insult is better at the very low doses that will usually characterize human exposures to chronic radiation than at the higher doses of the experimentalist, in which case the dose-rate factor for humans would be greater than 2, and our estimate conservative.

There is a moral in these developments to be borne in mind as the world struggles to evaluate the genetic implications of contaminants other than radiation. To obtain funding for their work experimentalists must develop test systems that produce results. It cannot automatically be assumed that the genetic test system yielding the most striking effects is any closer to truth than a recalcitrant system. By now much is known about locus differences in mutability; one set of loci can yield twice the response to radiation of another. Efforts to evaluate the risks from a potential mutagen must be multifaceted.

I would not like to close this chapter inadvertently leaving the reader with the impression that there is no need for concern about the genetic effects of radiation. The risks are less than assumed but still exist. With respect to the somatic (i.e., nontransmitted) risks of exposure to radiation, the principal concern is an increased rate of cancer in the exposed; the reanalysis of the Japanese data with the DS86 dose schedule *increases* the risk from what was inferred with the T65DR schedule.[15] Although one cannot equate the societal and familial impact of the development of cancer in an adult 20 years postexposure with the impact of a serious genetic defect present at birth, the present data suggest to me that guidelines that protect the public and the relevant work force against cancer will adequately protect against germ-line genetic effects.

There are times when, considering the fast-moving pace of modern human genetics, I have to wonder about the wisdom of my tremendous personal investment in the slowly moving question of the genetics effects of atomic radiation. I sometimes have felt a twinge of envy for those whose laboratory-based studies move so much more quickly. The twinge doesn't last long. The attempt to understand the genetic effects of radiation must be seen as one of the several most significant interfaces of genetics with societal concerns in this century, and the program in Japan is certainly the most extensive effort in genetic epidemiology ever undertaken. If one commits oneself to the field of human population genetics, one must be prepared to pay the price.

The best hope for refining still further the estimate of the genetic doubling dose of radiation for humans lies in still further studies on these Japanese cohorts. Of the various possible future studies, not surprisingly, thoughts have turned to bringing up a DNA-based system for the study of mutation. To this end, RERF is now establishing a large collection of lymphocyte-derived cell lines in familial configurations, the goal being at least 500 family constellations involving radiated parents and one or more children and another 500 involving unirradiated parents and their children. The staff for the previous biochemical studies at RERF have now all become familiar with the DNA technologies, and in Japan, here in Ann Arbor, and elsewhere, investigators are attempting to develop efficient techniques for detecting mutations that could be applied to these cell lines.

Figure C on page 210 illustrates the results of one of these new techniques. The entire genome of an individual has been enzymatically digested and a portion of the resulting DNA fragments distributed over two dimensions. Genetic differences between individuals (and mutations) can result in changes in spot position or intensity. The laborious task of examining gels for possible mutations is greatly eased by a computer algorithm our research group has developed.

As illustrated by the table on page 240, these DNA-oriented studies will, of course, not replace but rather supplement the earlier studies, specifically with reference to the frequency of point mutations. I personally cannot think of a more telling testimony to the dazzling advance of genetic science than that, in the lifetime of one of its practitioners, the study of the genetic effects of the atomic bombs had progressed from the grossest of morphological levels—congenital defects—to the ultimate reference, DNA.

As I completed this chapter and paused to look back on what I had written thus far, I had a sudden, stunning realization: the major genetic issues in my life were no longer accessible to others for study. We have to presume there will never be another atomic bomb explosion, and the disasters at Chernobyl, Semipalatinsk, and Chelyabinsk (in the former USSR) fortunately do not begin to replicate the circumstances of the Japanese explosions. With the great postwar mobility of the Japanese younger generation, the frequency of consanguineous marriage has fallen to well under 1% throughout Japan, and it would be extremely difficult ever again to assemble a series of consanguineous marriages as large as the sample we studied. The destruction and acculturation of tribal populations all over the world has accelerated so rapidly—as we foresaw—that while a variety of types of data can still be collected, the extent to which these data will reflect the precontact situation is increasingly debatable. Even the studies of the abnormal hemoglobins in Liberia would be impossible to repeat today, given the current civil chaos in that country. The genetic future may belong to studies at the DNA level, but our opportunities to conduct many kinds of population studies are rapidly vanishing.

A Personal Note

This is primarily a book about genetic science, but it is also intended to project some of the flavor of the conditioning and functioning of a scientist. I hope I have made apparent how fortunate I feel with respect to antecedents, wife, children, preceptors, and professional associates. There was one other important personal relationship I must mention. My wife is an identical twin. In our early years in Ann Arbor, Priscilla's twin and her neuropsychiatrist husband were in the Boston area, but in 1968 her husband, Gardner Quarton, assumed the position of Director of the University of Michigan Mental Health Research Institute. Suddenly, there were two "Mrs. Neels" in Ann Arbor, the confusion, because they were truly identical, enormous. A favorite family story involves the local fish-market. Buying us a fish one day, Priscilla overheard the counterman mutter to his associate: "That poor woman must be losing her mind; she bought the same thing just half an hour ago!"

I had approached my relationship with my brother-in-law in the wary fashion of affines cast together by circumstances beyond human control. I soon realized the lottery of life had in this one dealt me a winner. He had one of the three or four broadest, most penetrating minds I have ever encountered. Our intellectual joustings over dinner were often the high spot of the week. He died of cancer of the colon in 1989.

The incredible intricacy of the coiling of the DNA of a single chromosome. This chromosome spread also illustrates how the DNA strands of a chromosome attach to the protein scaffolding (bottom of picture) of that chromosome. Photograph reproduced from Cold Spring Harbor Symposium on Quantitative Biology 42, p. 355 (1978), by permission of U. K. Laemmli and the Cold Spring Harbor Library.

14

What Is a Gene?

The preceding chapters have been concerned with genetic developments in which I have been intimately involved. They have set the stage for the discussion of the current human genetic condition that is about to unfold. However, before I launch into this treatment of the human genetic condition, I feel it incumbent upon me to provide a précis on the molecular nature of the human gene pool, as it is understood today. Very simply, although this most certainly has not been a subject for personal research, I feel I must try to project the complexity of the genetic material with which evolution has endowed us and upon which, as we will discuss, some geneticists are now proposing to operate. These next two chapters are meant to supply only enough background to enable those unfamiliar with the subject to appreciate the essential features of the DNA that is our human heritage, on the thesis that if the upcoming treatment of a genetic program for humankind is to have the degree of precision desirable in so serious an argument, we really must consider briefly some of the amazing insights of recent years into the nature of the gene pool whose phenotypic manifestations we have been considering.

Only 40 years ago, the chromosomes were visualized as composed of proteinaceous beads (genes) on a string, and the gene pool as the sum total of all the homologous chromosomes of the individuals in the population under consideration. Now the definition of the gene pool has not substantially altered, but the chromosomes and their genes have assumed an enormous and puzzling complexity, as each stretch of DNA studied in detail is yielding its own set of surprises. From the standpoint of the student of human evolution and population genetics, the question of how this complexity is maintained—let alone how it evolves in an adaptive manner—becomes more challenging with each advance at the molecular level.

As an introduction to the complexity of the gene pool, how better to begin than to consider what has happened to our understanding of the hemoglobin genes since we left them in 1961? The unfolding of knowledge about these loci, a text for the coming of age of human genetics, is the work of many unusually able people, who have found the fact that hemoglobin comes in such convenient little packages devoted so predominantly to the manufacture of a single protein a great boon.[1] Lest, however, the gene pool of which I shall speak so frequently be visualized as comprised of multiple organizational units such as those encoding for the hemoglobin proteins, we must consider, more briefly, several other illustrative genetic complexes. Here, too, we will proceed by presenting the current understanding of the genetic basis for some of the traits we have already discussed, such as neurofibromatosis and retinoblastoma.

The Organization of the Human Hemoglobin Genes

The major hemoglobin of an adult human consists of four polypeptides, two of one type and two of another, i.e., the hemoglobin molecule is a tetramer. These two types of polypeptides are designated α and β; the structure of hemoglobin is thus $\alpha\alpha\beta\beta$ or $\alpha_2\beta_2$. The α-chain is 141 amino acids in length and so requires 423 nucleotide base pairs (bp) to specify the sequence, whereas the β-chain, with 146 amino acids, requires 438 bp. The two chains (and the underlying genetic fine structure; see below) show so many similarities in their amino acid composition that it is assumed they arose from a common ancestral gene. The type of thalassemia on which I worked 45 years ago is due to an inability to elaborate β-chains, but there is also a type of thalassemia due to the failure to produce α-chains. In the very early human embryo, the hemoglobin is also a tetramer, composed of two α-like chains (termed ζ) and two β-like chains (termed ϵ). At about eight weeks of gestation, the production of the ζ-chain is gradually superseded by the production of adult-type α-chains (for which there are two identical genes on the chromosome) and of the ϵ-chain by two β-like chains termed $^A\gamma$ and $^G\gamma$, each encoded by a separate gene. The two γ-polypeptides differ only in the presence of the amino acids glycine or alanine at the 136th position on the γ-chain. At this stage in development, five types of hemoglobin are being produced, namely $\alpha_2\epsilon_2$, $\zeta_2{}^G\gamma_2$, $\zeta_2{}^A\gamma_2$, $\alpha_2{}^G\gamma_2$, and $\alpha_2{}^A\gamma_2$. By six months of gestation, the production at ζ- and ϵ-chains has ceased and the production of γ-chains has decreased to a trace, but now two new chains are being made: the adult-type β and, in small quantities, a second β-like chain, termed the δ. This δ-chain is of the same length as the β but differs with respect to ten amino acids. The adult not only has hemoglobin $\alpha_2\beta_2$, but also a minor hemoglobin component, $\alpha_2\delta_2$, as well as a trace of the fetal hemoglobins $\alpha_2{}^G\gamma_2$ and $\alpha_2{}^A\gamma_2$.

The genes responsible for these hemoglobins are organized into two complexes, the α-and β-complexes. We will consider the β-locus and the complex of which it is a part in some detail, the α-locus and complex only incidentally. The structure of the protein-coding portion of the β-gene, located on chromosome 11, is shown in the upper half of the figure on this page. (The term "gene" is now applied to a contiguous stretch of DNA that functions as a unit and is responsible for a product.) Like most genes of higher organisms, the DNA that codes for the amino acids comprising the two hemoglobin polypeptides is interrupted by noncoding sequences, which are transcribed into RNA but then must almost immediately be edited out of this RNA to create the message that is translated into the amino acid sequence of hemoglobin. The DNA sequences whose RNA counterpart is edited out of the message are termed intervening sequences or, more usually, "introns," in contrast to

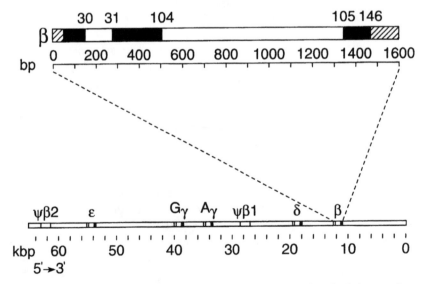

The structure of the gene encoding for the β-polypeptide of adult-type hemoglobin, and the structure of the larger complex of related genes of which the β-gene is a component. The black bands represent the exons, the light bands the introns. A full description will be found in the text.

the translated "exons." In the specific case of the β-globin gene, there are three exons specifying the hemoglobin polypeptide, divided by two introns. There are also short stretches of the DNA to the left and right, respectively, of the terminal exons (designated with oblique hatching), which stretches are also transcribed initially and seem to be essential to the functioning of the gene but do not contribute to the final gene product.

The discovery of these introns as a common feature of mammalian genes, in 1977, came as a major surprise. Their junction with the exons is always characterized by two specific base-pair sequences, at both ends of the intron, that supply a recognition signal to the mechanism responsible for removing the transcript of the introns from the processed messenger RNA. No really satisfactory biochemical justification for the presence of these inserts has yet been put forward, but there has been abundant speculation. They seem to be a needless complication to the system, since, as we will see, mutation in introns can have deleterious consequences; conventional genetic theory states that in the absence of a function to offset these deleterious consequences, they should be eliminated. The sequence of amino acids coded by an exon often corresponds more or less to a region of a complex protein associated with a specific activity of the protein, i.e., a functional domain, and it has been suggested that the presence of these introns facilitates the reshuffling of domains in the evolution of new proteins. They also provide a mechanism for obtaining several different RNA messages from a stretch of DNA, through a mechanism termed "alternative splicing." That the role of introns is more complex than this is suggested by the fact that within any given gene where they occur, intron length has usually been conserved (i.e., remains much the same), even in two species whose ancestors separated millions of years ago. Such conservation would seem unnecessary if the sole function of introns were to separate the "useful" stretches of DNA.

Beyond the head (5′) end of the β-globin gene, as for all functional genes, there is a stretch of DNA concerned with the regulation of gene activity. This sequence contains two groups of nucleotides, respectively about 25–30 and 70–80 bp above the point where transcription of the RNA message is initiated (the so-called TATA and CCAAT boxes), whose position and nature apparently must be maintained if transcription is to take place. Further 5′ to this region there are additional sequences that are involved in the DNA–protein interactions regulating the expression of each functional gene. For instance, for the ε gene of the β-globin complex, the 2000 bp 5′ to the protein-coding portion of the gene appear to have some four different types of regulatory sites, three of which occur in multiple copies, i.e., there is the "insurance" or redundancy for important control elements. Some genes have as many as 8 different protein-binding sequences in this 5' region, and there is even cross-talk between the regulatory proteins bound to the region, i.e., the level of gene activation when two different proteins are bound simultaneously is not just the sum of the activities when the two proteins are bound separately.

The first 30 amino acids of the β-globin polypeptide are encoded by 90 bp in exon 1, the next 74 by 222 bp in exon 2, and the final 42 amino acids by 126 bp in exon 3. From Ingram's previously quoted demonstration in 1956 that the sickle-cell allele resulted from the substitution of the amino acid valine for glutamine at the sixth amino acid position in the β-chain, it could be deduced, when the genetic code became known, and then later directly demonstrated by DNA sequencing, that in the sickle-cell allele a thymine nucleotide had been substituted for an adenine nucleotide at the second bp in the bp triplet constituting codon 6. By now it has become routine when a variant of normal hemoglobin is recognized, to determine the precise nucleotide substitution responsible for the variant.

The analysis of the thalassemias, which, it will be recalled, represent severe impairments of the ability to make hemoglobin, has also long since progressed to the DNA level. With reference to the type of thalassemia due to impairment of β-chain synthesis, some 90 different lesions in the DNA capable of seriously interfering with the production of β-chains have been identified. Approximately half are due to deletions of varying sizes in the β-locus DNA pictured in the figure on page 251, i.e., losses of nucleotides in intron and/or exon material. Several of these deletion-type thalassemias are apparently due to the fact that, because of their similarity, a β- and a δ-gene have mispaired during the meiotic process that precedes the formation of the germ cells and the chromosomes have then exchanged segments. The result is a poorly functional "hybrid molecule," part β-chain, part δ-chain, often called a Lepore-type hemoglobin from the name of the family in which this type of hemoglobin was first encountered. We will be returning again and again to the consequences for the gene pool of this kind of mistake.

Of the remaining lesions known to cause thalassemia, about half are due to changes in the genetic code in the exons, changes that either interrupt the translation of the genetic message at that point or result in a greatly impaired rate of hemoglobin synthesis. The remaining causes of thalassemia are of particular interest. They either involve the genetic code at the splice sites (the points at which intron-coded RNA is cut out of the RNA message) or are even within the introns. A very few involve the DNA preceding the first exon coding for hemoglobin. In these cases, mutation has either removed a splice site or created a new, "alternative" splice site, in consequence of which the processed RNA is abnormal. Here is unequivocal proof that the DNA in introns cannot simply be regarded as selectively neutral stuffing—"noise" in the system—but has a role to play, at least in the sense that a mutation in this region can be deleterious.

In the normal human, there are two identical functional α-genes on chromosome 16 (in contrast to only one β-gene on chromosome 11). These are very similar in their intron–exon structure to the β-gene, the chief difference being that the second intron of the α-gene is shorter. With respect to thalassemia due to a defect in the α-chain, the situation differs from that just described for the β-chain because of the occurrence of these two α-chain loci in close juxtaposition on the same chromosome. Thus a normal person has four functional α-chain genes, two on each homologous chromosome. Detailed studies of persons with defects in α-chain production (i.e., α-thalassemia) and their families have lead to the recognition of chromosomes with no or only one α-gene per chromosome, instead of the usual two. More or less by chance, individuals have also been encountered with three instead of the usual two α-genes on chromosome 16. Thus, when both chromosomes are considered, an individual might have 0, 1, 2, 3, 4, 5, or 6 α-globin genes. Those with none die before birth, while those with one, two, or three have defects in hemoglobin production varying from moderate to very minor. Persons with four such genes are the norm. Those with five also appear normal; to my knowledge, a six-pack has not yet been recognized. The explanation of this variation in α-chain number lies in the fact that the occurrence of two α-chain loci so close together seems at times to upset chromosome pairing, with unequal crossing-over resulting in chromosomes with either one or three α-gene loci, or even with no α-genes. In addition to these causes of thalassemia, many of the same kinds of defects mentioned earlier for the β-chain locus have been encountered in the α-chain, although, because there is a "spare" α-locus, the effects of these lesions are not so devastating as when only a single gene locus is present per chromosome, as in the case of the β-gene.

The four polypeptides that comprise the hemoglobin molecule are very tightly intertwined, the various polypeptides complexly convoluted. Studies of the genetically determined departures from the normal amino acid sequence reveal that many of them, by disturbing this complex configuration, alter the function of the molecule in some respect. Thus, when in the course of evolution one amino acid is substituted for another in the β-polypeptide, this may alter the configuration of that chain, which alteration may then have a very appreciable chance of impinging on the configuration of the α-polypeptide. Accordingly, at the molecular level, the complexity of evolution is enhanced by the fact that in a molecule such as hemoglobin evolution requires synchronized changes in both polypeptides.

The DNA complex to which the β-gene belongs is shown in the lower half of the figure on page 251. It extends over a sequence of some 60,000 nucleotides, or 60 kilobases (kb). As one proceeds from right to left (3' to 5'), one encounters, spaced approximately as pictured, the β-locus, the δ-locus, then a sequence with clear affinities to the other β-globin genes but nonfunctional (a pseudogene), then the two loci coding for γ-chains, and the ε-locus, followed by another pseudogene. The five functional genes all possess introns in the same position. The pseudogenes, which in this case also have introns, exhibit departures from the sequences of the functional genes of such a nature that they no longer code for a hemoglobin polypeptide. A gene cluster like this presumably results from successive gene-duplication events during the long course of evolution. Note that the sequence of activity of the five functional genes in development is in the order in which they occur on the chromosome, the δ- and β-loci becoming active at about the same time. The two γ-genes differ with respect to a single nucleotide, resulting in a different amino acid at the 46th position in the chain of 146 amino acids. The maintenance of such similarity in the γ-genes, despite the fact that this duplication occurred early in primate evolution, some 40 to 50 million years ago, has led to the suggestion that there exists some kind of "gene conversion" that keeps these two loci so similar.

This gene cluster presents in its most acute form the problem of defining a "gene." It now appears that genetic elements of vital importance to controlling the function of this complex, elements that control the transformation of undifferentiated cells into the red blood cell channel of differentiation, a step essential to the expression of the β-globin complex, are found at distances of approximately 6, 11, 15, and 18 kb upstream from the 5' end of the complex diagrammed on page 251. There are less well defined elements at similar distances from the 3' end of the β-globin gene cluster. Inclusion of these control elements in the definition of the β-globin gene cluster would expand the size assigned to this cluster by a factor of roughly 10. (There is evidence for even more remote control elements, but at some point these must be regarded as modifier genes rather than as components of the β-globin gene complex.

There is a similar cluster of six α-globin genes that, like the β-gene cluster, also contains pseudogenes (ψζ1, ψα1). The functional status of one of these genes (θ1) remains somewhat obscure. These pseudogenes, whose discovery occasioned as much surprise as the discovery of introns, usually contain introns very similar in position and length to those encountered in functional α- and β-genes. The most reasonable explanation for this type of pseudogene is that it results from a more or less direct insertion of a functional gene—but one that lacks the 5′, expression-controlling elements and therefore is silent—in an anomalous position in the chromosome. However, the silent gene is not immune to mutation, and in time, presumably released from selective pressure, accumulates nucleotide substitutions incompatible with coding for a normal hemoglobin molecule. Earlier I mentioned the possibility that if the β- and δ-genes mispair at meiosis when the germ cells are being formed (and crossing-over occurs), or if the pair of chromosomes bearing the two functional α-genes mispair, the result could be a deletion in the β-globin complex or loss or addition of an α-gene. To carry this suggestion one step further, it seems possible that a chromosome with no functional α-genes can result from a mispairing at germ-cell formation of a normal α-gene with an α-pseudogene, this mispairing followed by crossing over.

The discovery of pseudogenes has thus complicated our view of DNA dynamics no less than the discovery of introns. First, it seems clear that, especially when they are located in proximity to their gene of origin, they can at the time of meiosis confuse the pairing mechanism that normally brings homologous genes into precise apposition. As we have just seen, the result of failure of homologous genes to pair exactly can result in an appreciable genetic handicap. Second, some pseudogenes lack introns; this situation can only result from an insertion into a chromosome of DNA copied from processed messenger RNA. Such a read-back violates our concept of the inviolability of the germ line and, depending on where the read-back occurs, can interfere with functional DNA. Thus, the discovery of introns and pseudogenes has increased our appreciation of the potential for error in the system, without, as yet, a clear biological justification for their existence.

Because of the clinical significance of variations in the amount and functioning of hemoglobin molecules and the relative ease with which hemoglobin can be examined, the hemoglobin molecules have been more extensively surveyed for abnormalities than any other set of molecules. By this time it is probable that some ten million hemoglobin samples have been screened electrophoretically for variants. At last count, almost 600 different abnormalities of the α-, β-, δ-, and γ-globin chains had been recognized. This represents the result of mutation over many generations. It is a moot point what is more impressive: the documentation of the potential for error in the system, or the fact that with such complexity, so little seems to go awry. More or less this same complexity is present in all the other genetic loci, as we shall now see, but with some embellishments.

A homely analogy may be helpful at this point. Imagine a modern business, such as a printing shop, operating efficiently though the shop seems to be cluttered with debris consisting of nonfunctional replicas of many previously working machines, going back to the days when type was hand-set. To make matters worse, the working machines—small assembly lines—produce a primary product that contains seemingly extraneous padding that has to be excised before a finished product is possible. Something like this situation seems to characterize the majority of genetic loci. One can only wonder that the shop is functional at all, and reflect on how much is yet to be learned concerning how the frequency of error is held as low as the observed mutation rates suggest.

Neurofibromatosis and Factor VIII (Hemophilia)[2]

In considering the structure of some other genes, I will give precedence to the genes responsible for the inheritance of several traits discussed earlier in this book, but also introduce several supplementary examples. The first to be considered is the gene in which mutation results in neurofibromatosis (NF_1) (cf. Chapter 12). This gene has now been localized to a precise region of chromosome 17. Although the exact intron–exon structure has not yet been established in detail, the gene appears to be some 200 kb in length, with at least 40 introns. The gene is accordingly some 125 times as large as the genes encoding for the α- or β-globins, but the final gene product, a protein estimated to be 2818 amino acids in length, is only some 17 times longer than the α- and β-globin polypeptides. It is tempting to relate the high mutation rate exhibited by this gene to its size.

While size may play a role, that more than size is involved in mutation rates is demonstrated by the sex-linked dystrophin gene, encoding for a protein whose absence results in the fatal Duchenne's muscular dystrophy. The analysis of the dystrophin locus is somewhat further along than that of NF_1. It is a truly monstrous gene, some 2300 kb in all, in which there are 79 exons that yield a 14 kb processed mRNA, corresponding to a protein of some 3685 amino acids. It is thus approximately 1/1300 of the total human genome DNA content, or 1/100 of the X-chromosome. It is noteworthy that the mutation rate of the NF_1 gene, estimated to be about 1.0×10^{-4}/locus/generation, is similar to that of the dystrophin locus (about 0.9×10^{-4}/locus/generation), despite the much greater size of the dystrophin locus. Thus, factors other than size and intron number are clearly involved in mutation rates. (The dystrophin locus introduces an additional complexity into the concept of a gene: the transcriptional "read-out" of this gene may be initiated at different points on the gene, the site of transcriptional initiation depending on the cell type and the stage in embryonic development. These transcripts, when translated, result in a family of proteins with different functions.)

What renders the neurofibromatosis gene especially interesting is the discovery of three coding sequences of DNA "embedded" within a single intron of the NF_1 gene, sequences that by all criteria should encode for functional gene products. The term "embedded" is here used in a rather special way. These three sequences yield RNA that can be translated only if it has been transcribed in the opposite direction from the one in which the NF_1 gene is transcribed. Two of these sequences show many similarities to a gene associated with leukemia in mice. The other sequence appears to code for a glycoprotein of 433 amino acids that is present in the myelin sheath of all nerves and also in a type of nerve cell termed an oligodendrocyte. This last sequence (= gene) contains a single intron; so now there are genes within genes and introns within introns. The amino acid composition of this latter protein relates it to a family of similar proteins; the locations of the genes for this family have not yet been established.

A number of other genes have also been found to have a much more complex structure than the hemoglobin genes. We can consider only one more example, the gene responsible for a protein involved in blood clotting termed factor VIII, mutation in which can result in hemophilia. Factor VIII is also a large polypeptide, of 2332 amino acids. This requires some 6996 bp for its encoding. The total gene is 180 kb in length, comprising 0.1% of the X-chromosome. The gene is interrupted by 26 introns, one as large as 32,400 bp. The exons specifying the amino acid sequences range in length from 69 to 3,105 bp. There are three similar repeat sequences in the molecule, which suggests that the gene arose through a triplication of the gene for some primordial clotting protein but—surprisingly—there is also a DNA sequence that exhibits a very significant homology to the gene coding for the primary copper-binding protein of the blood plasma, ceruloplasmin.

The factor VIII gene is also of interest in that it currently provides the only other known example of a mammalian gene embedded within a gene. Within the largest intron (no. 22) there is an intronless gene oriented (as in the NF_1 gene) in the opposite direction to the main gene and associated with a 1.8 kb transcript. The protein for which this gene encodes has not yet been identified. There are two other copies of this gene outside the factor VIII gene but within 1,100 kb of it, i.e., closely linked. Interestingly, this inserted gene is expressed in many different cell types, including cell types in which the factor VIII gene is not expressed. Both the juxtaposition in the factor VIII gene of two apparently quite disparate types of building blocks and the insertion of extraneous genes in its introns are the types of events that might result from the activity of jumping genes.

It is noteworthy with respect to the genes for neurofibromatosis, factor VIII, and some other large genes, what a small proportion of the gene—roughly 1–4%—is involved in coding for the gene product. Thus, greater than 95% of these genes appears as of now to be "excess baggage." Earlier we mentioned the fact that some of the mutations resulting in thalassemia either inactivated the splice sites so important in the editing of the initial DNA transcript into its final (messenger) form or else created new splice sites within introns, thus resulting in an abnormal messenger DNA. In principle, all these larger genes should be much more prone to this type of mutational damage.

Retinoblastoma

The genes we have considered thus far usually encode for a familiar protein product, such as hemoglobin, or a blood-clotting factor. By contrast, it appears that the genetic locus at which mutation results in retinoblastoma (cf. Chapter 12) produces a protein involved in the regulation of cell function. The facts are these: In recent years a class of genes termed proto-oncogenes has been recognized. These genes first came to attention because mutations (of various types) involving one of them may result in the out-of-control cell proliferation that characterizes a malignant tumor. The mutation converts a proto-oncogene into an oncogene. Gradually it has become clear that these proto-oncogenes are involves in the regulation of many important "housekeeping" details of cellular metabolism, including cell division. A mutation in one of them can upset the delicate balance of events that characterizes the normal cell cycle. The term proto-oncogene is a real misnomer, reflecting the history of the discovery of these genes; in time each will acquire a name more specific to its true and usual function. The protein product of the retinoblastoma gene is confined to the cell nucleus and can be shown to bind to DNA, where the retinoblastoma gene product interacts very specifically with the regulatory region of a known proto-oncogene, *c-fos*, whose normal function is regulation of the cell cycle. The retinoblastoma locus normally down-regulates the activity of the *c-fos* locus; when this function of the retinoblastoma locus is lost through mutational inactivation, the expression of the *c-fos* locus is no longer normally modulated. Thus the study of a rare tumor has led to insights into the regulation of basic cell activities. More than 100 of these proto-oncogenes have now been recognized.

Another Example of Gene Complexity

Our final example of gene complexity is provided by the locus encoding for a cluster of hormones and hormone-like substances primarily elaborated in the pituitary, the so-called POMC locus. Although the definitive research on this locus involves cattle, the human locus must be very similar. This locus is responsible for the synthesis of a "brain hormone" (β-endorphin), a melanocyte-stimulating hormone, and a hormone active in the adrenal gland, as well as several other biologically active, relatively small, polypeptides. The primary gene product consists of a polypeptide with a molecular weight of approximately 28,500 daltons, not a large polypeptide. At one end is a "signal sequence," with a molecular weight of 2500 daltons. This is edited off as the primary gene product traverses the intracellular membranes. The remaining polypeptide is then further cleaved, the exact process depending on the tissue in which it has been produced. The first round of editing produces a piece of approximately 16,000 daltons molecular weight, and two smaller pieces, hormones, named adrenocorticotrophic hormone and β-lipotrophin. In various tissues there is then further cleavage, as follows: (a) the fragment of 16,000 daltons is cleaved to yield the so-called γ-melanocyte-stimulating hormone; (b) the adrenocorticotrophic hormone is cleaved to yield two smaller, biologically active polypeptides, one of them another melanocyte-stimulating hormone (the α-melanocyte hormone) and the other a corticotrophin-like peptide; and (c) the β-lipotrophin is cleaved to yield another lipotrophin (γ-lipotrophin) and the brain hormone β-endorphin. The γ-lipotrophin may be edited further, to yield another melanocyte-stimulating hormone, the β-melanocyte-stimulating hormone.

At first consideration, all this appears to be a remarkably "economical" process. Remember, however, that all this intracellular editing of that primary polypeptide begins only after the many steps preceding the elaboration of the polypeptide have been accomplished. Furthermore, specific enzymes must be involved in this editing of the primary polypeptide. These editing enzymes must be under genetic control. The process thus involves many possibilities for genetic error, some of which might result in cascade effects involving multiple hormones, a contingency which would not be a problem if the major products of this locus were synthesized independently. One must wonder whether this complexity, while not the ideal situation, reflects a fortuitous concatenation of evolutionary events with which an improbable evolutionary sequence has saddled the species.

The Coordination of Gene Activity

Thus far we have considered briefly several genes encoding for well-known protein products, generally termed structural genes, and one gene that appears to regulate the activity of another gene (the retinoblastoma gene). I presume that when the composition of the entire genome is known, this latter type of gene will be at least as frequent as the former type. Above both these gene-types in the functional genetic hierarchy is another set of loci, very poorly understood, responsible for the coordination of the functioning of the various genetic loci and the sequence of gene activity in development. The best-documented example is the case of the hemoglobin loci. Two kinds of coordination are involved. One is the successive activation of the various α-and β-globin genes during fetal development. The other is the coordination of the functioning of the α- and β-genes. The human adult elaborates equal numbers of α- and β-chains, even though there are two active α-loci and only one β. This coordination breaks down in the thalassemias, and the resulting hematological problems seem due as much to the relative overproduction of one kind of chain as to the underproduction of the other. How this coordinated activity is normally maintained is by no means clear. The problem of coordination of the functioning of the elements of the system extends to the entire cell. In general, each locus functions as is appropriate to the internal needs of the cell and the needs of the organism of which it is a part. The genes that are responsible for this coordination are sometimes termed "master" genes. They seem to work through coding for proteins, which then interact in complex ways directly with DNA. But how these proteins recognize the genes with which they should interact, and what regulates the activities of the master genes, is only now being worked out. This coordination occurs within the complexity of chromosomal structure shown on page 248! The task of defining the content of the gene pool—its anatomy—is coming along nicely, but the task of defining the regulation of gene function and the interrelationships between loci has barely begun.

The Repetitive Use of Successful "Genetic Building Blocks"

A fundamental principle that emerged early on, as the structure of various genes became known, was the repetitive use of certain genetic building blocks. Three different levels in this process may be recognized. The first level involves the use of repetitive segments within a single gene, as illustrated by hemophilia. The second level is illustrated by gene families, which may be clustered together, as for the hemoglobin genes, or more widely dispersed, as is the case for the collagen genes. The third level of building-block reuse is exemplified by genes within which a variety of functional "domains" of diverse origin may be recognized. An example of this has already been provided by the hemophilia (factor VIII) locus. There are numerous other examples of the latter type of gene. For instance, a receptor on the surface of some cells for a particular type of fat-complexed protein (the low density lipoprotein or LDL receptor) shows striking homologies in one region of the receptor with the C9 complement protein (a protein involved in antigen–antibody interaction), in another region homologies with three different proteins involved in blood clotting (factor IX, factor X, and protein C), and in still another region homology with a protein known to stimulate cell growth (the epidermal growth factor). Similarly, the gene for coagulation factor IX, one of the many proteins involved in the clotting of blood, involves two epidermal growth-factor domains, as in the previous example, a domain similar to one found in serine proteases (enzymes that break polypeptides down by inducing cleavage at the site of a particular amino acid, serine); an activation peptide domain (an amino acid sequence whose interaction with another protein results in the activation of factor IX). Also involved is a stretch coding for 12 glutamic acids (a sequence found in many other proteins).

As more and more genes are studied in detail, it should be possible, through homology studies, to decipher how many basic genetic building blocks have come into being during the course of evolution. It may be as few as 2000, most of which appeared relatively early in evolution. After this, the chief evolutionary advances resulted from combining various of these blocks to produce proteins with new functions. As far as geneticists are aware, these gene duplications and reshufflings of domains originally occurred at random. There is no known mechanism by which a cell can sense the need for a new protein and proceed to custom design the necessary gene. For each such random recombination of domains that represents a successful experiment and that survives the vicissitudes of getting established in the population, there must have been thousands upon thousands of unsuccessful experiments. If there were only 2,000 different building blocks, taken 2 at a time, they can be combined in 2,001,000 different ways, if we consider both homo- and heterodimers. The potential inherent in combinations of three and even four of these building blocks is even more astronomical. Most of these combinations would have no evolutionary future, but since humans seem to make do with "only" 50,000 genes, there are clearly ample numbers for evolutionary experimentation with domains. Unfortunately, the design of studies of mutation rates is such that only very incidentally would they detect recombinational events of this sort. We thus have no real estimate of how frequently they occur.

Developing the techniques that permit the molecular insights described in this chapter has obviously required considerable ingenuity, but, although they must be meticulously implemented, the techniques are not intrinsically highly demanding and difficult. They are thus finding extensive application all over the world, with discovery following discovery. The genetic loci whose fine structure has been so briefly touched upon in this chapter all encode for well-known proteins that are of special biological interest, but the loci were not chosen for study with any foreknowledge of the complexity of their fine structure. Clearly the surprises as to how the genetic material is organized and finds expression will continue for some time to come.

What is now abundantly clear is that there is no standard gene. Evolution represents a mind-boggling series of improvisations. For each successful "new" gene that in the cause of evolution becomes established in the genome, there were thousands, perhaps a million, unsuccessful experiments. When we come to discuss gene therapy, this is an important point: each gene is its own paradigm, this fact prohibiting any easy generalizations in this emerging field of inquiry.

Earth's gene pools from afar—perhaps the only gene pools in the entire universe (photo courtesy of NASA).

15

Defining the Gene Pool

The term "human gene pool" has been employed rather loosely thus far, as a synonym for the collective genetic material of our species. Now we are ready to lay out in some detail exactly what comprises this unique human resource. As we will see, the 3×10^9 nucleotides of the human chromosomes comprise a strangely admixed assemblage of many components, their functioning still very poorly understood. At literally every turn of the DNA, the prospect of genetic disaster looms, yet it all hangs together and manages to incorporate changes of adaptive value to the species. This requires of the gene pool a dynamic that is currently taxing our imagination.

When the genetic code was first elucidated, it appeared that only a few percent of the DNA was necessary for coding for all the proteins of the human body, and comments began to appear in the genetic literature about the frequency of "junk" and "selfish" DNA, the latter being DNA whose only function was to replicate with high efficiency. Now it is becoming apparent that when introns and control sequences are taken into consideration, many genes are much larger than the early proto-types were thought to be and that, in addition, the nongene DNA, rather than being random gobbledygook, is comprised of a variety of rather specific types. Although these types may vary in their quantitative rep-resentation from species to species, within any given species their pro-portions are relatively constant in the various members of the species. In view of how destabilizing to the character of DNA random mutation could be, I take this relative constancy to indicate that some kind of selection to maintain this constancy is at work. DNA that responds to selective pressures can scarcely be called "junk." This chapter has two purposes: first to indicate the emerging complexity of the genomes of which the gene pool is composed and second to provide the requisite background for when, in later chapters, we come to consider the subject of gene therapy.

The Kinds of DNA in the Gene Pool

A very simple qualitative and quantitative breakdown of the genomic material as now understood can proceed as follows.[1]

Functional Genes. Inasmuch as the preceding chapter was devoted to functional genes, not much more need be said here. At the moment, an estimate that, including introns and regulatory sequences, roughly 15% of the genome is comprised of such genes is tenable, using the term gene in the broad sense of an operative unit.

Pseudogenes. The frequency of pseudogenes is difficult to estimate because, by and large, their identification has been an incidental finding rather than the result of a systematic search. It is becoming apparent that for every functional gene there may be several corresponding pseudogenes. Since, however, these are often devoid of introns, and have usually lost their control sequences, pseudogenes will generally be less than half the size of functional genes. Thus as much as 10% of the genome could be comprised of pseudogenes. In the preceding chapter, we considered briefly the fact that many genes appear to be a juxtaposition of parts of previously existing genes, these parts often corresponding to what appear to be functional domains of the source genes. Inasmuch as mutation is customarily envisioned as random rather than directed, then pseudogenes could also serve as a source of potential evolutionary building blocks. However, pseudogenes are "pseudo," i.e., unexpressed, because they have lost control sequences or have accumulated in their exons mutations incompatible with a functional gene product. Therefore, only the portion of the pseudogene that preserves sufficient function to enter into the evolutionary process can serve as a building block. Pseudogenes will thus be a less efficient source of evolutionary material than active genes. On the other hand, the possibility of their participation in evolution exists, and it would be inappropriate to label them junk.

Potentially Mobile Genetic Elements. In experimental organisms, where they have been extensively studied, mobile elements are of two types. The first type, transposable elements, corresponding to the jumping genes described by Barbara McClintock, appear to move directly from position to position on the chromosome. The second type, termed retrotransposons or retrogenes, insert into new chromosomal locations in the fashion of retroviruses. Retroviruses, widespread in nature, are viruses that, unlike most viruses, possess only RNA in the infective particle. Once they have gained access to a cell, their RNA is transcribed into DNA by an enzyme known as reverse transcriptase, for whose independent discovery Howard Temin and David Baltimore, with Renato Dulbecco, received a Nobel Prize in 1975. This DNA then inserts into the chromosomes. Retrotransposons and retrogenes, structurally very similar to retroviruses but already inserted into the chromosomes as DNA, could move about by being transcribed into extrachromosomal RNA molecules that are then transcribed into DNA molecules by the action of reverse transcriptase and inserted back into the chromosome. Retrotransposons and retrogenes, and DNA sequences that bear unmistakable homologies to them, are surprisingly widespread throughout the human genome. Collectively, these sequences amount to perhaps 5–10% of the DNA.

Views as to the genetic significance of these germ-line retrogenes and retrotransposons and their traces in humans vary tremendously. By some estimates there are as many as an amazing million retroviral "footprints" (of various sizes) in the human genome. Since retroviral infection of various somatic tissues (as of lymphocytes by the AIDS virus) is common, we may ask whether these genomic retroviral footprints result primarily from retroviruses specifically targeted to germinal tissue or rather from retroviruses that strayed into the germ-line by accident. If the latter, it raises concerns for some kinds of gene therapy (see Chapter 21). Whatever their origin, what is their present function? One extreme view is that the great majority are simply relics of past infections that have lost their pathogenic significance through a variety of mechanisms and stay with us because the genome has no way of cleansing itself. They should thus be viewed in the same light as pseudogenes. The other extreme view holds that many of these sequences have incorporated into their structure foreign genes whose products are involved in important basic cellular processes such as differentiation and development. There is evidence that the presence of some of these sequences is required for the expression of specific human genes. Furthermore, as long as the specific RNA–DNA mechanism by which they inserted themselves into the chromosome persists intact, some of them are, as noted, potentially mobile genetic elements. These elements, inserting themselves throughout the genome, can be expected to disrupt established genetic functions, and there are already several examples of this.[2]

In Chapter 9 I described the grossly abnormal chromosomes first observed in cultured Amerindian lymphocytes but subsequently also encountered in English and Ukranian Caucasians and in Japanese. I also presented the evidence that cells exhibiting these abnormalities are not randomly distributed in time and space but come in waves, and speculated that this phenomenon might not be restricted to lymphocytes but could involve various cell types, including germ cells. In the latter, it would not be surprising if the phenomenon were associated with a concurrent increase in the mutation rate, resulting in mutations primarily characterized by gene rearrangements of the type discussed in the last chapter rather than changes in one or several nucleotides. I have speculated that this phenomenon might be due to the action of a virus. As discussed earlier, the direct action of an acquired virus is a possible explanation, but it is at this juncture tempting to return to the alternate hypothesis, that the finding is somehow related to these genomic retroviral footprints. Specifically, I have to wonder whether from time to time these slumbering retroviral ghosts are awakened *en masse*—reactivated—with the cytogenetic results we have recorded.

The extremely damaged cells pictured in the figure on page 167 are "dead end" cells: they cannot possibly divide successfully. But other, less damaged cells may survive but with some redistribution of their genetic material. If, as some have suggested on the basis of data from experimental organisms, especially *Drosophila,* these retroposons are a major factor in creating the DNA instability that leads to mutation and evolutionary change,[3] this DNA is anything but junk. If the case is ultimately made that "rogue" cells are due to retroposon (or other viral) activity, with all the implications I have suggested, then this thesis may well be the most important contribution of my scientific life. The seminal observations were completely unplanned, purely serendipitous. I have no qualms with this admission. Many more important discoveries are the product of serendipity than their authors will ever let on.

Repetitive DNA. A very abundant type of DNA consists of repetitive (i.e., frequently duplicated) sequences of various types. These are conventionally considered under two headings:[4]

(i) *Long interspersed nuclear elements* (LINEs). A LINE is defined as a "family" of homologous DNA sequences, each 6–7 kb in length, that are interspersed throughout the genome. Most of them fall into a single family of similar sequences, known as L-1, present in more than 10,000 copies. There are probably other, less numerous families. DNA sequence studies suggest that LINEs contain stretches of DNA that could code for one or more proteins, but the proteins have not been identified and the function of these (possible) proteins is unknown.

(ii) *Short interspersed nuclear elements* (SINEs). A SINE is a "family" of relatively short, homologous, interspersed DNA sequences, typically some 300–600 bp in length. Each family recognized thus far is represented by 100,000 or more copies. The best known of these is the *Alu* family, made up of about 1.5 million copies of an approximately 300-bp-long sequence, and comprising some 10% of the human genome. Other well-known SINEs are the variable-number-of-tandem-repeats (VNTRs) family, each generally characterized by a sequence of some 10–50 bp repeated in tandem from 20 to 50 times. Within any genetic lineage the number of repeat units is a specific VNTR is inherited, although mutational changes in repeat number are relatively common.

There may also be repetitive DNAs intermediate in size between the LINEs and the SINEs. These repetitive sequences so often show characteristics of mobile genetic elements that this category is not independent of the preceding category. We estimate that another 15% of the DNA consists of repetitive DNA that does not correspond to mobile genetic elements.

The function, if any, of LINEs and SINEs is also the subject of vigorous debate, and the jury is apt to be out for some time. However, whatever they do, two considerations suggest that they cannot be entirely neutral with respect to the functioning of the genome. First, the structure of the *Alu* sequences of different species suggest that they may originate from transcribed cellular RNA sequences that, because of their structure, are prone to be "read back" into the chromosome at various sites by a process akin to retroviral insertion. This characteristic should permit them to move throughout the genome, and there is already limited evidence to this effect.[5] This movement of DNA sequences has the potential to be disruptive to the functioning of the working genes. Furthermore, without some kind of selective restraint on the upper limit in which these sequences are represented, the manner of their spread suggests they should be much more numerous.

Telomeres and Centromeres. Two very specialized regions on each chromosome are the centromere and the two telomeres. The centromere, to which the spindle fiber attaches at the time of cell division, is essential to the orderly separation of homologous chromosomes. The telomere constitutes a "cap" on each end of the chromosome. If chromosomes are broken, as by radiation, the broken ends are "sticky"; broken ends, no matter what their origin, tend to unite with other broken ends. The telomeres, which consist of variable numbers of a six nucleotide repeat sequence, somehow ensure that the "normal" ends of chromosome are not sticky. The nature of both centromeres and telomeres is under active investigation. Perhaps 10% to 15% of the DNA is devoted to these purposes.

Genes for Which RNA Is the Functional Product. Although most of the various types of RNA molecules present in a cell will ultimately be translated into protein molecules, some RNA molecules function without translation. These for the most part are found in cellular organelles called ribosomes, where they are vital to the translation of other RNA messages into polypeptides. Other RNA molecules shuttle specific amino acids to ribosomes, for incorporation into the lengthening polypeptide. There are some miscellaneous smaller RNA molecules in the cell nucleus whose function is unclear. There can be no doubt about the functional significance of the majority of the DNA responsible for these various RNAs; it constitutes some 3–5% of the genome.

Mitochondrial DNA (mtDNA). For completeness, the extranuclear mitochondrial DNA must be mentioned. This is a circular, double-stranded DNA molecule, normally composed of 16,569 bp, found in small cytoplasmic organelles, the mitochondria. These mitochondria are believed to be the degenerate descendants of a bacterial species which early in the evolutionary process established on endosymbiotic existence in an ancestral eukaryote. [Recall that bacteria are haploid, not diploid like all (eukaryotic) plants and animals.] There are some ten copies of this DNA per mitochondrium, which, given some 800 mitochondria per cell, implies 8000 copies per cell. Since the mitochondria are located in the cytoplasm, inheritance of mitochondrial genes is largely maternal, through the egg rather than the sperm. By virtue of the absence of introns and the presence of only one regulatory region for the entire sequence, this relatively small sequence codes for 8 proteins, 22 RNAs which are involved in shuttling amino acids to the ribosomal polypeptide assembly line (tRNAs), and 2 species of RNA necessary to the functioning of the ribosomes (rRNAs). Several human diseases are associated with defects in mitochondrial proteins. Mitochondrial DNA comprises about 1% of the total cellular DNA, but in a highly repetitive form. There is no question that it is functional.

Other. The term "other" applies to whatever was not covered in the foregoing account—which adds up to almost half of the genome. However, I have no great confidence in the foregoing allocations, which, as noted, are somewhat overlapping (i.e., repetitive sequences may occur within introns). The allocations are certain to change with time, but they can serve as a point of departure for some important considerations. Is there any true "junk"? Mobile genetic elements insert at specific target sites in DNA. Each time a mobile genetic elements inserts itself into the DNA strand, short DNA duplications, usually of 5–15 bp, are formed at each end of the insertion. Should these duplications be termed "junk" DNA, even though the duplications create new target sites for the insertion of mobile genetic units?

The Potentiality for Error in this "System"

To understand the full impact of this complexity on our thinking, we need to go back to the day when a gene was visualized as a globular protein molecule and the genome was envisioned as strings of these beads, packaged as chromosomes. That gene had—and still has—two functions: to replicate itself, and to direct cellular metabolism and morphogenesis. Given what little was then known of enzymes, it was reasonable to think of genes primarily as enzymes that replicated in an inactive form but were also released in an active form, these active forms directing the synthesis of a wide variety of proteins (structural, enzymatic, and hormonal) as needed. By any reasonable standard, the complexity now revealed is several orders of magnitude greater than conceptualized only a generation ago.

To someone like myself, so oriented toward understanding the kind and frequencies of mishaps to the DNA that result in mutations, the DNA complexity we have just reviewed seems an open invitation to total disaster. These various kinds of DNA are all jumbled up together. The question of how this complexity of genetic fine structure is maintained, functions, and even evolves, especially in the face of the mutational erosion we discussed earlier, is the central problem of biology in general and genetics in particular. Except for the telomeres and centromeres, the various kinds of chromosomal DNA are intermingled in such a fashion that if there is some over-arching organizational principal, it has yet to be discovered. There is the same air of improvisation to genome structure as to gene structure.There is no standard blueprint from which the attempts at gene therapy we will be discussing later can proceed. Rather, the approximately 3×10^9 nucleotides which are to be found in a human egg or sperm often appear to have been assembled by a committee that could not agree on first principles, other than to make do with whatever constellation of DNA was functional and to conserve as much from the past as possible, on the thesis that it might someday be useful. Let us consider briefly the "overhead costs" created by six aspects of the structure of human DNA.

(i) **The Packaging of DNA in the Nucleus.** The amount of DNA in the 46 chromosomes packed into a human nucleus would, if unraveled and laid out end-to-end, extend two meters. The length of the DNA is thus approximately 200,000 times greater than the diameter of the nucleus within which it is contained. How this miracle of packaging is accomplished (see the figure on page 248)–and in a way that discrete sections of DNA may be functional in the nondividing nucleus—and why the molecular strands that constitute each chromosome do not become hopelessly entangled at mitosis is unknown. How often, given the angulations of the DNA strands that must exist under these circumstances and the tensions on the strands that must arise as the DNA condenses into the chromosomes seen at mitosis, do the strands snap, especially at sites of imperfect bonding?

(ii) **The Inherent Instability of DNA.** In each human cell there are some 6×10^9 nucleotide pairs, for each of which, as discussed earlier, a purine or pyrimidine molecule is an integral part. These molecules are subject to spontaneous loss from the DNA chain. If lost purines or pyrimidines are not correctly replaced, the result can be a change in the code, i.e., a mutation. The probability of such a spontaneous loss at any site is very low—in a mammalian cell about 3×10^{-11} every second—but T. Lindahl[6] has calculated that because there are so many nucleotides, a mammalian cell in an actively dividing cell line should spontaneously lose some 10,000 of these elements during the usual 20-hour generation period.

(iii) **Mitotic Risk.** At each cell division, unless the DNA reproduces perfectly, a mutation of some type will occur someplace in the genome. Prior to the meiotic cell divisions that result in eggs and sperm, there have been on average some 21 cell divisions in the precursor cells in the ovary but, by age 28 (the average age of male reproduction), an estimated 380 cell divisions in the testes. At each of these rounds of DNA replication there is the possibility for error. Let us round the numbers off, to an average of 200 cell divisions in the germ-line, sexes combined. Error could creep in at each of these replications. Earlier we estimated that the mutation rate detected in germ cells was approximately 10^{-8}/nucleotide/generation. But since that is the cumulative error in 200 cell divisions, then the error per cell division is less than 10^{-10}. No process humans have devised approaches this in accuracy. How is such accuracy achieved?

Furthermore, at the stage when individual chromosomes are replicating themselves in preparation for the next cell division, there is often an interchange of segments between the original and the nascent chromosome (the so-called sister-chromatid exchanges). Although these exchanges have not been studied in the early germ-line cells, they must be presumed to occur there as well. If these exchanges were not precise, the result again would be a mutation.

(iv) **Meiotic Risk.** In addition, at the meiotic divisions, which result in the formation of the functional sperm and eggs, there is crossing over between specific paternally and maternally derived (homologous) chromosomes. About 50 cross-over events are realized in the average human germ cell undergoing meiosis. With our current understanding of the ultimate basis of mutation, it is clear that these exchanges between members of a pair of homologous chromosomes must be excruciatingly precise if crossing over is not to be accompanied by a mutation. Even a mispairing by one or two nucleotides at the time of the pairing that accompanies these exchanges can result in germ-line mutation. The mechanism whereby the corresponding portions of two homologous chromosomes are so precisely attracted to each other is unknown. Consider, now, those aspects of the genome that have the potentiality to throw this pairing off. First, members of a gene family may "mismate." We have already documented the ill effects of this at the α- and β-globin locus. Second, some pseudogenes are sufficiently similar to the gene from which they originated that they might be expected to be a cause of confusion in the pairing process. Finally, highly repetitive sequences of DNA should impair the precision of the pairing process; how does a given sequence in chromosome A establish its precise match-up with the right sequence on its homologue when there are so many other similar sequences in the genome?

(v) **The Presence of Introns.** For at least two reasons, the presence of introns should be a drag on the organism. First, they are transcribed into the nascent RNA messenger only to be almost immediately discarded. As we saw in the preceding chapter, for many genes, the superfluous "cuttings" from that first transcription of introns-plus-exons will far exceed the extent of the final message. [For instance, the gene responsible for the cellular protein that binds to certain fat-transporting molecules (the LDL receptor) is 45,000 nucleotides in length, within which there are 18 widely spaced exons, totaling 2517 nucleotides, that code for the 839 amino acids in the protein. About 94% of the initial RNA transcript is discarded. The enzyme phosphoglycerate kinase is encoded by a gene 23,000 nucleotides in length, in which there are 10 exons, totaling 1248 nucleotides that code for the actual enzyme. Again about 95% of the first transcript is edited out and discarded.] These "cuttings" must be disposed of, and rather quickly, before they clog the cells' vital functions. Try to visualize running a business this way!

A second nuisance value of introns is that, through mutation, new "splice sites" may arise within the introns. These confuse the mechanism whereby the introns are edited out of the primary RNA transcript, resulting in the inclusion of extraneous RNA in the RNA that is to be translated into a protein, and, since the intron DNA is usually gobbledygook as regards a genetic message consistent with the function of the protein being synthesized, the result is a disaster for protein synthesis.

Thus far, as we have noted, the strongest suggestion for the function of introns has been that they divide the DNA into convenient stretches (the exons) which, recombined, can be the building blocks of the code for a new protein. To be sure, as discussed in the preceding chapter, several proteins are known that seem to be hybrids—one portion of the protein coded by exons derived from the gene responsible for protein A, another portion coded by exons found in the gene responsible for protein B. There is, however, a good deal of the *"post hoc, ergo propter hoc"* in this line of reasoning, i.e., since we see several examples where the exons of different genes seem to have been combined to create a new gene, therefore introns exist to break the genetic material up so it can be combined in this fashion. It is worth nothing in this connection that an intron is often inserted between the members of a 3-nucleotide codon for an amino acid. Otherwise stated, one of the three nucleotides that comprise the codon that specifies a particular amino acid may be adjacent to one end of the intron, the other two adjacent to the other end. This implies that these exon building blocks will sometimes not code for a consecutive string of amino acids, still another aspect of DNA organization that appears to be of dubious functional value.

Our knowledge of the molecular basis for mutation, limited as it is, already raises some problems for the "building block" justification of the presence of exons, especially in higher organisms. As far as we know, evolution does not look ahead, storing in some protected place goodies for a rainy day, like a squirrel. The genomic material is "exposed" to natural selection at all times. Given the documented ill effects of mutation in introns, we could postulate that natural selection should, in the short run, in many respects favor the organism and the species with the minimum number of introns, which is to say, no introns. On top of this, the probability of a functional new protein emerging from a reshuffling of exons must be very low. The appearance of a "new" gene in that convolution of 3×10^9 nucleotides requires not only that the domains brought together by chance result in a protein with a tertiary (e.g., coiling) structure permitting function, but that they acquire a suitable 5' controlling sequence and that all this be inserted in a region of the DNA where it does not create a lethal effect! Finally, the new gene, at first an isolated event, must, if it is to become the norm, spread throughout all the members of the species.

The *a priori* odds against the emergence of a new gene through domain shuffling are incalculable but must be enormous. There must be many, many unsuccessful experiments in the creation of the genes for new proteins, the impact of which should be generally unfavorable to the organisms in which they occur. In popular parlance, the species could be nickled and dimed to death while waiting to hit the jackpot. It now appears that the basic structure of many of these proteins emerged during those first several billion years of evolution, when life forms were small and simple with a tremendous reproductive capability. Evolutionary experiments were much easier and less dangerous for the species then. One could infer that now we are left with a genetic fine structure the logic and dynamics of which must be sought in the past rather than the present. But the present is left to maintain this structure.

(vi) **Back to the Retroviruses, Their Footprints, and Other Possible Intruders.** We mention again the danger to the genome of the widespread insertion of retroviruses during primary infections but, perhaps worse, the possibility that these retroviruses, once inserted, sporadically give rise to copies which, again inserting at numerous sites, should multiply the potential genetic damage of the primary infection.[7]

The Additional Potential for Damage from Chemicals and Background Ionizing Radiation

In addition to these readily identified reasons for spontaneous instability, our DNA is constantly exposed to potentially disruptive extraneous influences. During the time span from birth to the middle of the reproductive period, background radiation is now estimated to result in an effective dose equivalent to human gonadal tissue of approximately .09 Sv equivalent.[8] Since perhaps 5% of ions generated by this exposure escape immediate recombination and function briefly as free ions, that translates to as many as 3×10^{11} free ions released per gram of gonadal tissue over this period. Furthermore, in the "natural" environment, which I will equate with the nonindustrialized environment, there are many mutagenic compounds, such as certain chemicals produced by molds and leafy plants, which, reaching our germinal tissue and interacting with DNA, could act as natural mutagens.[9]

Genetic Repair

The fact that errors in the maintenance of "resting" DNA and further errors at the time of DNA replication are no more common than suggested by the mutation rates I have cited is apparently due to a complex of highly efficient and still very poorly understood repair enzymes. The number of such enzymes is unknown, but, at present, some 30 have already been recognized, and the number awaiting discovery is surely at least another 30. They have been chiefly studied in bacteria. In the four billion years of evolution, the ability living organisms have evolved to protect their DNA is truly astonishing, but perhaps not so astonishing when we reflect that living organisms have no more basic challenge.[10] These repair enzymes have two kinds of functions. On the one hand, they must survey the nonreplicating DNA for lost purines and pyrimidines and chemical kinks (adducts) or breaks in the chain and take appropriate action when these are encountered. On the other hand, they must ride shotgun on the replicating DNA strand, correcting any infidelity from the original in the new strand. Specific functions have been assigned to many of these enzymes. A description of these functions would take us too far afield. Since the frequency with which damage to the DNA or primary errors in replication occur is unknown, there is no way to assess the efficiency of the repair system, but quite possibly upward of 99% of all the genetic damage that occurs during the cell cycle is accurately repaired.

When most people think of the protection of a population from genetic damage, they think in terms of protection from mutagenic agents. But another kind of protection would involve ensuring the optimal functioning of the enzymes responsible for DNA repair. Perhaps in severe malnutrition or during a febrile episode the functioning of these genes is impaired. A decrease in efficiency from 99% to 98% could double the spontaneous mutation rate. A high-priority research item for the future is to understand what external factors can be brought to bear on the functioning of these systems. What vitamins, minerals, or energy balance will keep them functioning at maximum efficiency?

The Amount of Genetic Variation in the Gene Pool

The development of biochemical methods of identifying genetic variation resulted, beginning with hemoglobin some 40 years ago, in the realization of just how much hidden genetic variation in proteins there was beneath our relatively similar exteriors. Then, 10–20 years ago, insights into how much genetic variation existed at the DNA level began to emerge. This latter variation can be considered in two categories. One is qualitative variation in the DNA code, i.e., variation in the precise nucleotide present at a specific site in the DNA. A current estimate is that variation will be encountered at about 1 in every 200 sites examined in nuclear DNA. This must be considered a very conservative estimate, since the kinds of laborious, large-scale surveys that would reveal the true amount of variation are yet to be performed, and the data are largely for exons, in which variation is presumably subjected to stronger stabilizing selection than elsewhere in the genome. But even if the estimate of 1 in 200 were correct, it would imply that there were $(1/200) \times 3 \times 10^9$ or 1.5×10^7 variable nucleotide sites in the human genome. Let us assume there are only two alternatives at each variable site. The number of *potential* genotypes created by this finding is $2^{1.5 \times 10^7}$, or $10^{4,515,450}$. This number is a *tour de force*, since some of these genotypes have a negligible probability of occurring, but the example makes its point.

A second kind of variation, more quantitative in nature, consists of small deletions, duplications, or inversions in the DNA. We have already suggested that very small lesions of this type may be almost as common as variable nucleotide sites. In addition, there are larger lesions, involving perhaps 200 to 200,000 bp. Here our own preliminary results suggest that within every sequence of 2.5 million nucleotides, one can expect to encounter at least one variant of this type.[11] If we consider the typical germ cell, it might carry 1000 such variants. Calculations of the amount of variation in the human genome are in many respects egregiously premature. Nevertheless, the realization of how enormous the store of genetic variation within the human (and other) species is must be considered one of the genetic revelations of recent years.

The Turmoil in Our Genome

The foregoing considerations, coupled with what we have discovered about the genetic dynamics of tribal populations as discussed in Chapter 11, are currently leading me to a very different feeling for the genetic dynamics of human populations, especially during the years before civilization, than that which I held during my student days or as recently as 20 years ago. Then I thought of the gene pool much in the Websterian sense of pool: "a small body of standing liquid." Now, however, I see a state of agitation at both the cellular (nuclear) and population levels that, at best, deserves the appellation "controlled turmoil," and at worst, "organized chaos." In short, the view of gene pool genetics at which I have arrived drives me into the arms of so-called "complexity" and/or "chaos" theory, but to pursue this subject at this juncture would lead us far from the purpose of this book.[12]

Despite all the possibilities for mishap that have just been documented, the system obviously works, or else we wouldn't be here to analyze it. The fact that it seems to work so well when we can see so many possibilities for error to me underscores our present imperfect understanding of DNA dynamics. To one who considers the DNA from the standpoint of mutational possibilities, the potential for error is staggering. But, if our calculations are correct, the average gamete contains "only" some 30 mutations involving nucleotide substitutions (plus the other types of mutational change I have mentioned). "Only" is in quotes because while it is to me a marvel that number is so low, still, 30 random nucleotide changes per generation per gamete is a large number to be dealt with by natural selection. Some of these changes, occurring for instance in the repetitive DNA, might be neutral in their effects, but some must involve genetic material discharging important functions. Even in this case some mutations could be neutral in their effects, but others must be deleterious. This is not speculation but a fact supported by many experimental studies.

Although the application of biochemical and molecular techniques has revealed—and continues to reveal—much more inherited variation than was previously suspected, our studies on the effects of inbreeding reveal that relatively little of the variation encountered in DNA is associated with unfavorable phenotypic effects. Somehow the genetic system is doing a remarkable job of cleansing itself of such deleterious mutations as do occur. Calculations cited in Chapter 6 suggested that the average Japanese germ cell contains only 2 or 3 genes that if rendered homozygous would be incompatible with survival during the period between birth and maturity (or 4 to 6 genes with 50% impairment of survival, or 8 to 12 genes with 25% impairment, etc.). This seems a rather low figure considering the mutational input. The implication is that most of the variation revealed by the biochemical studies is not very important to survival and reproduction. On the other hand, there are still so many examples demonstrating that mutations in proteins can have deleterious consequences that we are forced to wonder how the accumulation of these deleterious mutations is kept so relatively low. What the foregoing adds up to is a very imperfect understanding of the functioning of the genome and the manner in which "natural" selection operates to eliminate genes that do not meet the organism's needs.

Finally, now, we know that this system, which seems so delicately poised, is really quite resistant to external perturbation. The gene pool is tough, possessed of remarkable homeostatic powers. Earlier we suggested that it requires something like an instantaneous exposure to 2 Sv equivalents of radiation to increase the mutation rate by 100%. This amount of radiation should result in the release of perhaps 1×10^{13} very reactive, free ions per gram of gonadal tissue! This is really a very strong pulse of energy. From experimental studies, we can postulate that most of the damage is repaired, the exact amount depending on the kind of damage. To a first approximation, perhaps 95–98% of the radiation-induced damage is repaired. We do not understand these repair mechanisms any better than those repairing "spontaneous" damage. Ionizing

278

radiation is perhaps the worst kind of damage a cell can receive. But whereas Muller, a generation ago, could view radiation as uniquely disruptive, now it appears rather as the extreme in a spectrum of disruptive forces for which evolution has prepared the DNA. One has to wonder whether the ability to repair less disruptive kinds of damage than that caused by radiation is even greater. As I contemplate the potentiality for chaos with which we now know the nucleus must "normally" contend, the ability of the genome to cope with the added insult of ionizing radiation as well as it seems to is much less surprising than it would have been 40 years ago.

Not only does our DNA each generation survive its various hazards relatively unscathed, but in our hominid line adaptive evolution has apparently occurred at a relatively rapid rate. This is to say, not only has the species kept even with the deleterious effects of at least some mutations (which requires negative selection), but adaptive evolution has apparently occurred, in respect to the genes influencing pelvic structure, brain size, and life span, to mention only three of the more important attributes in which humans differ from other primates. These developments must be based on the incorporation into the genome of mutations contributing to these evolutionary developments. A fish, a frog, or a fly on the evolutionary trail has hundreds or even thousands of potential offspring, from among which natural selection each generation can choose the best adapted. For our species, the reproductive potential is, of course, much less. This implies for humans a much more restricted range of genetic variation in their offspring upon which natural selection can operate. Yet, per generation, we seem to have done very well indeed.

The insights of the past 20 years into the nature of the human gene pool boggle the mind, and are coming at such a pace that no description can be up-to-date. The resulting redefinition of humans and their relationship to other living forms will ultimately have all the impact on our self-image that accompanied the Darwinian revolution. But, despite my great admiration for these advances, I hope to have made it clear how much we have yet to learn concerning how the DNA manages its extensive business. This position may come as a surprise to those accustomed to reading in the daily press and weekly magazines about the triumphs of the new genetics. There are indeed triumphs, but it is one level of complexity to identify a new gene or transfer a single gene from one organism to another under circumstances in which it will function, and several levels of complexity greater to understand the total system of which that gene is a very small part, and the long-term consequences of such gene transfer.

The environmental degradation of acid rain (courtesy of P. B. Neel, 1989).

16

Just Too Many People

ach time I returned from a field trip to the rain forest, the resulting
reminder of the extent of humankind's recent transmogrification
of the world in which we have evolved has had sledge hammer
emotional impact. It would be impossible for me to conclude this account without some discussion of the human condition today as viewed
by a population geneticist. In the next three chapters, I inspect the
genetic problems of contemporary society through lenses tinged by the
experiences laid out in past chapters. Then, in the final three chapters,
I will ask what genetic considerations should be kept in mind as we look
to the future.

The human condition at the end of this century, viewed broadly, is
a cause of deep concern to many scientists, including this one. In the
upcoming discussion, building on the previous chapters, I propose not
to attempt to deal with all the profound social issues at hand but to
restrict my comments to developments with clear genetic implications.
Not only has *Homo sapiens sapiens,* step by step, abandoned the tribal-
type population structure that was essential to our past amazing evolution, but in its explosive numerical growth and accompanying exploi-
tative environmental policies, humankind is rapidly imperiling the
prospects for a comfortable existence and a full and accurate expression
of the genetic potential of the species. It is of course not necessary to
travel to the interior of South America or Africa to appreciate this point,
but the resulting contrast sharpens perspectives difficult to maintain in
the present U.S. environment. To the population geneticist, the only real
biological asset of the human species is its gene pool, in all its glorious
and poorly understood diversity. There are serious threats to the integrity of that gene pool and its optimal phenotypic expressions.

"Threats to the integrity" in this context are developments that carry the potential of altering the gene pool *qualitatively*, with poorly understood consequences. Simple *quantitative* changes, which preserve the gene pool essentially as it is but scale it down in size, are not regarded as a threat and indeed, as we will see, are highly desirable. In what follows, I address threats to the human gene pool wherever it is represented—no matter how packaged by national or ethnic considerations, the gene pool is all one resource.

The genetic problems of the human species can be categorized as "more pressing" and "less pressing," the terms being more or less synonymous with "shorter range" and "longer range." The dominant short-range genetic problem of our times, to whose characterization the entirety of this chapter is devoted, is the rapidly increasing discrepancy between population numbers and the resource base that must support that population, with all of the implications thereof. "Short range" to the population geneticist implies the next two or three generations—not the two or four year cycle of the politician. In the next chapter, we consider some of the longer-range issues.

If in the following chapters I draw frequent contrasts between present circumstances and what we encountered in tribal groups, it is not to be interpreted as nostalgia for the genetic "good old days." These references are simply a way of rather forcibly addressing the issue of how far humankind's current population structure departs from that which characterized human evolution.

Too Many People, Too Few Resources

Burgeoning Populations

Within my lifetime, which now encompasses three generations, the world's population has gone from 1.8 to 5.3 billion persons. In one more generation, by the year 2020, if birth and death rates follow the United Nations' "middle" projection of 1992, it should reach approximately 8 billion, and finally plateau in 2150 at 11.6 billion! (Such projections of course assume that the conditions for human existence continue as they now are, an assumption necessary to the forecast but by no means assured.) Somewhere along the way from a Yanomama-type existence to modern civilization, there was the loss of the commitment to restraints on population growth exhibited by most so-called "primitive" populations. While much has been made in Western culture of the motivation (justification) provided by the passage in Genesis giving man dominion

Just Too Many People

over the earth and enjoining him to be fruitful and multiply, when the opportunity arose, Muslims, Buddhists, and Confucians have been equally fruitful.

The results of this loss in so many parts of the world of any deep cultural commitment to attempt to balance population numbers against perceived resources have been exacerbated by the developments of the past century with respect to sanitation, immunization, and antibiotics. The face of disease, especially in the Western world, has undergone a revolutionary revision, of which, if we are to keep our thinking straight, we need constantly to be reminded. The table on this page contrasts the ten leading causes of death in the United States in 1900 (a landmark year to geneticists because of the rediscovery of Mendel's Laws) and

Death rates and percent of total deaths for the ten leading causes of death, 1900 and 1988. (From Monthly Vital Statistics Report, National Center for Health Statistics, v. 39, no. 7, supplement, Nov. 28, 1990.)

1900		1988	
Cause of Death	Deaths per 100,000	Cause of Death	Deaths per 100,000
Cardiovascular-renal disease	345.2	Diseases of heart	311.3
Influenza and pneumonia	202.2	Malignant neoplasms, including neoplasms of lymphatic and hemato-poietic tissues	197.3
Tuberculosis	194.4	Cerebrovascular diseases	61.2
Gastrointestinal disease	142.7	Accidents and adverse effects	39.5
"Senility"	117.5	Chronic obstructive pul-monary disease and al-lied conditions	33.7
Accidents	72.3	Pneumonia and influenza	31.6
Malignant neoplasms	64.0	Diabetes mellitus	16.4
Diphtheria	40.3	Suicide	12.4
Typhoid fever	31.3	Chronic liver disease and cirrhosis	10.7
Complications of preg-nancy	13.4	Nephritis, nephrotic syn-drome, and nephrosis	9.1
All causes	1,719.1	All causes	882.0

283

1988. Note the disappearance from the list during those 88 years of such infectious diseases as tuberculosis, gastrointestinal diseases, diphtheria, and typhoid fever, and the striking decrease in mortality from influenza and pneumonia. Conversely, malignant neoplasms, cardiovascular disease, and diabetes mellitus are now absolutely and relatively much more important. These latter are diseases in which genetic predisposition often appears to play a role. Since these are crude rather than age-adjusted death rates, they reflect not only the medical and sanitary developments of the past century but also the changing age composition of the population. Crude death rates per 100,000 population in 1900 were approximately twice those in 1988. The result is that whereas in 1900 average life expectancy in the U.S. was 47 years, in 1988 it was 74. The mortality structure of the developing nations remains somewhat closer to that of the U.S. in 1900 than in 1988, but to the extent that these nations succeed in closing the economic gap that separates them from the developed nations, and this development is reflected in their fertility rates, their age structure will increasingly approximate that of the United States and other developed nations, raising for them the same set of difficult issues we will be discussing shortly.

It is noteworthy that the scheduled population growth is expected to swell urban much more than rural areas. Of a world population estimated at 4.8 billion in 1985, 2.0 billion (42%) was urban. But of the estimated population of 8.0 billion in 2020, perhaps 4.6 billion (58%) will be urban. The slums of Mexico City, Rio de Janeiro, Bombay, New York City, and Calcutta can be expected to double in size.

Shrinking Resources

Concern with the implications of unbridled population growth is an issue going back to Malthus' "Essay on the Principle of Population" in 1798, if not further. What has given a new urgency to the concern is the more recent realization of the extent of the accelerating depletion of the non-renewable (and renewable) resource base on which this population growth has been dependent, accompanied by environmental degradation so well documented in recent years that I need present only a smorgasbord of the salient facts.[1] The impoverishment of the environment can be expressed in many ways. In keeping with my farm background, let me first consider the current rate of loss in the agricultural base needed to feed this expanding population. All figures are, of course, somewhat approximate, and carry a large error. But the trend is clear.

Thus, it has been estimated that globally, soil losses each year are 24 to 26 *billion* tons in excess of new soil formation. By some accounts, more than half the topsoil of the U.S. at the time of the discovery of the Americas had already been lost. It has further been suggested that by the year 2000, at which time the world's population will be some 22% greater than at present, continuing erosion, plus urbanization, desertification, and saline build-up, will, worldwide, have claimed 20–30% of the arable land available 50 years earlier. This loss is in the face of estimates that in this same year 2000, the world will require agricultural production 60% greater than in the early 1980s. Acid rain in Europe and North America will have resulted in the loss of about the same expanse of nonarable but formerly timbered land as has been lost to these other developments.

In an effort to offset this loss and provide for population increases an area of humid tropical forest equal to the size of Great Britain is every year being converted to other uses. Not only can this not continue indefinitely, but it is now clear that these tropical areas are very delicate and unless managed with a care not yet in evidence will relatively soon be agriculturally worthless. By one estimate, for instance, each year in Latin America some 30 million acres, an area equal to the U.S. state of Ohio, are lost to agriculture because of erosion and the rapid loss of soil fertility and other desirable characteristics.

Much of the world's agriculture now depends on irrigation. Because of deforestation, rivers and streams are becoming more seasonal in their flow. In addition, the water table, representing the underground resources that can be drawn upon, is falling at an alarming rate in some parts of the world. The United States supplies as worrisome a case study in the problems now arising from excessive or unwise irrigation policies as can be found anyplace in the world. The High Plains region of the United States, a broad band extending from parts of Wyoming, South Dakota, and Nebraska in the north to Texas and New Mexico in the south, is underlain by a vast reservoir of water impounded in sand–gravel termed the High Plains or Ogallala aquifer. Until about 50 years ago, this aquifer was in reasonable balance between input from seepage and loss through springs and underground rivers. Much of the irrigation of the region draws on this aquifer. The saturated thickness of the aquifer averages 200 feet, but there are marked local variations in thickness. A U.S. Geological Survey in 1983 found that "The rapid increase in pumpage since 1949 has resulted in extensive water-level declines locally. . . . Since irrigation began, water levels have declined more than ten feet in 50,000 square miles and more than 50 feet in 12,000 square miles of the aquifer. Water levels have declined locally more than 100 feet in parts of Kansas, New Mexico, and Oklahoma, and declines as great as 200 feet have been reported in Texas."[2] The depletion has continued since that authoritative report was issued. Decline in well yields and increases in the costs of pumping water have already, given crop prices of the past several decades, caused many irrigators to cease operations, especially in Texas and New Mexico, where irrigation developed earliest. The water supply problems of the U.S., serious though they are becoming, are minuscule compared with those of, for instance, the Middle East countries.

In the 1960s, 1970s, and early 1980s, this loss of agricultural base was largely offset by the Green Revolution. Agricultural productivity increased in pace with population growth, the result of three factors: 1) the intensive cultivation of short-stemmed rice and wheat strains and hybrid corn, all selected to be highly responsive to the very liberal use of fertilizer; 2) nearly a tripling of irrigation in some areas; and 3) the increasing use of chemical insecticides, herbicides, rodenticides, and fungicides. Unfortunately, over the past five to ten years, the thrust represented by the Green Revolution seems to have peaked. The optimists point out that, in principle, agriculture and food technology have thus kept far ahead of the growing population, and that hunger and famine today are usually the result of political upheaval, administrative breakdowns in food distribution, or inability of the poor to buy food. The pessimist points out that these causes of hunger and famine will persist and be intensified for the foreseeable future, and that, in addition, the store of genetic variability to be drawn upon in the breeding of new strains resistant to plant pathogens and insect depredation is diminishing due to ecosystem destruction and, as just discussed, the agricultural base is being eroded at an alarming rate.

To what extent the "genetic engineering" we will discuss later can spark a significant further round of crop improvements is moot. For now, society is left with grains whose yield is highly dependent on heavy fertilization. The run-off from fields treated with large amounts of fertilizer is a major source of stream, river, lake, and ground-water pollution. In addition, the dense stands of similar plants, i.e., crop monoculture, invite fungal and bacterial disease and insect depredation. The latter threat is met by pesticides in the amount of almost 500 million kilograms in the U.S. annually (plus a liberal use of herbicides to control ground cover). On the face of it, this is a good investment; dollar returns for the direct benefits to farmers have been estimated to range from $3 to $5 for every $1 invested in the use of pesticides. However, these cost–benefit figures do not reflect the indirect costs of this use, such as reduction of fish and wildlife populations, livestock losses, inadvertent destruction of susceptible crops and natural vegetation, honeybee losses, evolution of pesticide resistance (requiring ever heavier use of pesticides), and destruction of the natural enemies of the pests and the creation of secondary pest problems. It is a telling commentary that the share of crop yields lost to insects has nearly doubled during the past 40 years, despite more than a tenfold increase in both the amount and toxicity of synthetic insecticide used.[3] Recognition that this strategy of pest control is in trouble has led to recent emphasis on the use of biopesticides; now a recent alarming development is the appearance of resistance on the part of insect crop pests to an important number of these newly developed biopesticides.[4]

About 78% of the energy that drives modern civilization comes from fossil fuels—coal, oil and gas—the chief of which is currently oil. Presumably objective studies suggest that *at present rates of consumption* all of the known oil producing areas will have exhausted their resources by the year 2020, except for Latin America and the Middle East. At current rates of consumption, the resources of Latin America are good for 50 years and the resources of the Middle East for 110 years. While new discoveries and new extractive procedures will undoubtedly push those dates forward in time, increasing demand will push the dates backward. No one can doubt that, at the very least, within the next 50 years this nonrenewable resource, whose exploitation is now so vital to supplying the energy upon which some populations thrive, will increasingly be restricted to the Middle East and, with the passage of time, become a source of contention—as will access to the remaining supplies of natural gas and coal. Just as it has taken several generations for the industrial nations to become so dependent on oil, so it will take several generations for a smooth transition away from oil, at the end of which the number of people and/or the standard of living will decline unless another major source of cheap energy is discovered. It is a little noticed irony that while the developed nations maneuver around the fossil fuels, in parts of the Third World there is already, due to deforestation, a crisis in the supply of the firewood necessary to prepare a simple meal. In step with this consumption of its oil reserves, the world has exploited more of its mineral resources in the last 20 years than in all of preceding history.

Finally, two other developments that must be mentioned are the much discussed greenhouse effect[5] and polar ozone holes,[6] the one involving a worldwide warming trend, the other a change in the spectrum of radiant energy reaching the earth from the sun. Both of these effects result from human activities, the first from the high level of energy consumption, the second from the release of certain humanly engineered chemicals, notably chlorofluorocarbons, into the atmosphere. Their relevance to the present situation is that both of these developments imply a further reduction in the resource base necessary to support the world's population.

Facing up to the Problem

In a now voluminous literature, the Ehrlichs' books of 1968 and 1990,[1] The Limits to Growth Study of 1972,[7] and the Global 2000 Report to the President in 1980[8] have stressed the issues I have touched upon especially well, emphasizing the immediate need for profound socioeconomic adjustments. More recently, the Union of Concerned Scientists has sounded a global warning, endorsed by 101 Nobel Prize winners—a majority of the living recipients of the prize in the sciences—plus some 1600 additional prominent scientists.[9] The *apparent* failure of some of the dire predictions of these treatments to materialize on cue have given considerable comfort to those, especially economists, who feel that human ingenuity can improvise solutions to impending scarcities almost indefinitely. Unfortunately, as a recent follow-up to the Limits of Growth Study documents, the deterioration in the world's situation appears to be proceeding about on the schedule envisioned in the earlier report, with the per capita GNP falling in 49 countries during the 1980s.[10] It seems clear that our wanton pillage of the earth's natural resources has, on a generation scale, almost run its course, even though the exact timetable on which the bills come due if present levels of resource degradation persist remains obscure. The immensity of the implied readjustments is such that even if the real crunch is two or three generations off, the time to initiate the transition to a more sustainable lifestyle for the world's population is at hand.

However, although national and international studies have thoroughly documented these causes for apprehension, the recent record of the political establishment, especially in the United States, in assuming a *real* leadership role in defining the dimensions of the problem as the facts have become apparent is sorry. It is not as if there has been no discussion of the subject. In 1969, I participated in excellent hearings, chaired by Senator Edmund Muskie, as to whether the U.S. Senate should establish a Select Committee on Technology and the Human Environment. The problems I have just enumerated were already emergent; their potential magnitude was already frightening.[11] The Committee never became a legislative force. In the U.S. it is not good politics to be perceived as providing a less than abundant view of the future, nor as challenging the Western world's model of development, with its emphasis on having rather than being. Furthermore—and coming closer to home—it is to me surprising to what a limited extent these facts and their genetic implications have elicited discussion in a genetic community now preoccupied with the intricacies of the human and other genomes.

The blindly oblivious pronatalist policies of the recent Reagan and

288

Bush administrations in the U.S.—reversing a concern for population growth enunciated by Presidents Truman, Eisenhower, Kennedy, Johnson, and Nixon—have not only impeded a rational approach to the regulation of fertility but, on the international level, substantially diminished the U.S. emphasis on birth control at a time the Third World was in a more receptive mood than ever before.[12] The President Bush who continued Reagan's policies in apparent response to relatively small but superbly organized, conservative antiabortion groups who supported both his campaigns, is the same person who as Representative Bush said: "I am convinced that we can never come to grips with the problems of poverty and hunger without a really enlightened family planning effort both in this country and abroad." Those same two administrations toned down the impetus toward the development of more fuel-efficient automobiles, which followed the oil scarcities of the 1970s, and dismantled the various programs that they had inherited designed to develop in the U.S. greater energy efficiency (approaching that of Japan and Europe) and alternative energy resources. Unfortunately, the National Energy Policy of the Bush Administration, announced in February, 1991, in the opinion of most who have considered it, showed no more understanding of the finiteness of resources and the need to begin the transition to a lower level of energy consumption than did that of the Reagan Administration. Notable in the light of recent developments is that, under this policy, the U.S. will be consuming more Persian Gulf oil in 20 years than it does today.

Given the worldwide situation, it is all too easy for the average American to believe that priority should be given to controlling population growth in such regions as India, China, or Central America. However, as the Ehrlichs have pointed out in their 1990 book (p. 134), "a baby born in the United States represents twice the destructive impact on earth's ecosystems and the services they provide as one born in Sweden, 3 times one born in Italy, 13 times one born in Brazil, 35 times one in India, 140 times one born in Bangladesh or Kenya, and 280 times one born in Chad, Rwanda, Haiti, or Nepal."[1] In this matter of reconciling population with consumption of the earth's resources, the U.S. is in no position to moralize. What renders the conspicuous consumption of the United States so stridently obscene is that in recent years much of it has been sustained by deficit financing. If the U.S. is to exercise any kind of leadership role in the world, it must balance any espousal of population control on the part of others by vigorous steps to cut back on its own energy consumption.

In writing an account such as this book, one tends to emphasize the positive aspects of one's life. But along the way there were lost battles, lots of them. In 1971, I found myself on a National Academy of Sciences committee charged with nominating the next president of the Academy, the nomination being tantamount to election. Our eventual selection was Philip Handler. Concurrently with this development, I was asked to stand for election to the Council that advises the Academy's president. Having had a hand in getting Phil into this position, I thought I could do no less than agree to make myself available. I was elected, and this set up another educational interlude. It was a somewhat frustrating experience. There were major issues that I felt the Academy, as the nation's most prestigious scientific body, should be addressing. They included many of those we have just discussed. Phil was at the outset of his presidency quite concerned with an administrative reorganization of the Academy, to make it and the associated National Research Council more functional. Council meetings were predominantly the nuts and bolts of operation and reorganization—we didn't seem to get to those big issues. Finally, in an effort to crystallize action, I wrote a letter, which is to be found in the Appendix of this chapter, urging that the Academy undertake a White Paper directing the attention of the government and the nation to the problems facing the next generation. Obviously—because no such paper was ever issued—I failed to make my case. Phil received the suggestion with reasonable enthusiasm, but somewhere, in winding its way through the Academy's important Committee on Science and Public Policy, plus several other "sounding boards," the project bogged down. The reasons are complex, but undoubtedly there were good minds who felt mine was too extreme a position to bear the Academy's imprimatur.[13]

My own previous most personal immersion in facing up to the unpleasant facts ahead began in 1978, when the Smithsonian Institution approached me for assistance in organizing one of the symposia it sponsors every three or four years. After some negotiations we settled on the theme, "How Humans Adapt: A Biocultural Odyssey," an effort to appraise where humankind now found itself in its improbable journey through time and space. The symposium was duly held in 1981, and the papers, edited by D.J. Ortner, published in 1983.[14] The 18 chapters, by such distinguished authors as Kenneth Boulding, Lord Asa Briggs, L.L. Cavalli-Sforza, James Gustafson, Mary Midgley, Nevin Scrimshaw, and Stephen Toulmin, constitute one of the most depressing documents with which I have ever been associated. We were (almost) united in seeing grave troubles ahead. A sentence in my own essay perhaps caught the feeling: "Like Odysseus, whose ordeals returning from the siege of Troy will forever be associated with the title of our Symposium, we can anticipate enormous challenges ahead, but unlike Odysseus, there is no return to familiar shores once we have met our challenges."[15] The Smithsonian tells me the book sold 1800 copies. Never did the combined work of a distinguished group slip more quietly beneath the waves of human indifference.

The Specific Genetic Implications of these Developments

Many of the current generation of geneticists will have difficulty in identifying with the foregoing issues. To the contrary, I suggest that the potential magnitude of the coming crunch is such that the genetic implications of these developments on a worldwide basis overshadow all other genetic considerations concerning humankind's genetic future. Three implications stand out.

1. *Famine and epidemic disease.* The first implication of the situation just described is that, beginning within the next several decades, the human population will very likely increasingly be subject to localized "die offs." The impact of these disasters will not be equally distributed across the earth's populations but will be selective, precipitated by famine and/or epidemic disease, often compounded by a breakdown in government and governmental services for any of a variety of reasons. Three examples of famine at this writing involve the populations of Ethiopia, Somalia, and southern Sudan. The press coverage has been so graphic that little more need be said concerning the impact of famine.

Coupled with the prospects of localized famines is an increased vulnerability of populations, especially malnourished populations, to epidemics of disease. As emphasized in Chapter 9, the disease pressures with which humankind evolved were principally endemic. While these diseases exacted a toll in morbidity and mortality, usually only a few percent of the population were ill at any one time. As population density increased, epidemic disease became increasingly important. The impact of various plagues on human populations has been well reviewed.[16] One much-quoted example suffices: in Europe, the bubonic plague is estimated to have killed about one-third of the total population between 1346–1350. The mortality in Russia and the Balkans cannot be estimated, but there is no reason to believe it was any lower. There were lesser recurrences of the plague in the 1360s and 1370s. This is but one of some half a dozen major epidemics stretching back to the dawn of European history. Cities have been hell-holes in time of plague—and all the large cities of the world are destined to become larger—some under appalling conditions. In this century, vaccination and immunization programs have been extraordinarily effective in controlling many epidemic diseases (and contributing to the population glut), and there is a tendency to believe that major epidemics are no longer to be feared. This viewpoint overlooks the fact that a pandemic of "mutant" influenza killed 20 million people in 1917–1918.

The latest major player on the epidemic disease scene is, of course, the Acquired Immunodeficiency Syndrome (AIDS), caused by the human immunodeficiency viruses (HIV-1, HIV-2), a disease unrecognized some ten years ago, but now, thanks to the ease of travel, worldwide. The World Health Organization estimates that already some 30–40 million persons are infected with the virus, of whom the vast majority will develop AIDS, and the infection is spreading. With AIDS now predominantly a heterosexually transmitted disease, the limits of this epidemic are difficult to define. It is now predicted to kill as many persons as the bubonic plague in Europe. In the absence of an effective therapy or vaccine, the public-health response to the disease requires the much more difficult path of behavior modification—"safe" sex, "safe" drug use, "safe" prostitution—in groups resistant to change, groups some segments of society prefer to put out of mind. It is unfortunately this unthinking denial, driven by hostility and excessive fear, that has resulted in so much lost time in facing up to the gravity of the situation. Society cannot afford the luxury of moralizing in the face of any potential epidemic. In the past, the spread of epidemic disease has been geared to the relative slowness of human travel; currently the traffic out of any of the airports of a major city will deposit individuals all over the world within 24 hours.

The diseases emerging as major threats need not be new. Worldwide, infectious diseases remain the major cause of death. New, more malignant strains of infectious agents may emerge at any time from this ever-present base. In the 20 years immediately following World War II, the World Health Organization mounted a major campaign to eradicate malaria. The initial successes of the campaign soon faded as the mosquitoes that carry malaria developed resistance to the insecticides that at first were so effective, and the malaria parasite developed resistance to the drugs to which it was formerly susceptible. In many regions where malaria had been almost eliminated, the disease has made a vigorous comeback, sometimes surpassing previous levels of infections, and greater population densities can only accelerate this trend.[17] Tuberculosis is a second alarming example of the resurgence of an "old" disease that at one time appeared to be under satisfactory control. With the advent, beginning midcentury, of such effective drugs as isoniazid, rifampin, and pyrazinamide, the medical importance of the disease declined to the point where in the United States the previously extensive network of tuberculosis sanitoria has virtually disappeared. Meanwhile, the disease had made a striking comeback in such groups as AIDS patients, intravenous-drug users, the homeless, and migratory workers, all agents of spread for this airborne disease as they move through the populations of the world. Of especial concern is the fact that drug-resistant strains of the bacillus are emerging, so that it is by no means clear that the established measures of control will remain effective. Many other agents of disease are now exhibiting a very disturbing resistance to previously effective drugs.[18] New, useful drugs will come along much more slowly than in the past. Once again, nature is exhibiting the depths of her complexity.

A recent study [19] lists some 55 diseases, with causes ranging from viruses to fungi, that may be labeled emergent. Some represent recrudescences of well-known pathogens, others must be considered new. Their impact ranges from minor and highly localized to the major catastrophe represented by AIDS. The majority of the "new" diseases appear (like AIDS) to have originated in the tropics, as, in the course of destroying the tropical rain forest and its animal inhabitants, humans come into contact with unfamiliar agents of disease.

Control of epidemics is, of course, a humanitarian imperative, to which we all subscribe. In this context, however, it is important to address their genetic implications. As with famine, were the impact of these future epidemics evenly distributed throughout the world, this would be a cruel but effective method of reducing population size but maintaining all the present human variety. But this is not the way epidemics work. Already the impact of AIDS on Africa has been horrifyingly disproportionate. Thus:

> Based on several models developed to assess future scenarios, in the Sub-Saharan region, the total population could be reduced by as much as 50 million persons. AIDS typically increases national crude death rates by nearly 50 percent. In the next 15 years, AIDS mortality will reduce urban life expectancy by approximately 19 years.
>
> Because of pattern II (i.e., heterosexual) transmission and increasing numbers of infected women, nearly two-fifths of all births will result in HIV infection of the newborn. As a result infant mortality levels will increase by more than 20 percent, reversing the hard-fought gains of the past decade. . . . The reversal in child mortality will begin in approximately 4 years.
>
> As a direct result of adult mortality, WHO estimates that this decade will bring more than 10 million AIDS-related orphans.

Both HIV-infected and non-infected orphans have the same short life span, the former because of their disease, the latter because of the loss of both parents in their lives due to AIDS.[20]

Unfortunately, there are already indications that the ravages of AIDS in Asia will soon equal the havoc in Africa. Those who find the idealism of espousing a program for equal protection of all the gene pools hopelessly naive in the face of statistics like this need to be reminded of the self-interest in the control of any (unpredictable) epidemic.

Selective and localized die-offs in the trail of the four Horsemen of the Apocalypse (or the contacts of the New World with the Old) are an established feature of human history. But whereas, previously, pestilence or famine or even the depredations of war could be seen as manifestations of a God's mysterious will, it is now increasingly difficult to maintain that position, and they must be perceived in large measure as a human responsibility. We have (see p. 31) the reins of our destiny in our own unprepared hands. Only in recent years have we begun to appreciate the genetic consequences of *localized* die-offs. Given that the only true biological asset of the species is its gene pool, which is locally differentiated into ethnic groups with distinctive gene frequencies, the loss of any substantial portion of any one of these local groups impoverishes the whole. In remote times, localized extinctions of population groups may have been one vehicle of natural selection; now it is increasingly unlikely that such losses serve that role. True, there are very substantial overlaps between the genetic content of the ethnic gene pools, but our experience with private polymorphisms in Amerindian tribes suggests there may be some absolute ethnic differences, the significance of which is unknown. Early humans, in their ignorance, could rub their hands in satisfaction after polishing off an obnoxious neighbor; we, with this new knowledge, should wring our hands in some despair each time a neighbor population, no matter how seemingly obnoxious, is decimated—the species might, in the future, profit from some of those lost genes, and the mutational process will replenish the store of genetic variability only very slowly.

2. *Chronic malnutrition.* The second implication of this population–resource trend is that, for increasing numbers of people, the expression of their innate capabilities will be thwarted by malnutrition. Of the present world population of some five billion persons, already about one billion are estimated to be on a seriously suboptimal diet, inconsistent with the realization of their genetic potential. This number can only increase with present population trends. The resultant stunting (and increased disease susceptibility) is unfortunate for the body but probably

even more tragic for the mind; severe pre- and postnatal malnutrition may impair brain development, and postnatal malnutrition also results in an apathy to environmental stimuli during critical periods when essential sequences of experience must be acquired to provide for continued orderly development. No "primitive" Amerindian tribe that was culturally intact would in lean times continue to bring into the world new lives that could not be adequately nourished. As I wrote in my Smithsonian essay: "The success of the human odyssey is not to be measured by the numbers in which man overruns the earth, but by an equation which must include the quality of life as well."[21] It is ironical that there are also parts of the world—and the U.S. is one—where *over*-nutrition also curtails life expectancy and productivity. Whether this results from self-indulgence or frustration (as may be the case for reservation Amerindians), the genotype is malexpressed.

3. *Ecosystem impoverishment.* The third implication of the people–resources conflict is the continuing impoverishment of the richness of the ecosystem. The most obvious aspect of this is the loss of plant and animal species.[22] The number of distinct species in the world has been estimated at between 4 million and 30 million, the majority living in the still poorly characterized tropical rain forests. A conservative estimate is that 4000 to 6000 (some authors estimate as high as 17,000) species are being lost each year. The more spectacular of these losses or near losses are noted in the newspapers (as endangered species). In Kenya in the last two decades, due primarily to the activities of poachers (driven by failing local economies), the estimate of the rhinoceros population has decreased by a staggering 97%, from 20,000 animals to less than 600, and the estimated number of elephants has fallen 83%, from 130,600 to fewer than 20,000. Many of the losses of small, inconspicuous species in the tropical rain forests can only be surmised. In the oceans, periodic profuse blooms of algae, fueled by pollution, increasingly decimate marine life. The plant losses include the disappearance of the ancestors and relatives of many cultivated crops, an irreplaceable reservoir of useful genes for breeding pest- and disease-resistant plants. The agricultural gene banks of the world are not keeping pace with the losses. The processes of evolution are such that some species loss is to be expected, but the current rate of loss appears to be about 1000 to 10,000 times more rapid than before the human activities of the past two centuries. At present rates of destruction, the tropical rain forest, the principal reservoir of species, will have virtually disappeared by mid-21st century.

Now is a period of some reappraisal concerning the ethical basis of our relationship with other life forms. Mary Midgley, Tom Reagan, and Paul Taylor have been writing with especial sensitivity concerning the responsibilities that come with man's dominion over the animals of land, sea, and air.[23–25] I find it fascinating how the moral position of these ethicists is being extended by reductionist science, which is providing the basis for a quantification of our relationship to other life forms such as never existed before. The old apologia "There but for the grace of God go I" now translates into "There but for the grace of a few nucleotides go I," but the latter statement now cuts across species lines.

Several years ago I undertook from a literature survey to determine just how similar the proteins of man and chimpanzee were. I could find 11 proteins for which the exact amino acid sequence was known for both species. Of the total of 1442 amino acids in these 11 proteins, man and chimpanzee differed in only 6. Given that each amino acid requires 3 nucleotides for its specification, and assuming that each of these amino acid differences depended on a single nucleotide difference, this implies 6 nucleotide differences in the total of 4326 nucleotides encoding these proteins. Man and the chimp are in these respects 99.9% the same. This is something of a tour de force—the critics will say we haven't looked at the right genes—but this was all the protein material on which to base an opinion at that time.[26] The fact is that DNA biochemistry has not only brought new insights into the unity of life but raised in many minds profound questions as to why an animal whose DNA differs so little from that of many other animals should have license to behave in such an untrammeled fashion. What renders the license to the far-reaching environmental harm caused by modern man the more striking—as emphasized by Simons[27]—is the fact that only five to seven million years ago, 0.15% of evolutionary time, his ancestors were such insignificant members of the ecosystem. When humans were free to explain their origin with a creation legend appropriate to the culture, a little hubris with regard to other life forms may have been excusable. Now, with the DNA-based certainty of just how intimate our relationship to other life forms is, it's a singularly closed mind that doesn't find it a time of reevaluation of the nature of our interactions with the rest of the living world.

In consequence of the amazing insights of molecular genetics, the ethical dilemma identified by Midgley and Reagan is mounting. For millennia so much done in and to the world has been justified as a god-given right—the gods changing and differing between groups. As the life-form whose DNA confers upon us a special ability to impact on the DNA of other species, our stewardship of the gene pool must transcend narrow species boundaries—all we species are in it together. The advances of molecular genetics force us to perceive ourselves as genetically related in a completely definable way to every other life form.

These other life forms are an enormous family of cousins. And one member of that family, humankind, is making it absolutely miserable for most of the rest of the family. At the same time, that obnoxious family member is so degrading the environment that even if he were to make a Herculean effort to set matters right, it is not now clear how soon and how well the complex system can respond.

The argument extends beyond ethical considerations: the full extent of our self-interest in the protection of the ecosystem is easily over-looked. Our life-support system ultimately depends on the domestica-tion of various plant and animal forms plucked out of this complex skein of life. It is by no means clear what yet remains "out there" to be domes-ticated, as a source of food or medicine. In this age of molecular genet-ics, the term domestication takes on an entirely new meaning. Each species represents a special solution to the problem of survival. It does this through the medium of genes and their products. The loss of any life form impoverishes our ultimate ability to understand the ingenuity of nature to evolve solutions to the challenges of successful living. What modern ecology does is to demonstrate how these solutions are inter-related. Our (temporarily) dominant position in the hierarchy is com-promised when we contribute to the extinction of other life forms, be-cause it was only in the presence of many of these forms that our position evolved.

This view of the interrelatedness of life's myriad species is scarcely new. J.Z. Young put it very succinctly in 1971: "The entity that is main-tained intact, and of which we all form a part, is not the life of one of us, but in the end the whole of life upon the planet."[28] In the extreme, this view merges with the sometimes controversial concept of Gaia, equating the earth to a living organism whose components, as in any successful organism, are constantly but harmoniously readjusting to changing cir-cumstances. A less florid statement of the concept is that the biota regulate the abiotic domain in a manner that resembles homeostasis. The concept has a long history, culminating in recent decades in the writings of J.E. Lovelock and L. Margulis.[29] To a medically trained mind like my own, the logical pursuit of this concept drives one to analogize the recent behavior of the human species with the natural history of a malignancy that, originating in a single cell of a specific tissue, even-tually threatens the whole of which it was once a very insignificant part.[30] Much has been written about the sense of alienation that has overtaken modern man as he realizes he may be the product of blind evolutionary forces rather than occupying a preordained position as a result of divine intervention.[31] To me, any sense of alienation stemming from all our insights into the evolutionary process is compensated for by the realization of the kinship of life forms, and the need to strive for harmony within this complex nexus is no less than the need religious man has had in the past to establish harmony with his God.

When we humans arrogantly "take charge" of the total ecosystem— as we have—we pit our strand of DNA and what it has enabled us to create against the collective wisdom encoded in the DNA of perhaps twenty million other life forms, plus the complex relations that have evolved between them. It's a match in which in the long run we cannot hope to prevail. All that other DNA is freely evolving, to meet changing conditions. Ours is probably not. Planning for the future must involve an attempt at ecocompatibility difficult to visualize today.

Appendix

February 14, 1972

Dr. Philip Handler
President
National Academy of Sciences
2101 Constitution Avenue
Washington, D.C. 20418

Dear Phil:

This is an effort to expand somewhat on my remarks at the last Council Meeting, concerning the desirability of a *brief* statement, a "white paper," if you will, on the *possible* magnitude of the *new-type* problems facing the U.S. and the world. I repeat my convictions, that this should be sober but not alarmist, and that the "right" statement at this particular moment might have as much impact as anything the Academy does in this decade.

The obvious practical problem is how to place bounds on this statement. Do we limit ourselves to the strictly scientific, or do we stray over into such issues as, for instance, the consequences of nuclear war (which, after all, the Academy proposes to study). Furthermore, since the issues all interrelate, what's the logical structure to a statement?

Perhaps the best first step is to agree on an outline for a presentation, an outline which in effect is the strategy of the paper. If this could be agreed to, then one can better select the people to flesh out the statement. How's this for openers?

I. Cognizance of the recent spate of dire predictions and concerns regarding the next 20–30 years.

II. The principal identified problems:
   1. Increasing populations–which surely must level off at some point.
   2. Diminishing agricultural base to feed this population, due to urban sprawl, erosion, etc.
   3. Possibility that certain key crops are near the ceiling of energy conversion, i.e., the "green revolution" may be running its course.
   4. Dwindling water supplies.
   5. Pollution of all types—industrial, agricultural, human waste—with the hazards to health and productivity.
   6. "Energy crisis," and growing energy consumption gap between have and have-nots. Probability that have-nots will curtail (or mark-up) energy sources available to haves.
   7. Deterioration of cities—does modern transportation make the megapolis concept outmoded; physiological price of urbanization.
   8. The burden of armaments.
   9. The product of all these: unrest, nationalism, conflict.
III. The spectrum of *valid* opinion.
IV. The consequences if extreme point of view correct.
V. The prudent course.
VI. Necessary steps to place problem in perspective, and to manage transition.
VII. SOME SUGGESTED PRIORITIES.
VIII. Issue to be decided in next 30 years?
IX. Judgement of history (our children) if we fail.
X. The high order of dedication and statesmanship required.

If a proper group of wise men (*Council*) could agree this is the way to go, then comes a session in which the appropriate custodians of special knowledge go to work.

I was impressed by the relative consensus with which this suggestion sailed through, and the offer of help from such diverse personalities as Sinsheimer, Brown, and Bronk. Your Council seems to have a sense that now is the time.

Regards,

James V. Neel, M.D.
Lee R. Dice University Professor
of Human Genetics

The adult males of a Yanomama village lined up shortly after dawn, preparing to leave on a hit-and-run raid on a neighboring village (1966). This represents the level of man's ability to inflict harm on his fellows until, in evolutionary terms, only yesterday.

17

Some Longer-Range Problems for the Gene Pool

The issue of reconciling population numbers with sustainable resources in a genetically equitable manner throughout the world transcends in its acuteness all other issues confronting our species. There are, however, also some longer-range genetic issues faced by the human species, which, on the usual scale of political and social planning, are in general very long-range indeed, with nothing like the urgency of the issue just discussed. They can be grouped under three headings.

Departures from the Population Structure and Bases for Survival and Reproduction of Prehistoric Tribal Cultures

Throughout most of history, man has aggregated in very small groups, at most amounting to a few hundred individuals. I have presented each of these small pockets of humanity as an experiment in evolution. Because of the small size of the pockets and the kinship-dominated social structure, this structure carries the potential for relatively rapid genetic change. The reconstruction of human evolution requires that when a better adapted human type arose, it replaced or assimilated the existing types in, geologically speaking, relatively short order.

It is reasonable to surmise that humankind has disrupted the complex genetic structure that resulted in its rapid evolution no less than humankind has disrupted the ecosystem in which it functions. We do not yet understand that genetic structure well enough to perceive precisely what is implied, but this is scarcely surprising: even with respect to the ecosystem, where the results of disruption were more immediately obvious, recognition of the magnitude of the disruption has been slow in coming. Nevertheless, a number of developments render it unlikely that human evolution can proceed in the future as it has in the past.

First, population isolates, as represented by some of the surviving tribal populations, have virtually disappeared. Otherwise stated, the unusual combinations of genes represented in localized ethnic groups— experiments in evolution—are being broken up by migration and inter- marriage at an ever-increasing rate. Second, the sheer increase in hu- man numbers means that a genetic "innovation" of adaptive value will require a longer span of time to become predominant in a population than previously. Third, given the accelerated mobility of people, the trend is toward one large, increasingly interrelated, homogeneous gene pool, in which gene pool, because of its new size and structure, the appearance and spread of pockets of genetic innovation is less likely than previously. There is now more inertia to change in the gene pool. Finally, the genetic basis for survival and reproduction appears to have altered dramatically. This is not to say that human evolution should be regarded as completely finished, but that, if it continues, it will acquire a different dynamic. We may have tamed some of the turmoil in our gene pool to which I have referred in earlier chapters but at an unknown evolutionary price.

The first three points in the preceding paragraph are self-explan- atory, but the last point requires discussion. The Darwinian natural selection to which biologists attribute our divergence from other primat- es can operate in only two ways, namely, through the differential fertility or the differential survival of the genetically more adapted (and adapt- able) organisms. Consider, first, the changing bases for differences in individual fertility. It is a fact that among the nations of the world, almost without exception, fertility differences are negatively correlated with socioeconomic status. To what extent these differentials imply a relaxa- tion in biological selection has been debated for years. Without entering into that debate, I would, on the basis of studies of minimally contacted tribal populations, suggest that during the long span of human evolution there was a clear association, at least for males, between "ability" and reproductive performance, a result of the greater fertility of leaders or headmen. Today this association appears to be substantially weakened. Here I do draw very heavily on our Amerindian experience, and re- cognize the relative lack of data from other cultures. I also recognize that I am very subjective in the opinion that headmen are superior individ- uals. One of the major disappointments of our fieldwork was that, de- spite much brainstorming, we could never devise a field test of Yanomama "smarts"—and if we had devised one, the Yanomama would have no motivation to take it seriously.

The possible genetic implications of headmanship are obvious. Let us consider that we have at our disposal an Index of Innate Ability (IIA), which some will be tempted to equate to intelligence. It is a quantitative trait certainly related to intelligence, based on the additive effects of alleles at many loci, but since the quality we call intelligence has been validated only as a predictor of school performance, we had best not allow ourselves to be ensnared by that word. Let us assume that the average Index within a village which contains 50 reproducing adults is 100, but that the headman has an Index of 120, in which case his 49 peers will average 99.6. We will assume that in this egalitarian society, where the educational opportunities are remarkably uniform, the Index really measures an innate difference. On a very simplistic, additive model for the inheritance of this Index, assuming the IIA values of children are the average of the two parents, it follows that if the headman marries women of average IIA but has twice as many children as other men, then the average IIA of the replacement population of 50 should be 100.4. The selective implications of such a mating system projected over hundreds of generations is obvious. Let us quickly note that the IIA is relative, so that one does not make long-range predictions by adding generational gains! There are many unreal and simplistic assumptions of this model, but even so, the potential this population structure offers for positive selection for the IIA seems incontrovertible.

These probable changes in the genetic basis for differential fertility may be matched in importance by changes in the basis for survival up to and through the age of reproduction. Three points will suffice.

First, at present in the Western world only a few percent of liveborn infants fail to survive to reproductive age. Among the Yanomama, that percent (exclusive of infanticide) is 25 to 30. With the transition to agriculture and urban life, prereproductive mortality probably reached 50% before, with vaccination and sanitation, it began to fall to present levels. The most striking reductions in prereproductive age mortality have occurred in the past two centuries. We have argued that the pressure of disease on children as a potential agent of selection was greater during the agricultural transition than previously. However, our knowledge of precise causes of death in both unacculturated Amerindians and early agriculturalists is so imprecise that it is impossible to cast those higher losses of earlier times into any precise genetic framework. Our intuition would be that the "genetically stronger infants" (a very imprecise term) would survive better than the "genetically weaker," but once we depart from a consideration of a few relatively rare inherited diseases of childhood, we have difficulty giving substance to that statement. A point that should not be overlooked is that infant survival under difficult conditions may be as much a reflection of the mother's genotype—her ability to cope—as the infant's. Since, however, the infant has received half its genes from its mother, infant deaths because of maternal inadequacy still have genetic significance. On a more concrete level, the treatment of children with specific genetic diseases is improving. Children with such genetic disorders as hemophilia, childhood diabetes, and phenylketonuria are ow surviving and reproducing to a much greater extent than previously. Genes which would formerly have been lost are now being transmitted to the next generation.

Second, the role of diet in survival and reproduction is also changing. Humans are surely the most omnivorous of all species. The range of diets upon which, throughout the world, humans are subsisting exceeds that of any other animal. But recently in many respects the breadth and pattern of our diet has in some parts of the world altered dramatically, and that we have adapted to this change does not mean it is without biological cost.

The change in diet is in two diametrically opposed directions. On the one hand, as we have already discussed, simple *chronic* malnutrition was probably quite uncommon in tribal populations, rarely thwarting genetic potential. It is now increasing, especially in the developing nations and the Third World, threatening to undo the recent improvements in public-health statistics. While extreme malnutrition will result in the death of any child, it is not clear to what extent there is a borderline zone of chronic malnutrition in which genotype plays a role in survival. If there is, then chronic malnutrition introduces a type of genetic selection uncommon in the past.

At the other nutritional extreme, in some countries a combination of overalimentation and malalimentation, coupled with physical inactivity, is now emerging as a significant cause of disease, sometimes at a relatively early age.[1] The difference between the diet of precivilization humans and modern western humans is succinctly summarized in the table on page 305. The diet-related "diseases of civilization" include hypertension; high cholesterol levels followed by atherosclerosis with its involvement in heart, cerebral, and renal disease; diabetes mellitus of the adult-onset type; and the complications of obesity. As documented in Chapter 9, we did not recognize these conditions in our medical studies on the Shavante and Yanomama. Although these diseases are primarily dietary in origin, there is also evidence for a genetic component in the predisposition to each of these, so that their emergence has

Estimated Energy Sources in Paleolithic, Third-World, and Modern Western· Diets. After D. Burkitt, *The Pharos,* Winter, 1991, p. 19, reproduced by permission. See also: Eaton, S.B., Shostak, M., and Konner, M. 1988. The Paleolithic Prescription. New York: Harper & Row.

	Paleolithic	Third-World	Modern Western
Starch	60%	60%	20%
Sugar	Minimal	Minimal	20%
Protein	20%	15%	15%
Fat	20%	15%	45%
	Mostly saturated	Mostly saturated	Mostly unsaturated
Fiber	50–100 g/d	50–100 g/d	<20 g/d
Energy density	Dilute	Dilute	Concentrated

the potential of creating new selective forces. The serious medical impact of these diseases is, however, predominantly—dare I say 90%—postreproductive. While, therefore, they are societally of great importance, and will be considered at some length in the genetic program I will discuss in the final chapters, the advent of these diseases does not have anything like the potential for altering Darwinian selection inherent in the changing bases for successful reproduction or the control of infectious disease or the advent of mass malnutrition.

A third alteration in survival patterns that may have genetic implications is the difference in the basis of traumatic death, including infanticide, between populations such as the Yanomama and civilized populations. In most tribal societies, children recognized as malformed are killed immediately after birth. Once admitted into the society, so to speak, they are not killed later. Most children with severe congenital defect would, even if not killed at birth, soon die in tribal societies. We do not know how discriminating the cullings of infanticide were, i.e., to what extent they extended to lesser degrees of congenital defect, including such of the inherited genetic syndromes as could be recognized at birth. The role of genetic factors in the etiology of congenital defect compatible with survival ranges from clear and simple in some of the skeletal abnormalities to complex and ill-defined in such entities as the hare-lip/cleft plate syndromes. Thus while it is difficult not to see some genetic significance in the killing of malformed infants, it is not feasible to put the subject in a quantitative perspective. On the other hand, with regard to the preferential female infanticide practiced by many tribal groups, no genetic (as contrasted to sociological) significance can be attached to the practice.

The issue becomes more complex when we consider traumatic death later in life. Chagnon's studies suggest that perhaps 23% of all Indians, primarily males, die traumatic deaths from such causes as snakebite, falls, and, especially, wounds sustained in duels or warfare.[2] By comparison, in our society accidents rank fourth as a cause of death, homicides, twelfth, and suicide (virtually unheard of among tribal populations), eighth. Combined, these amount to 5% of deaths. It is extremely difficult to attach a genetic price tag to traumatic death and even more difficult to contrast the genetic cost of traumatic death in Yanomama-type and civilized societies. I would suggest, however, that among Amerindians the outcomes of the personal combat of two antagonists or the hit-and-run raids of one village on another have more genetic overtones than do the highway accidents in which one party is drunk and the other party a "random" victim, or do the homicides occurring during robbery attempts or drug-triggered shoot-outs.

J.F. Crow has developed a simple Population Index of Potential Selection, based on the frequency of death prior to the age of reproduction plus the variability in number of children per male or female.[3] On the basis of the data we collected on the Yanomama, we estimate their Index to be 4.24, as contrasted with an Index of 0.95 for U.S. Caucasians.[4] That Index expresses the *potential* for selection if all deaths and all fertility differences were genetically determined, which of course is not the case. Nevertheless, I would have to suggest that relatively more of the deaths and reproductive differentials among Yanomama have genetic overtones than in the U.S. today. This implies that whatever the true Index of Selection is, the differential between Yanomama and U.S. Caucasians is greater than the 4.24:0.95 ratio we have calculated. This is a thought-provoking statistic.

Increasing Exposure to Mutagens

The second of the developments with possible long-range implications for the gene pool to be mentioned is a possible increasing exposure to mutagens, as, in desperation with a deteriorating world situation, standards of exposure to mutagens in the environment and work place are lowered. While on general principles this development should be resisted, I find this prospect less worrisome than 40 years ago. The human genetic material, honed by the evolutionary heritage of four billion years of contending with potential mutagens, is more resilient than was earlier thought to be the case. This conclusion emerges from the growing knowledge of genetic repair mechanisms and the empirical observations from Hiroshima and Nagasaki. Likewise, our studies of inbreeding effects in Japan reveal a lower accumulation in the gene pool of deleterious genes than was earlier projected to be the case. Just how the ill effects of mutation are eliminated is unclear. The gene pool is resourceful in ways still not well understood. This does not mean that it can be mutagenized with impunity—only that this issue is less pressing than seemed to be the case 40 years ago.

Nevertheless, concern over the genetic effects of exposure to ionizing radiation continues to run high. Short of nuclear war, the principal man-made exposures of populations to radiation, in order of importance, are from 1) the diagnostic use of sources of radiation in the practice of medicine; 2) occupational exposures, including the operation of the nuclear-energy cycle, from mine to ultimate waste disposal; and 3) fallout from the testing of nuclear weapons (now *very* minor). (Therapeutic exposures to radiation, as in the treatment of cancer, are predominantly postreproductive and so of little genetic significance.) From the genetic standpoint, we are interested only in dose to the gonads prior to the end of reproduction. The total average individual dose in the U.S. from all the above-mentioned sources probably does not exceed, up to the average age of reproduction (25–30 years), about .02 Sv. This is a lesser exposure than is received from the natural radioactivity of the environment over that same period (about .09 Sv).[5] (This latter figure includes the exposures to radon which, while a natural phenomenon, are intensified in certain regions by residence in well-insulated homes.) On the basis of the studies in Japan and the knowledge that chronic radiation is less mutagenic than radiation acutely delivered, .02 Sv is well under 1% of the exposure necessary to result in a 100% increase over the spontaneous mutation rate. Thus, for the populations of the industrialized countries, radiation exposures appear to remain well within the limits set by responsible committees and agencies as prudent, as a reasonable trade-off for benefits received. On the other hand, even these small exposures do mean in principle some increase in mutation rates, at a time when selective pressures appear to be relaxing.

The situation with respect to the chemical mutagens is much less clear. In recent years, there has been increasing recognition that many plants elaborate for their protection against herbivores potent chemicals that are mutagenic in experimental test systems. The charring of meat, or of any protein, produces breakdown products that can be shown to be mutagenic. On top of this "background," industrial and agricultural technology is introducing into the workplace scores of chemicals that, under suitable conditions, can be shown to be mutagenic. Since some of these chemicals are highly important to success in industry or agriculture, the question is not one of eliminating them so much as regulating their use. Bruce Ames and colleagues,[6] in particular, have been arguing that the same set of demonstrable biochemical defenses that animals, including man, have evolved to detoxify naturally occurring noxious compounds should be effective in countering the effect of many man-made chemicals. The challenge of understanding the genetic implications of these chemical exposures presents one further enormous complication not encountered with radiation: whereas the physical properties of radiation mean that gonad doses can usually be relatively accurately calculated, the chemical mutagens—even if they escape detoxification—will reach the human gonad through complex channels that make calculating actual doses to the germ cells very difficult. The demonstration that a given chemical is in high doses mutagenic in bacteria, which lack the layered defenses of a mammal, is of limited relevance to the human situation, but yet when reported such demonstrations often elicit public alarm.

The future potential for genetic misadventure inherent in nuclear weapons deserves special consideration. This is a potential with which I consider myself to have lived intimately since 1946. In addition to the constant reminder of the effects of nuclear weapons growing out of my long-term association with the follow-up studies in Japan, periodically my level of consciousness has been raised by "resource person" performances in Washington. Two of these stand out. In 1959, Representative Chester Hollifield, then Chairman of the Subcommittee on Radiation of the Congressional Joint Committee on Atomic Energy, held hearings on the biological and environmental effects of nuclear war. These hearings[7] were really the first comprehensive, unvarnished consideration of the potential impact of nuclear war on the United States—or, by extrapolation, on any adversary in a nuclear exchange. I was there to discuss the genetic consequences. This introduction to the gruesome consequences of any large-scale nuclear exchange paled by comparison with a second experience, service in 1978 as the biomedical resource person on an Office of Technology Assessment (OTA) Advisory Panel on the Effects of Nuclear War. Out of this came a monograph, "The Effects of Nuclear War," published in 1979, which may be considered an update to the earlier Hollifield hearings.[8] By now bombs of megaton potential were available. (A megaton is the explosive power of one million tons of TNT.) The OTA study considered what a single such bomb would do to a major city. The Russian example was Leningrad, the U.S., Detroit. The choice of a city only 40 miles from Ann Arbor, a city with which I am quite familiar and whose panic-stricken evacuees would quickly overwhelm Ann Arbor, gave the exercise extra poignancy.

A single such weapon would utterly devastate or inflict severe damage on an area of 50 square miles. Detonated over the approximate heart of Detroit, it would result in some 200,000 immediate casualties and 500,000 persons with significant injuries, a considerable fraction of whom would die within the next several months. A large number of the people who retreated to the comparative safety of air raid shelters or the basements of reinforced-concrete buildings in anticipation of the bombing would find themselves trapped. But unlike the resources that are quickly mobilized following disastrous earthquakes, if the bombings were widespread there would be no rescue squads. The breakdown of our complex system of food distribution would undoubtedly result in mass starvation among the survivors. Depending on how the bomb was detonated and the meteorological conditions, radioactive fallout at lethal levels could extend several hundred miles downwind from the detonation.

The United States and the Soviets at peak nuclear strength together had about 50,000 nuclear weapons, the average weapon being 10 to 20 times more powerful than those exploded over Hiroshima and Nagasaki. Several other nations have lesser numbers. With the dissolution of the U.S.S.R., both the U.S. and the former Soviet states have announced they are no longer producing atomic weapons, and stockpiles are being reduced. Unfortunately, this hopeful development is to some extent offset by the discovery of the capability of a determined Third World country such as Iraq to develop a strong nuclear program. There is also a growing concern that in the current turmoil in the former Soviet Union, some of the present Soviet nuclear weapons and/or the Soviet expertise in their manufacture may make their way to would-be nuclear powers.[9] Superpowers fully aware of the implications of "mutual assured destruction" are surely less inclined to nuclear brinkmanship than a country intent on regional domination.

Since this is a genetically oriented book, we need to consider the genetic implications of full-scale, local or widespread, nuclear war for the gene pool. Even with the worst scenario, it seems unlikely the *average* survivor would have received more radiation to the gonads than would double the mutation rate. I can say categorically that the species can absorb this increased mutation rate—at, of course, a price. The much greater impact on the gene pool is from the great and selective mortality of all-out nuclear war. Depending on the countries involved in the exchange, the different subdivisions of the human gene pool would not be equally decimated. It is difficult to escape the conclusion that a major nuclear exchange between well-armed adversaries could alter the qualitative composition of humanity's gene pool as dramatically, and in a shorter time span, as any previous event in human history.

In the raids of tribal bands or villages on one another, all except the really handicapped or disabled men participate. There are no occupational deferments. Such encounters may have been instruments of genetic selection. At some point in human history, the fighting began to be relegated to the physically and intellectually more fit. The military deferments for physical/mental reasons which obtain when armies are conscripted, as in the recent World Wars, render war an agent of negative selection. Nuclear war will be neither an instrument of positive nor negative selection—just mass slaughter. If there was a redeeming feature of war as previously fought, it was the sense of purpose, camaraderie, and group loyalty it engendered in those who fought together and survived. I have seen the males of an Indian village leave on a raid—the shared danger and its outcome will be a point of departure for around-the-fire exchanges for years. The atomic bombs take even this away. Someone can push a button and carnage will overwhelm the world before the sense of purpose and loyalties that enable one to cope with suffering can ever emerge.

As I briefly described life in a Yanomama village, some of my readers undoubtedly thought, how stressful. From the evolutionary standpoint, however, it is we, not the Yanomama, who should regard ourselves as the stressed. Our ancestors have survived, albeit for most of the time as a very minor species, for some five million years under Yanomama-like (or much more primitive) conditions. How long and how well we can survive under present conditions remains to be seen. Fortunately or unfortunately, the compartmentalization of responsibility in modern society plus failure to recognize that it really is one world, helps us put the full gamut of our stresses out of mind.

Emergence of a Gerontocracy

A final point of concern with the current human genetic condition is much more subtle than those already treated. It involves the emergence of a gerontocracy. In the spring of 1984, quite unexpectedly, I received an invitation from the Secretary of Health and Human Services to join, for a four-year term, the Advisory Council to the National Institute on Aging. This is an 18-person group that has rather broad advisory responsibilities to the Institute. I was selected to function as a genetic-resource person. The subsequent experience has been thought-provoking and sobering. I had not previously given much professional thought to the panoply of problems associated with the phenomena of aging. Now I had a special license to devote some attention to both the social and biomedical aspects of aging, a license that rapidly let me into some unexpected genetic considerations. These considerations have been of two types, namely, on the nature of the still very mysterious process of aging, and on the long-term genetic implications of the increase in the proportion of the population we refer to as senior citizens (to which I have, of course, long belonged). We will come to the former subject in Chapter 20; here I propose to discuss only the latter topic.

The fact concerning the emergence of a gerontocracy in the industrialized nations are spectacular. Thus, in the U.S., the percent of the population over age 65, the nominal age of retirement, has already (1990) reached 12.4%, and is predicted to reach 16.1% by the year 2020 and 19.2% by 2050. This projection is based on current actuarial statistics, i.e., it assumes no major prolongation of life expectancy. In the Yanomama we estimated that proportion at 0.3%! There are 19 nations in which an even greater proportion of the population is over 65 than in the U.S. An equally impressive statistic is that in the U.S. between 1940 and 1980, the number of oldest old (aged 85 and over) increased from 370,000 to 2,271,000, and if current trends continue is estimated to reach 6,000,000 to 8,000,000 persons by the year 2020.[10] It seems clear that more and more of the voting and political power in this world is passing into the hands of its senior citizens.

By 2020, if current trends toward prolonged education for youth and early retirement for seniors persist, of each 10 persons in the U.S. about three will be of working age, but of the seven not working, only 2.7 will be children. These statistics concerning the gerontological trend raise a troubling question, meticulously discussed by Preston[11] and by Richman and Stagner[12]: how will efforts to meet the real and perceived needs of this older population impact, in a world of shrinking resources, on meeting the health and educational needs of (nonvoting) children. The relevance of this issue to the full expression of the genetic potential of the young is obvious; the record thus far is not good, as in the past several decades benefits and privileges increase for seniors but decrease for children. For instance, in the U.S. the proportion of elderly defined as living in poverty *decreased* by 50% between 1970 and 1989, primarily as the result of the linking of Social Security benefits to inflation, whereas during the same period the proportion of children living in poverty, under circumstances not conducive to realizing their genetic potential, *increased* 26%. A major reason is that during this period state and federal governments cut the principal income-support program for poor children—Aid to Families with Dependent Children—by 43% in inflation-adjusted dollars.[13] In Chapter 2 I commented on the increasing preoccupation of medicine with life's terminal ramifications. It is not only general benefits for the elderly that have increased disproportionately vis-à-vis the children, it is also the elderly's share of medical services, especially when the latter term is extended to embrace the nonmedical support side of these medical services. How far society should move to meet the enormous social and medical needs of this aging population is emerging as one of the prime ethical issues of our times.[14]

The Time Scale for Genetic Change

To place the rather somber enumeration of potentially adverse genetic developments of the past two chapters in perspective, we must now turn briefly to a consideration of the time scale for genetic change. There are, in fact, two time scales. To the extent that genetic change is brought about by gradually changing selective pressures or increased mutation rates, the rate of change is slow. To the extent that change it is due to cataclysmic events—famines, epidemics, or nuclear war—it is of course very rapid.

The usually slow rate of genetic change through what is conventionally envisioned as natural selection is treated with appropriate formulae in all genetic texts. The rate depends on the type of inheritance underlying a particular trait. It is sufficient for our purposes to say that the "appropriate formulae" suggest that even with the relaxation of previously important selective pressure or the advent of rather strong new selective pressures, it would take many generations, covering hundreds of years, for the frequencies of the genes responsible for major human traits to change dramatically. There is an inertia in the genetic process in large populations, inherent in the many pairs of genes involved in determining most of the important attributes of humans. Even those of us most concerned about the genetic future of our species would not argue that the etiolation of our species is imminent. Here the contemporary human geneticist differs from the eugenicist of the 1920s and 1930s, who, without today's understanding of the principles of population genetics, believed deterioration could occur very rapidly. Genetic deterioration has a relatively long fuse, in a world whose attention seems riveted on tomorrow.

The Primitive Mind of Modern Man

It is a sobering thought that the mental equipment with which we face the foregoing (and other) major issues has not altered substantially with the advent of civilization. Human beings with brains the size of those of modern man—the so-called Neanderthal man—appeared about 100,000 years ago. About 40,000 years ago, this type was replaced by a type more like us in physical appearance—Cro-Magnon man—but with essentially the same-sized brain as Neanderthal. Although there is no one-to-one correspondence between brain size and intelligence (however that be measured), overall the correlation must be assumed to be high. Thus the Neanderthal mind that spent thousands of years perfecting the art of making stone tools and developing the Levallois technique for flaking flint is, by anatomical indications, not very different from the mind that is dealing with the emergent technologies of the past century.

There are those who find this concept difficult to accept. Surely, it is argued, there must have been some changes in brain structure to cope with this new complexity. There is not the slightest shred of evidence for this. Furthermore, as I just pointed out, genetic change, short of that due to great catastrophes, is really rather slow; there is no way biological evolution can have been commensurate with the cultural evolution of the past several hundred years.

There is an unfortunate tendency to confuse material advance with intellectual advance. A Yanomama must master all aspects of his culture, whereas in these times of specialization we of civilization individually must master only a small portion of ours. Technical developments free us to spend more and more time digesting (and extending) the portion of the culture with which we become involved. I would argue, however, that the decisions to be made on the way to headmanship are intrinsically as delicate as those involved in emerging as a business tycoon or a political figure, but perhaps a little more dangerous. The would-be headman who miscalculates may lose his life, whereas the tycoon or politician who fails has only to declare bankruptcy or wait for the next election.

Anthropologists have often commented on the "harshness" and "cruelty" of life in many tribal cultures. There is little privacy under these conditions. Everything is out in the open, in full view of both the villagers and the anthropologist. As now, in the U.S., statistics begin to accumulate on the sad frequency of child abuse, as it is recognized that battery is the most significant cause of injury to women, as homicide ranks number 12 among the causes of death in 1988, as children continue to disappear at a frightening rate, and as modern war becomes increasingly indiscriminate in its victims, it is difficult to argue that our minds have been gentled to any significant degree in recent centuries. Indeed, we have lost an important check on the impulsiveness of these minds. As pointed out in Chapter 8, the extensive kinship structure of most tribal villages provides relatives to intercede and protect if intrafamily violence is deemed excessive. Our "privacy," plus a mobility destructive to familial associations, have together largely removed this check.

It may seem to some that I am belaboring this point. Critics would do well to recall that in the first 30 years of the Darwinian revolution, in the full flush of social as well as biological Darwinism, many prominent scientists held that the world's primitive cultures were the biological losers in the evolutionary progression that culminated in the advanced Europeans. To A.R. Wallace, who independently developed concepts so similar to Darwin's, the problem was of a somewhat different nature. He wrote:[15]

> We see, then, that whether we compare the savage with the higher development of man, or with the brutes around him, we are alike driven to the conclusion that in his large and well-developed brain he possesses an organ quite disproportionate to his actual requirements—an organ that seems prepared in advance, only to be fully utilised as he progresses in civilisation. A brain one-half larger than that of the gorilla would, according to the evidence before us, fully have sufficed for the limited mental development of the savage; and we must therefore admit that the large brain he actually possesses could never have been solely developed by any of those laws of evolution, whose essence is, that they lead to a degree of organisation exactly proportionate to the wants of each species, never beyond those wants—that no preparation can be made for the future development of the race—that one part of the body can never increase in size or complexity, except in strict coordination of the pressing wants of the whole. The brain of prehistoric and of savage man seems to me to prove the existence of some power distinct from that which has guided the development of the lower animals through their ever-varying forms of being. (pp. 193–194).

The implication that the brain of primitive man had somehow evolved over and beyond the stimuli of natural selection did not sit well with Darwin and became perhaps the chief point of difference between the two, but both, viewing Indian culture through nineteenth century

English eyes, agreed on its simplicity and retardedness. The alternative, that there was a complexity to those tribal societies which had nurtured evolutionary developments, a complexity they had not penetrated, seems not to have occurred to either. What Wallace and Darwin both overlooked was that the autochthonous civilizations of the Mayas, Aztecs, and Incas at first contact stunned the conquistadors, who demolished them as rapidly as possible. Furthermore, as regards agriculture, the foundation of civilization, it has been estimated that at the time of the Spanish conquest, the Andean Indians had domesticated and were cultivating as many as 70 separate crop species, almost as many species as had been brought under cultivation by the farmers of all Asia or Europe. It is particularly noteworthy, in contrast to the hand-to-mouth existence of Europe, that because of the Inca's agricultural productivity and remarkable organization, it was usual to have three to seven years' supply of food in storage.[16] What the conquistadors encountered were often only remnants of a previously much greater culture.

I am certainly not arguing the equivalence of Amerindian to European and Chinese civilization (both the latter the products of extensive borrowing from each other), but rather that the complexity of Amerindian culture too often has been underestimated. The "savage" had not only in many respects demonstrated the potential of that mind, but both Darwin and Wallace—influenced by the lack of understanding of genetic inheritance—failed to realize the equivalence of the "savage" to their own ancestors 2000 years earlier, and the impossibility that biological evolution could have bridged that gap in so short a period.

By now the repeated examples of the ability of "savages" under favorable conditions to master in a generation or two the ground rules of our civilization should have gone far to put to rest those sweeping concepts of inferiority; I repeat the earlier assertion that, viewed through eyes not clouded by the material trappings of civilization, the intellectual life of tribal man is on average in many respects as complicated as our own. When the hunting and gathering ancestors of the readers of this book began to take up agriculture some 3000 to 10,000 years ago (depending on location), they were already essentially the same biological material as ourselves today—and as the surviving Amerindian societies. Step by uncomprehending step, the minds that evolved under tribal and pretribal circumstances have modified the population structure and the world in which that evolution occurred to the point where only resolute and carefully considered counter action can avert tragedy on an unprecedented scale. Indeed, there are gloomy days when I feel major tragedy cannot be averted—the more realistic challenge is to minimize the pain in this readjustment.

The Royal Courts of Justice, London (1993).

18

Assessment of Genetic Risk

A s we have seen, the detonation of the atomic bombs over Hiroshima and Nagasaki and the subsequent testing of more powerful weapons, plus the advent of nuclear energy, inevitably created immediate and serious concerns over the genetic implications of these developments. Likewise, with Auerbach and Robson's demonstration in 1944 of chemical mutagenesis, it was quickly realized that occupational exposures to the large numbers of chemical agents used as pesticides and herbicides, incidental exposures resulting from residues of these agents in the environment, and exposures resulting from the disposal of useless residues of the manufacturing process all had possible genetic, not to mention carcinogenic, implications. But although exposures to man-made sources of radiation and chemicals are sometimes accidental, and in some instances even wanton, increasingly they are the result of some type of risk–benefit analysis, sometimes inchoate, sometimes the result of the best effort society can command.

Thus far, despite the recent amazing developments in genetics, these risk–benefit efforts remain beset by controversy. It is easy to clobber microorganisms and mice with large doses of radiation or a chemical and demonstrate that mutations are produced, A not-so-small industry devoted to testing chemicals for mutagenicity has come into being in the past 40 years, the usually positive results of each such testing raising yet another alarm. It is quite another matter to extrapolate from the effects of a mutagenic agent on an experimental organism to the effects of very low doses of this same agent on humans, with a different metabolism and conditions of exposure, or to obtain the necessary data on which to base decisions directly from studies on humans.

Developing a proper perspective toward the implications of these readily identifiable chemical and ionizing mutagens has recently been greatly complicated by the realization of the extent to which mutagens and carcinogens have been and are a feature of the "natural" environment. All of us, on average, during the first 30 years of our lifetime receive about .09 Sv equivalent of radiation to the germ cells from our "natural" environment. Furthermore, many plants used as foods synthesize toxic mutagens and carcinogens in relatively large amounts, apparently as a defense against bacterial and fungal invasion, and insect and other animal predators. Finally, subjecting foodstuffs to high temperatures, i.e., some types of cooking, can convert amino acids and other natural food components to potent mutagens.

In this chapter I propose to review some of my own experiences with efforts to set guidelines for human exposures to these mutagens. These efforts provide a textbook study of the difficulties in taking intelligent actions in the face of limited knowledge but societal pressures to set guidelines. They also underscore the need for improving the methodology of epidemiological studies of human populations thought to have sustained a mutagenic experience.

The BEAR Report of 1956

In October of 1955 I received a letter from Detlev W. Bronk, then president of the National Academy of Sciences, stating that the Academy proposed "to study and report on the biological effects of atomic radiations resulting from scientific, military, and industrial uses and to recommend needed research in related fields," and inviting me to join a committee being organized to this end. As the activity unfolded, a number of separate committees were formed, whose combined efforts resulted in a report of the Academy in 1956, entitled, "The Biological Effects of Atomic Radiation";[1] I served on the Committee on Genetic Effects. The senior members of this committee were an unusually distinguished group, almost without exception members of the Academy, many of whom—Glass, Beadle, Wright, Demerec, Hollander, Kaufman, Muller, and Sturtevant—I had met during my *Drosophila* days. Three of us were distinctly the juniors of the group: W.L. Russell, then in the midst of his extensive studies on the radiation genetics of the mouse; J.F. Crow, a statistically oriented geneticist adept at the types of calculations our activity required; and myself, fresh from our first major summary of the study in Japan. The Committee's chairman was Warren Weaver, a mathematician by training who in 1932 had joined the Rockefeller Foundation, where—to some significant disgruntlement in the biological community—he had shortly assumed responsibility for programs in the natural sciences. By 1955, he had mastered a great deal of biology, but as committee chairman, he played his background to the hilt, constantly pressing the committee members to "make it clear to me; remember, I'm not a biologist." The result (I am prejudiced) was a document which for clarity at all levels has not been superseded by any of the many subsequent reports. This committee activity was the first time the subject of human genetics had been brought into the picture on an important policy issue since the heyday of eugenics, when Harry Laughlin, of Eugenic Records Office notoriety, in the 1920s, had advised the Congressional House Committee on Immigration and Naturalization in a fashion most geneticists would prefer to forget (see Chapter 1). I, for one, was very much aware of the symbolism of genetic principles being brought back into considerations of national interest.

In the mid-1950s, both the U.S.S.R. and the U.S. were engaged in above-ground testing of nuclear weapons, and the initial principal focus of concern was on the effects of fall-out. The Committee, however, soon came to the realization that to discharge its commitment properly and develop perspective it had to consider all types of exposures to ionizing radiation. The Committee meetings rapidly became an exercise in conscience and risk–benefit analysis on the part of a group of very intelligent and highly concerned citizens. As we did our homework, we assembled all the information on human radiation exposures, natural and man-made. Over the range of radiation doses at which genetic experiments with mice have been carried out, the relationship between dose and effect has been linear, i.e., the effect has been directly proportional to the dose. It must be recognized that at that time (and at present) experiments have never been carried out at the relatively very low doses that (short of an atomic or other radiation disaster) will characterize human exposures. To yield significant results, such low-dose experiments would require what have seemed to be prohibitively large samples of mice. Accordingly, in treating the implications of the additional low-level exposures which, it was thought, humans would receive in the future, we were forced to assume no threshold to the genetic effect of radiation, i.e., any radiation exposure, however small, would have some genetic effect. In the absence of information to the contrary it was a necessary (and tenable) assumption. But it has also been, right down to the present, an arguable assumption. In order to make some of our calculations, we had to extrapolate from the laboratory data to the effects at doses only 1/100 to 1/1000 of those at which the seminal experimental observations had been made. The potential for error was apparent to everyone, but the need for even the roughest of guidelines was also apparent. No one liked the prospect of any increase in inherited genetic defects, but no one felt the various radiation exposures to which members of society would be exposed in the future were capricious—each arose from what most considered a well-defined need, and we had to balance those needs against the probable biomedical effects of the radiation.

This was the first time any group had considered the total population exposure to ionizing radiation. The findings at that time (1955), slightly different from the figures presented in the preceding chapter for current exposures, were somewhat surprising. The best estimates of gonadal exposures over a 30-year period (i.e., up to midreproductive age) that our consultant physicist could generate were: 1) background radiation: about 4.3 roentgens (principally from cosmic rays, geological radioactivity, and radioactive elements incorporated into the body); 2) medical exposures to radiation: 3.0 roentgens; and 3) fall-out from weapons tests: 0.1 roentgen. (Note that at that time we were still using roentgens as the unit of exposure.) The exposure to radon characteristic of some homes was not yet recognized. Exposure to fall-out from weapons testing, in some respects the proximate reason for convening the Committee, was almost negligible (and now, of course, is even less). Background radiation was higher than anticipated, and the relative contribution of medical practice to human exposures was also greater than anticipated.

With Weaver's constant prodding, the key final recommendations of the group, issued in 1956, emerged simple and clear (we also had no legal counsel), namely:[1]

That the medical authorities of this country initiate a vigorous movement to reduce the radiation exposure from X-rays to the lowest limit consistent with medical necessity; and in particular that they take steps to assure that proper safeguards always be taken to minimize the radiation dose to the reproductive cells.

That for the present it be accepted as a uniform national standard that X-ray installations (medical and nonmedical), power installations, disposal of radioactive wastes, experimental installations, testing of weapons, and all other humanly controllable sources of radiations be so restricted that members of our general population shall not receive from such sources an average of more than 10 roentgens, in addition to background, of ionizing radiation as a total accumulated dose to the reproductive cells from conception to age 30.

The previous recommendation shall be reconsidered periodically with the view to keeping the reproductive cell dose at the lowest practicable level. If it is feasible to reduce medical exposures, industrial exposures, or both, then the total should be reduced accordingly.

That individual persons not receive more than a total accumulated dose to the reproductive cells of 50 roentgens up to age 30 years (by which age, on the average, over half of their children will have been born), and not more than 50 roentgens additional up to age 40 (by which time about nine tenths of their children will have been born.) (pp. 28–29)

Those recommendations have often been misinterpreted. The recommendations of the last paragraph were meant to cover the occupational needs of industry or the military, but were not meant to imply that industry and the military routinely permit exposures right up to that limit. The recommendation was population- rather than individual-oriented; it was anticipated that if some persons exceeded a ten roentgen exposure, others would fall well below it, so that the average to the total population would not exceed ten roentgens. Elsewhere the report was careful to spell out the probable impact of this exposure, as it was then understood. It was not negligible, but neither were the benefits, including, as many perceived the situation at the time, maintenance of a precarious peace through mutual assured destruction. The Report was generally received for what it was, an honest effort to deal with a difficult problem. I do not recall any significant criticism to the effect we had been too lax.

Our Committee was reconvened in 1959. There had been one extremely important genetic development since our original charge. The principal sources of data for these recommendations were the mouse experiments of W.L. Russell, the data from human studies playing at that time—properly—a minor role. All of the early experiments on mice had involved single exposures to relatively high doses of radiation. Now Russell and colleagues had looked at the genetic effect of dividing a given amount of acute radiation into multiple, small doses, or of reducing the rate at which a given dose was administered. The findings, reported in 1958 and referred to in Chapter 13, were dramatic: the mutational yield under these circumstances decreased by a factor of at least three.[2] Since, short of nuclear war, most human exposures will come in the form of small, intermittent exposures, this observation had many implications for the human problem, i.e., the observation meant that our guidelines, based primarily on the genetic effects of acute ionizing radiation on the mouse, were on the conservative side. (A second development with possible equal implications for the human situation was to come six years later. All of the early work in mouse radiation genetics was based on experiments with male mice, since females became sterile at relatively low doses of radiation. In due time it became absolutely essential to study the genetic effects of radiation on the other half of the species, and, in 1965, W.L. Russell reported that although the first litters from irradiated female mice showed as many mutations as corresponding litters sired by equally irradiated male mice, the later litters from these same females provided no evidence for any increase in the mutation rate.[3] The basis for this finding is still poorly understood, as is its applicability to the human female, but the fact itself has not been challenged. The net effect of this finding, assuming some relevance to the human situation, is also to make the earlier recommendation appear conservative, i.e., carry a higher margin of protection than originally thought to be the case.)

There was one other important development relative to the reconvening of the Committee: T.H. Dobzhansky was added to its membership. This set up a classic scientific confrontation. Muller's view, which had dominated the first round of Committee meetings, was that essentially all mutation was deleterious, with individuals heterozygous or homozygous for a mutant gene in some way handicapped. More than that, each new mutation, unless (very rarely) favorable, eventually resulted in one genetic death. This led to the further view that a population would be best off in the short run if all that mutationally derived variation could be eliminated and the species represented by a single "best" genotype. An analysis of the then very limited results of studies on human inbreeding, in collaboration with J.F. Crow and N. Morton, had persuaded him that the variation encountered in human populations was mostly detrimental and maintained by mutation pressure.[4] Our studies in Japan, showing a substantially smaller inbreeding effect than the studies on which Morton, Crow, and Muller relied, had not yet been completed. Dobzhansky, on the other hand, from his extensive studies of variation in wild *Drosophila* populations, had come to the view that a large measure of diversity was desirable, so that the species had a reservoir of genetic traits on which to draw as it moved into various ecological niches or encountered changing conditions. He was also impressed by the example of sickle-cell anemia, where a single dose of the gene is advantageous in malarious environments but the double dose distinctly deleterious. How many more such balanced polymorphisms were waiting to be discovered?

Muller had come into the initial round of meetings with a well-thought-through position and firmly held views that exerted great influence on Committee thinking. Now these were challenged. I found myself increasingly siding with Dobzhansky, on the grounds that we knew too little about human genetics (a view certainly reinforced by the later developments we have discussed) to be as definitive in our treatment as Muller would have wished.

The exchange assumed a new dimension as Sewall Wright more actively entered the fray. Wright, and the Englishmen J.B.S. Haldane and R.A. Fisher, were the founding fathers of modern statistical genetics. Wright—always a rather quiet person—had clearly not been happy with Muller's certainty in the first round of Committee meetings, but he had not pressed the issue to an impasse. Now he was ready. He began to challenge Muller more and more. The contrast between the mathematically rigorous thinking of Wright and the brilliantly intuitive thinking of Muller was fascinating. I recall one memorable exchange when Wright pushed Muller further and further on a point of difference. The "J" which was Muller's middle initial was an abbreviation for Joseph, and many of his friends called him Joe. "How can you be so sure of this, Joe?", asked Wright. Muller thought for an instant and then, drawing himself to his full 5'4", he replied, "I just know, Sewall, I just know." That terminated that particular discussion.

Any committee report represents a compromise, and, in due time, a second Committee report that rather glossed over some of these strong differences of opinion was hammered out (and appeared in 1960).[5] But when the time came for individual members of the Committee to sign off the new document, a surprise: Wright would sign off only if he could add an Appendix, entitled, "On the Appraisal of the Genetic Effects of Radiation in Man," in which he would lay out the complexities as he saw them. More discussion but eventual agreement. One small quotation will convey the thrust of Wright's Appendix:[5]

There is one point of view under which the appraisal of genetic damage from increased radiation is a relatively simple matter. If we assume that there is one best genotype and that this is homozygous in all type genes, it follows that all mutational changes from this are injurious and selected against. For each mutation there will be on the average one elimination (or "genetic death") to restore the status quo (in a static population; more than one in a growing population). If we define damage in terms of number of genetic deaths, it follows that all mutations produce equal damage in the long run and it merely becomes necessary to estimate the number of mutations produced by a given amount of radiation to appraise the damage. There are, however, several considerations that make this point of view unsatisfactory. (p. 18)

There followed an essay that laid out the value of diversity to a population and the ways, some advantageous, some disadvantageous, that mutation contributed to this diversity. The essay, written before biochemical techniques had revealed the enormous store of genetic variation in proteins and DNA, was a polite rebuke to overly simplistic approaches to the problem of the genetic effects of radiation. Wright, of course, believed in the generally deleterious effect of induced mutations but, given the complexity of their impact, was having greater difficulty with his balance sheet than Muller, as was I. Be this as it may, in its 1960 report, the Committee basically reaffirmed its earlier recommendations, in language that recognized the problem had grown no less complex in the interval since the Committee had last been convened. Unfortunately, the 1956 report received far more attention than the 1960 report, and Wright's thoughtful essay went relatively unnoticed.

Time passes, allowing for more meetings of national and international advisory committees than I care to recall. In general, the recommendations of our Committee have held up surprisingly well in the hands of those bodies that are somewhat closer to the regulatory stream, such as the U.S. Federal Radiation Council, the National Council on Radiation Protection and Measurements, and the International Commission on Radiation Protection. Along the way, an important development was to subdivide that permissible additional 10 r to the general population into an average dose limit to the gonads from nonmedical radiation sources of 5 r in 30 years, and a similar amount from medical sources. It was always clearly recognized that these were *maximum permissible* exposures, and that, in practice, society should strive to keep exposures As Low As Reasonably Achievable (the so-called ALARA principle). There is little room for argument with the principle, but, as we shall see, what is reasonable to one group may not be so to another.

Our original Committee had agonized over a recommendation that would strike a balance between societal needs and benefits and genetic risk. When in doubt, we tended to be conservative. The best evidence of that conservatism—and the conservatism of the regulatory bodies as well—is that when in 1958 the data came in on the lesser genetic effect of divided doses of radiation, the permissible limit was not extended upward. Moreover, when, as mentioned earlier, W.L. Russell in 1965 reported the very important observation concerning the failure to recover mutations among the later litters born to X-ray-treated female mice, again there was no tendency on the part of regulatory bodies to assume this situation also obtained for the human female and to relax the maximum permissible population exposure to radiation. Neither was there any such movement as the data from Japan, indicating humans might be less sensitive to the genetic effects of radiation than mice, began to accumulate.

In describing some of the additions to the data base on the genetic effects of radiation that have occurred since the maximum permissible limit was first suggested, it is not my purpose to suggest that the limit should have been increased in the light of these new discoveries. Rather, it *is* my purpose to indicate that this limit can now be recognized to be extremely conservative, a fact that, as we shall see next, seems to be lost sight of by regulatory agencies as they pursue the ALARA concept.

Translating Principle to Policy: The EPA and High-Level Radioactive Waste

During and since the 1960s, here in the United States the public mind, reflecting widespread anxiety over the complex and poorly understood dangers of many technical developments, has had a fixation on the precise attribution of responsibility for all the risks of modern living. The legal profession has been quick to respond to these concerns. I quote from a distinguished legal scholar:[6]

Serious efforts to impose legal controls on the sources of societal risk date only from the mid-1960s. Most prominent in the public mind has been the creation of specialized federal agencies to directly regulate particular societal dangers—the Environmental Protection Agency, the Occupational Health and Safety Administration, the National Highway Traffic Safety Administration, and the Consumer Product Safety Commission as examples. But far more pervasive in scope and far more effective in impact than agency regulation has been the expansion of civil law to control risk. Since the late 1960s, our civil justice system has adopted the premise that civil damage awards enforcing liability rules and statutory rights can optimally regulate *every* source of societal risk. Building from this premise, our civil courts have become the most powerful institution of the modern state for regulating risk. (p. 207)

My most informative experience with the response of the regulatory agencies to the perceived need for precision in setting risks created by this ambiance came through the Environmental Protection Agency (EPA), established in 1970. From the outset, EPA had an Office of Radiation Programs (ORP). Congress, in the Toxic Substances Control Act of 1971 and the Comprehensive Environmental Response, Compensation, and Liability Act of 1980, better known as "Superfund," instructed the EPA to regulate exposures to mutagens (as well as agents that cause cancer and birth defects and other adverse health effects). Then in 1982, Congress enacted the Nuclear Waste Policy Act, requiring a plan for the disposal of high-level radioactive waste by 1985. High-level waste results from the operation of nuclear power and weapons plant and from outmoded nuclear weapons. The perceived health effects of the stored waste, both somatic and genetic, were clearly to be a driving consideration in the preparation of this proposal. The efforts of ORP to deal with this highly emotional issue provide the material for this section.

ORP published criteria for the disposal of high-level waste in the Congressional Record in late 1982.[7] In connection with this activity, they established a nongovernmental subcommittee to review their proposal. This is where I became involved, as a member of this subcommittee. Later, as a member of EPA's Advisory Committee to ORP, I also became involved in EPA's attempt to develop regulations for exposure to airborne radioactivity and also to the radioactivity resulting from disposal of the kinds of radioactive compounds resulting from some medical research and treatment. However, it will be quite enough to consider the problem of high-level waste disposal in some detail.

It is difficult to select a single quotation from a complex federal document that will transmit the biomedical essence of the document, but, with some explanation, the following excerpt from EPA's published criteria for a high-level waste repository comes close:[7]

Our assessments of repository performance gave estimates of the possible health effects expected from releases after disposal. These estimates can vary considerably depending upon the assumptions used and the geologic media considered. For well-designed 100,000 MTHM model repositories in salt and granite—using engineering controls that we believe are readily achievable—we estimate that the health risks over 10,000 years would be no greater than the risks from an equivalent amount of unmined uranium ore. To provide a specific basis for our proposed standards, we selected a limit on long term risks of 1,000 health effects over 10,000 years for a 100,000 MTHM repository. Our assessments show that a wide variety of repository designs and sites can reduce risks below this level. (p. 58196)

The reader must appreciate that the term "health effects" is a euphemism for serious cancers and genetic effects, and "MTHM" an abbreviation for "metric ton of heavy metal." The reader must also appreciate that at that time, in calculating risks in the face of uncertainty, EPA customarily presented the highest risk yielded by any calculation, and then did not transmit to the public any of the uncertainty in the estimate, nor the fact that this was a maximum risk. Finally, for the sake of perspective, the reader should know that at that time it was estimated that the storage of this high-level waste would require about eight to ten of these repositories, and, under the proposed standards, in 1982 each would cost at least ten billion dollars.

There are three aspects of that quotation that quickly elicited comment from our Committee as it settled into its work. If, back in the ground, this waste results in "health risks . . . no greater than the risks from an equivalent amount of unmined uranium ore" (which this once was), then we are not dealing with an additive but a substitutional risk. No one gave the government credit for reducing risk when the ore was mined, why proceed as if the government had created a risk when the ore is returned to the earth from which it came—but now, because of careful site selection, with less risk of contaminating water supplies than previously? The public needs assurances the waste will not be treated carelessly and even that it will be a lesser source of radiation after disposal than before, but not a count of health effects which presumably would have occurred even in the absence of the manufacture of weapons or nuclear power.

The second notable aspect of this paragraph concerns the "1,000 health effects." Over that 10,000 year period, this amounts to one effect every ten years. Later in the document we learn that by the same formula used to calculate those excess health effects, during that ten-year period in which the repository results in one health effect, there will in the U.S. be 40,000 effects from background radiation, and that—presumably to provide perspective—if natural background radiation increased from the reference figure of $0.100\ r$ per year which EPA employed to 0.101, there would be with the formula they employed 400 of these extra health effects annually. (Remember, that at sea level cosmic radiation amounts to about $0.06\ r$ per year but at the high altitude of some Colorado towns, to about $0.20\ r$). First, the uncertainty surrounding this calculation can be appreciated by anyone who read Chapter 13. (The uncertainties surrounding cancer risks are less than those surrounding genetic risks, but still very major.) The extrapolation involved requires going from observations on experimental animals receiving several hundred *r* and Japanese with exposure of ten to several hundred *r* down to doses less than 0.001 *r*, and, as we have seen, even the estimates of the effects of the high doses have considerable uncertainty. Second, the attempt at the regulation of a risk down to 1/40,000 of the "natural risk" from environmental (i.e., background) radiation is without precedent. Third, the writing of regulations based on protection over a 10,000 year period is also without precedent. The behavior of any geologic repository over such an extended period of time cannot be predicted, nor can population densities in the region of these repositories, or above the aquifers and along the watercourses into which some of this radioactivity might find its way.

It seemed rather clear to me (and I believe to the majority of our Committee) that in formulating this plan, EPA had not only been overzealous but had put the regulation in an unfortunate framework. As the figure of 1000 health effects was presented, it appeared certain to arouse concern over the dangers of radiation, and arouse concern it did until put into the perspective that the regulation was directed at less than 1/5000 of the 5 *r* deemed the maximum permissible population exposure by our NAS and subsequent Committees.

Our EPA Committee—no less conscientious than the first NAS Committee—was virtually unanimous that this was regulatory overkill. We felt that the permissible "releases after disposal" had arbitrarily been set too low, with the magic figure of 1000 health effects badly distorting perspective, and that, to begin to bring these risks in line with those society is accustomed to accepting, the release constraints could be relaxed by a factor of ten. Here, however, we found ourselves forced to observe a very fine line. Regulatory agencies such as EPA have in recent years attempted to maintain a clear distinction between "risk assessment" and "risk management." Our charge was not the latter but the former: had ORP used the proper models and the relevant data in arriving at this figure of 1000 health effects? It was made clear to us by staff that it was not our role to comment on the appropriateness of the figure; that fell under "risk management." Some of us were much less than happy at this instruction, boxing us as it did into a strictly technical corner. In the end, we decided we could not in good conscience function in this narrowly technical context: the first recommendation of our final report, submitted in January of 1984, read: "The Subcommittee recommends that the release limits specified in Table 2 of the proposed standards be increased by a factor of ten, thereby causing a related ten fold relaxation of the proposed societal objective. . . . " We may have blurred the lines between risk assessment and risk management, but I do believe we made our point.

We (and I assume others) did not labor in vain. The document finally passed by the Congress (50 FR 38084 Sept. 19, 1985) recast the permissible exposure resulting from the presence of a repository to exclude for 1000 years an annual exposure to any individual living in the vicinity of the repository of 0.025 *r* to the whole body or 0.075 *r* to any critical organ, with no specification of the resulting health effects. Earlier, I mentioned the emergence of the ALARA concept. I believe that, in retrospect, the proposed, original regulation on waste disposal will be seen as the high-water mark (or low-, depending on one's viewpoint) in attempts to push the ALARA concept to its limits. Exposure standards had been set at ridiculously low levels, with the risk management bearing precious little relation to the risk assessment.

The efforts to site these repositories are telling testimony to the public near-hysteria over the effects of radiation. Even after the identification of potential sites that appear to meet all specifications, none of the various states with the requisite geology and the arid climate for such a site have been willing to accept one. The inability of the U.S. to reach a solution to this problem remains a significant obstacle to the further development of nuclear energy, at a time when such energy could play a significant role in the transition from the high dependence on fossil fuel to the new energy consumption patterns of the future.

The EPA and Low-Level Radioactive Waste

Similar sorts of issues arose in the writing of the proposed regulations regarding airborne radioactivity and low-level waste. In general, the exposure standards were far more stringent than the standards regulating other risks in our industrialized society. If such standards had come "for free," the Committee would probably not have again demurred. However, there was an appreciable cost, to industry and to government, in meeting these standards. For instance, one of our Committee members, Merril Eisenbud, calculated that in regulating the risks due to the airborne radioactive element polonium-210, emitted from industrial facilities producing elemental phosphorous, the EPA proposal would require of the facilities producing this substance an expenditure over a 20-year period of $300 million to avert a single estimated "health effect." The contrast between this and other "public health" expenditures speaks for itself.[8] It was not difficult to perceive a variety of societal needs upon which the money released by a relaxation of these regulations could be spent to better purpose—albeit the system does not work that way.

For myself, during this EPA involvement I was already well along in the appraisal of the future expressed in the last two chapters. By that time, it was already becoming clear, as discussed in Chapter 13, that the genetic effects of radiation on humans were probably even less than were implied by the guidelines from which our Committee was operating at that time. The zealousness with which the Office of Radiation Protection of EPA pursued this regulatory overkill, while the official establishment of which it was a component (i.e., the government) was treating some of the other major issues bearing down upon its citizenry with such benign neglect, was to me a classic example of distorted perspectives. Yes, I believe in protection against radiation exposures, but protection as a component of a much more balanced program of risk evaluation than exists in the U.S. at present.

In reviewing the regulations proposed later concerning air-borne radioactivity and low-level waste, another sort of issue came to the fore. Although, as documented in the earlier quote,[7] 40 CFR Part 191 (Proposed) paid lip service to the uncertainties in the calculation of the 1000 health effects, the range of these uncertainties was never indicated. Later documents, presenting proposed regulations regarding exposures to radionuclides and low-level waste, also presented probable effects as a single figure, despite the great uncertainties as to just what the release of radioactivity would be following storage and the biological effects of these releases. Without wishing to impugn the motivations of persons I respect, I suspect that in a regulatory setting there was a reluctance to reveal just how much uncertainty there was concerning the factual basis of the regulation. We scientists on the reviewing panels looked at the matter quite differently. We felt these uncertainties should be brought out at every turn. To present a certitude of outcome well beyond the facts could only choke off the necessary additional research.

We members of the advisory committee to ORP transmitted our concerns about the need to indicate uncertainty in rather strong language to the Science Advisory Board of the EPA, of which Board we were an appendage. The Board—which already had its own misgivings—endorsed our position equally strongly. The result was a 1985 report by the Carcinogenicity Guidelines Group of the Science Advisory Board to the then Administrator of the EPA, William Ruckelshaus, a report strongly recommending disclosure of uncertainties in all projections of risk. There was minor resistance within EPA, but the policy now prevails.[9] I regard this as one of the most worthwhile developments to which I have ever contributed.

I do not mean to wax caustic at the expense of EPA. Conscientious public servants, under enormous political pressure, were doing their best to deal with a complex and controversial issue. Zealots within some environmentalist groups would maintain that any adverse effect from an environmental contaminant was too much. The problem is in the application of the ALARA concept. Who is to decide what is "reasonably achievable"? There is no governing principle.

Radiation Genetics Goes to Court

Of all my various other entanglements in the past 30 years in society's efforts to adjust to the risks of exposure to ionizing radiation, one, just completed, stands out. In 1983, Yorkshire Television broadcast a program drawing attention a cluster of cases of childhood leukemia in the village of Seascale, West Cumbria, England. This report was subsequently confirmed by a government-appointed Independent Advisory Group: although the numbers were small, the rate of leukemia in children resident in Seascale was approximately ten times that in children in the surrounding areas. The Advisory Group recommended a case-control study of all leukemias and other related tumors (lymphomas) diagnosed since 1950 in persons under 25 years of age resident in the West Cumbria district. (A case control study matches affected persons with one or more unaffected persons on the basis of age, sex, and residence, and then compares these two sets of individuals with respect to numerous circumstances that might be relevant to the problem under investigation.) The responsibility for directing such a study was placed in the hands of a highly respected epidemiologist, Martin Gardner. The publication of the results of this study in 1990 created a mild uproar: of some 60 possible associations investigated, the statistically most significant to emerge was the long-time employment of fathers living in the village of Seascale in the nearby Sellafield nuclear reprocessing installation, operated by British Nuclear Fuel (BNFL).[10] On the basis of the size of the sample studied, only 0.6 Seascale children with leukemia should have been fathered by employees at Sellafield; the actual number was 4. Dividing this latter number by 0.6 yielded a relative risk of 6.7. (The figure of 0.6 children is a statistical expectation, fulfilled by observing either 0 or 1 leukemic child.) With one less child (3 affected), the finding would have lost statistical significance. The authors of the study favored a genetic explanation of their finding, i.e., the induction of leukemogenic mutations in these fathers, despite the implication of this correlation that human sensitivities to the genetic effects of ionizing radiation were far greater than was consistent with the by now voluminous literature on this subject.

A British law firm, Leigh, Day, and Co., advertised its willingness to file claims against BNFL, and, not unexpectedly, relatives of two of the children soon initiated legal actions against BNFL, claiming damages for personal injuries and, in one of the two cases, death. According to the London Times, the resulting trial, before the Royal High Courts of Justice of England, was expected "to be a record breaking legal action costing up to £10 million." I became involved as witness for the defendant (BNFL). The epidemiological study meets all scientific standards. However, the exposures to the small sample of fathers involved were relatively low, in the region of 0.2 Sv equivalent of chronic ionizing radiation to the body surface prior to the conception of the affected child, with no father receiving more than 0.4 Sv equivalent of chronic radiation.[11] The exposures were within the current occupational limits for workers in the nuclear industries. The apparent finding was in flat contradiction with the failure to observe an increase in leukemia in the children born to atomic bomb survivors discussed in Chapter 13, for which reason the Japanese data cast a long shadow over the trial.

The attribution of these findings to the workplace induction of mutations in the fathers of these children was immediately suspect on several grounds. 1) As just stated, the observation suggested sensitivities at least a thousand times greater than the data from Japan would support. 2) When so many statistical tests are conducted (about 60), just by chance several should yield a "significant" result, i.e., the finding could be a statistical artifact. 3) Human leukemia in past studies has never appeared to have the simple genetic basis of, say, retinoblastoma, which basis might provide the expectation of a response to radiation. 4) Although an increase in leukemia in the offspring of male mice receiving relatively large doses of X-rays to spermatogonia has been reported in one of three mouse strains, in the sensitive strain the baseline frequency was about eight times higher than in humans, and the increase was only 1/15 as great as would be predicted from the Seascale observations.[12]

Not surprisingly, the results of the study by Gardner and colleagues quickly stimulated a number of new, related studies, with results also germane to the trial. 1) Since the genetic effects of radiation are not gene-specific but shotgun in nature, a mutagenic insult capable of creating such an increase in leukemia should have increased the probability of other unfortunate mutational events in Seascale to an extent that could scarcely have gone unnoticed, but thus far such an increase has not been observed.[13] 2) Since the publication of the Gardner study, three other similar studies, two (overlapping) on a cluster of leukemia in children in the vicinity of the Dounreay Nuclear Reprocessing Plant in Scotland, and one of leukemia in the children of employees at five nuclear facilities in the Province of Ontario, Canada, have failed to duplicate the findings of the Gardner study.[14] 3) A study on the children of additional workers in the Sellafield plant who do not live in Seascale has failed to reveal any increase in leukemia in these children.[15] 4) Finally, a systematic search for other leukemia clusters in the north of England brought to light one in the village of Egremont North, just 7 kilometers north of the Sellafield plant, but now there was no association between the occurrence of the disease and father's employment in the Sellafield facility.[16]

The case, tried before a single judge (no jury), has elicited an enormous volume of highly technical testimony. Our Japanese data on cancer in the children of atomic bomb survivors have been completely reanalyzed by two independent statisticians—with results identical to our own. On February 25–26, 1993, after some eight months of preparing depositions, analyzing the depositions of others, and careful reading of the voluminous transcripts of testimony, it was time for my first court appearance. (The second came on May 24, 1993.) The trial was held in London, at the Royal High Courts of Justice of England. This is a beautifully Gothic edifice, constructed in 1874 (see page 316). The various courts themselves, mostly rather small (some 30' × 30' but with great height), open off a soaring Great Hall. Entering that Hall, I was suddenly very conscious of the weight of history. The building, which houses the highest courts dealing with civil (as opposed to criminal) justice, as well as the (ultimate) Court of Appeals, have been the scene of England's most famous libel trials. Since by this time some of the allegations of plaintiffs' witnesses concerning the inadequacies of our Japanese study bordered on the libelous, this setting seemed entirely appropriate. What an unpredictable web of circumstance had led sometime Lt. Neel, attached as an afterthought to that survey in 1946, from his first glimpse of a devastated Hiroshima to a role—maybe even a key role—in yet another historic trial in this setting.

As one aspect of making their case, the barristers for the plaintiffs had somehow to demonstrate 1) that our study of leukemia was of poor quality although (patronizingly) the best to be expected under the circumstances, and 2) that prior to the initiation of the study in 1948, there had been a virtual epidemic of leukemias, which our study had missed. In addition, inasmuch as our other studies in Japan had not yielded evidence for genetic effects of a magnitude consistent with expectation based on the Seascale findings, these too had to be discredited. It was all very civilized, but there is, however, a certain amount of tension inherent in the confrontation of two men, one, the barrister for the plaintiffs, intent on denigrating the life's work of the other. I admit to being taken aback by the innuendo, obfuscation, and distortion that characterized the tendentious efforts of counsel for the plaintiffs to cast doubt upon both the experimental (murine) and Japanese data, to create an aura of plausibility for a genetic etiology for the findings of the Gardner study. It was a sad commentary on the adversarial ambience that characterizes society's approach to dealing with the type of complex scientific evidence with which it is increasingly confronted in environmental issues.

The suit required the court to decide whether the large body of negative evidence just mentioned overrode the results of the one small but apparently adequate, positive study on which the legal action was based. From the first it was clear that the outcome of the trial would be a landmark decision: If the Court found for the plaintiff, there would be a flood of similar actions brought wherever there are nuclear installations, as well as challenges to the adequacy of current government regulations concerning permissible exposures to low-level ionizing radiation. If the court did not find for the plaintiff, there would be a legal precedent that, if adequately publicized, could play a major role in allaying unwarranted fears over the effects of low-level ionizing radiation. As the case has unfolded, the plaintiffs have altered their original position, suggesting that there is at work in Seascale a "factor X," possibly a virus, that interacts synergistically with the effects of radiation, apparently not realizing that their position still requires postulating a very specific predisposing mutation in each of the fathers, a position against which the odds are overwhelming. Furthermore, given the demonstration of a second focus of leukemia so close to Seascale, a focus in which radiation cannot be implicated, the scientific principle of parsimony demanded that the case for radiation playing any role whatsoever in the findings had collapsed. The issue for the Court, given the rules of evidence, was, however, how to move from an apparently proximal convincing relationship (the original study) to a much larger body of data that give no support to the Sellafield findings. Both sides have been extremely thorough in their review of all possible aspects of the situation; the transcript of the trial will be a "bible" and guideline for any future actions alleging genetic damage from potentially mutagenic exposures.

I have gone into such detail to project to the reader some understanding of the full complexity the issues concerning the genetic effects of radiation may assume when they move into the courts. An equally important point is to illustrate the unfortunate consequences of precipitous action based on a single unusual study. As we look into the future, with more such actions undoubtedly in the offing, it is already clear that when studies of this nature yield a most unusual result, the cause of science (and human affairs) is best served by a very cautious interpretation of the data until confirmation is at hand.

NOTE ADDED IN PROOF: On October 8, 1993, the Court found for the defendants.

A point of real concern to one who has consulted for various agencies in Washington involved in the regulation of, or compensation for, genetic risks, is that each of the many agencies involved in these complex matters, acting under its particular Congressional Mandate, in risk setting and management proceeds independently of, and with little cross-reference to, the activities of other agencies involved in risk evaluation. The result is a complex set of conflicting actions and standards. There is a screaming need for a more uniform approach, a need recognized in various ways by the Carter, Reagan, and Bush Administrations. The current annual cost of compliance with antipollution laws is estimated at 115 billion, more than 2% of the gross national product. It is important that this enormous sum be spent as effectively as possible. Congress, also recognizing this need, mandated in 1981 through P.L. 96-528 that the National Research Council review the process of risk assessment in the Federal Government. A strong recommendation of the 1983 report of the Committee of the National Research Council which implemented the study was that: "Uniform inference guidelines should be developed for the use of federal regulatory agencies in the risk assessment process" (p. 162).[17] I would endorse a recurrent suggestion [18] that a Presidential Executive Order be issued to ensure that such guidelines be developed.

Once Again, the Special Problem of Evaluating the Genetic Effects of Chemical Exposures

As brought out in the preceding chapter, the evaluation of genetic risk is at present far more difficult with respect to chemical exposures than with respect to radiation exposures.[19] As is the case with radiation, there is a "background" exposure to the chemical mutagens, since they occur naturally in many foodstuffs and their fungal contaminants.[20] On top of this, a variety of industrial and agricultural practices result in exposure to chemical mutagens. At this point, however, the similarity of the situation to that posed by radiation exposures stops. Because of the physical (penetrating) properties of ionizing radiation, the estimation of gonadal doses from radiation exposures is relatively straightforward. This is not at all the case for the chemicals that are potentially mutagenic, which are absorbed, transported, and broken down by the body in many different ways. Thus, while the theoretical possibility exists of small amounts of many mutagenic chemicals reaching the germ cells, where they may (or may not) interact with each other in a synergistic fashion, there are virtually no data on the extent to which any of the potentially mutagenic chemical compounds actually do reach human gonadal tissue following the types of exposures currently encountered in the workplace.

On the basis of the experimental studies with chemical mutagens thus far conducted in many laboratories, plus a demonstration by our group of very similar mutation rates per cell generation in cultured somatic and germ-line cells,[21] I doubt that the current chemical exposures of the general public are, except under most unusual circumstances, any more of a genetic threat than the radiation exposures. However, I cannot speak with the same assurance to the risks of chemical exposures increasing mutation rates that I can bring to bear on the corresponding radiation risks. Segments of the public—those living near chemical dumps, those whose water supplies are chemically contaminated, even in trace amounts—are greatly concerned. This understandable concern will not be met by easy reassurances, and unless met with evidence, will persist and intensify for the foreseeable future.

This continuing concern over the genetic effects of chemical exposures leads to a very concrete suggestion. At the end of Chapter 13, I discussed the new technologies becoming available for the study of mutation, technologies the RERF proposes to apply to further the understanding of the genetic effects of the atomic bombs. I suggest that as soon as perfected, these same technologies, in concert with such established procedures as a register of congenital defects and a survey for cytogenetic abnormality, be applied to very major studies of the offspring of the presumably most chemically mutagenized individuals who can be identified in the various countries in which these exposures occur, plus a study of a suitable comparison (control) group. It is impractical to make the appropriate studies of the effect of *all* the potential chemical mutagens; let us select, at least for a beginning, certain outstanding candidate mutagens. This is a "worst case" approach. If studies on an appropriate scale indicate no significant increase in mutation rates in the children of these highly exposed individuals, an upper limit to any possible effect can be set. If an effect is demonstrated, then a legitimate cause for concern has been identified, and further studies can be undertaken regarding the effects of lesser exposures. Without some such studies, I suspect the present inchoate fears will persist indefinitely, resulting in expensive and protracted legal challenges and unwise regulations.

The most appropriate subjects for such "worst case" studies are the children of those individuals who have survived childhood malignancies because of intensive chemotherapy (sometimes accompanied by radiation) and have now reached the age of reproduction and are, in fact, reproducing. Some of the chemotherapeutic agents used in treating childhood malignancies, such as nitrogen mustard, procarbazine, cyclophosphamide, or doxorubicin, are highly mutagenic in an experimental setting, but the generally fatal course of childhood malignancies has justified their use in a desperate situation. That mutations have been induced in the somatic cells of some of these persons is indicated by the fact that, ten or more years following treatment, the relative risk of a (second) malignancy (presumably initiated by a mutation in a somatic cell) is some 10- to 20-fold above normal, the precise figure depending on the duration of follow-up, the type of the primary tumor, and the treatment modality.[22] It is a reasonable hypothesis that some mutations have also been induced in the germ cells of these patients, but the precise risk cannot be estimated from the cancer data. The number of children born to these cancer survivors and available for study will be small, but with the DNA technologies, each child in principle should be highly informative.

A proper study would be expensive by biomedical standards. On the other hand, the B-2 bomber that the Pentagon has proposed developing as the next level in deterrent technology will cost at least $865 million per plane. Given the dimensions which fear of chemical carcinogenesis and mutagenesis have assumed in the public mind, it would seem sound public policy as we enter this time of ecosystem reappraisal to initiate a study of the type described above, at a fraction of the cost for a single B-2.

Once again, it must be stressed that even if a mutagenic effect of certain chemical exposures were detected in humans, it does not follow these chemicals must be banned. As in the case of radiation, a risk–benefit analysis must be conducted. This risk–benefit analysis becomes increasingly tense as the need to protect crops in the face of the needs of a mounting population intensifies the pressure to use herbicides, insecticides, and fungicides, some potentially mutagenic.

If further major genetic studies are undertaken on the children of putatively chemically mutagenized parents, particular attention must be given to arranging a source of funding and an operating group that is, so to speak, "beyond reproach." With respect to sources of funding in the U.S., surely the National Cancer Institute, the organization that has sponsored the development and testing of so many of the drugs to treat the childhood malignancies, can do no less than fund an effort to define the genetic legacy of these drugs. With respect to the conduct of such studies, the level of concern over the potential biases of industry and even some governmental agencies is running so high that neither industry nor government should conduct the study. While it seems inevitable that funding for such studies come from federal sources, a consortium of universities would perhaps be the most appropriate organization for their conduct. Finally, the study must maintain constant liaison with appropriate lay groups.

Toward a More Mature Approach to Genetic Risk Evaluation

I begin this section by reminding the reader that there is no such thing as an absolutely free lunch in this complicated world. It is difficult to think of a benefit that does not carry a cost. If we are to enjoy the benefits of the diagnoses resulting from the use of X-rays and the use of radioactive isotopes, if we are to meet a critical percentage of our power needs with nuclear energy, if the chemical industries are to play a role in our lives, and if pesticides and herbicides are used in food production, there is from each of these some probable cost in health effects. Given the population numbers now directly or indirectly dependent upon these processes, there would also be a cost if these technologies were immediately withdrawn. The precise evaluation of these costs, as a basis for more informed judgments, is proving to be an extremely difficult problem. This is especially true with respect to genetic effects, which involve the descendants of exposed individuals. Under any reasonable scenario, the effects should be relatively small in proportion to the background of genetic disease already in the population, and, since there are not known to be any unique genetic effects of chemical or radiation exposures, these effects will manifest as a generally small increase over background values of many, diverse endpoints.

From the scientific standpoint, the most appropriate investigations of potential genetic damage are what epidemiologists term *prospective cohort studies,* such as the studies of the children of atomic bomb survivors exemplify. These are relatively expensive by the standards of biomedical research, but yet, given the high societal relevance of the genetic question, it is not unreasonable to spend as much money on such a study as on the construction of a single bomber. The sponsoring agencies, usually governmental, must not look for ways to cut costs when an adequate study design has been submitted. If this means fewer but more authoritative studies, so much the better for the public.

Given the many difficulties in mounting comprehensive genetic studies on the children of groups of persons thought to have been subjected to a mutagenic experience, I suggest that once the appropriate intellectual and observational bridge has been built between mutagenesis and carcinogenesis, serious consideration must be given to an increasing reliance on the occurrence of cancer in the exposees as a guide to genetic damage in their offspring. With respect to radiation, that bridge is well along: the extensive studies in Japan on the occurrence of cancer in the survivors of the atomic bombings permit estimates that the organ doubling dose of acute, whole-body, ionizing radiation for fatal cancers of all types through 1985 is about 2.5 Sv equivalents.[23] (There are a number of ways this calculation can proceed, and I will not insist on this precise estimate, but none of the other calculations yield strikingly different results.) Our estimate of the genetic doubling dose, also through 1985, was about 2.0 Sv equivalents. For now, without delving into mechanism, let us accept this correspondence as an empirical fact. Data on cancer frequency are more easily collected than on the panoply of genetic effects. One, of course, cannot strictly equate the societal impact of a cancer with onset in the later years of life with a genetic effect present at birth. Nevertheless, I am moved to suggest that the regulations which society has and will develop, on the basis of a cost–benefit analysis, to protect against cancer in individuals exposed to ionizing radiation, should, on the basis of present knowledge, be adequate to protect future generations against genetic damage: cancer can serve as a "pit canary" for genetic damage.

With respect to chemical mutagens, the semiequivalence of the cancer–germ line mutation axis is by no means as well established as for ionizing radiation. Earlier I urged a major study of germ-line mutation in the children of individuals receiving chemotherapy for childhood malignancy. This study should have the additional value of contributing to the bridge which for chemical agents needs to be built between the frequency of somatic cell events resulting in cancer and germ-line events manifesting as genetic effects in the children of exposees.

Finally, as society continues to attempt to engage in risk assessment and containment, let us not be seen by future historians as picking at the minutiae of risk regulation while the children and grandchildren soon to come may be battered by major societal risks the current generation was too timid to face. A corollary to this viewpoint is that in the straitened circumstances ahead, it may become even more necessary than in the past to prioritize the recognized environmental genetic risks and concentrate on the control of those which have the greatest potential for harm and/or can be most expeditiously controlled.

A modern "gathering." By contrast, prior to recent outside contacts, the average Shavante or Yanomama would meet about 1000 people—perhaps 1500—during the course of his/her entire life. Did evolution prepare us for this degree of crowding?

19

Genetic Medicine for Populations

I n these three final chapters, we will consider some possible next developments in the human odyssey from the viewpoint of a population geneticist. The seriousness of the population–resource issue sketched out in Chapter 16 seems—with occasional exceptions—either not to have penetrated the collective political mind, or to be considered by politicians too depressing to discuss with the electorate. It may require several large-scale famines that come close to exhausting the food reserves of the world to drive the message home and precipitate the necessary readjustments. For humankind to extricate itself from the complex array of ominous situations with which it is now confronted will require on the part of nations a coordinated, concerted effort heretofore only seen in the face of a major military challenge, one perceived as threatening national survival. With particular reference to the United States, a great paradox is abroad in the land: having on general principles supported the scientific establishment more bountifully than any other nation, we now are so reluctant to accept and act upon its message when, instead of promising the sun, moon, and stars to all, it is unpleasant. We keep hoping for a quick technological fix. In the present situation, the appropriate scientific/learned societies of the various nations could serve a useful purpose by developing carefully crafted resolutions expressing the gravity of the situation.

Each nation must now develop an appropriate Task Force to cope with the situation, a Force whose composition and objectives will be determined by the nation's resources.[1] The only means of coordination at hand is the United Nations. The Task Force of a nation such as the United States will oversee the activities of a number of Working Groups,

each dealing with an appropriate aspect of the problems I have discussed (and related problems left undiscussed), but, because of the
interrelatedness of the issues, these Working Groups must have a degree
of communication and coordination in their activities that has seldom
been achieved in the past. At the international level, the model for
coordinated Working Group activity may be the protocol that grew out
of an international conference in Montreal in 1987, at which a treaty was
signed to slow down the global production of the chlorofluorocarbons
threatening the atmospheric ozone layer.[2] Unfortunately, the treaty has
no provision to prevent an unscrupulous entrepreneur who wishes to
evade the constraints in industrialized nations against manufacturing
chlorofluorocarbons from moving his operation to a desperately poor
(and more compliant) Third World nation.

The alternative to a concerted effort is a drifting, worldwide deterioration in the ecosystem, accelerating as population numbers mount,
until the intra- and international squabbling over the remaining resources results in a tragic parody of the world that could have been.
Even with the most prompt action the nations can generate, the situation will worsen and the tension increase; it is imperative some Task
Force be in place quickly to provide tangible evidence that a policy
exists and is being pursued. It is true that the dire forebodings of many
scientists cannot be rigorously proven, and this could be a convenient
excuse for inaction by "practical" decision makers of all stripes who
realize the immensity of the implied readjustment. Unfortunately, by
the time it is completely clear to the electorate what is transpiring, the
delicate equipose of the complex ecosystem may be catastrophically
disturbed.

I would anticipate the Task Force would include a Genetics Working
Group. What considerations should guide the activities of such a Group,
as society attempts to deal with the complex of issues so briefly mentioned? In suggesting some objectives, it is inconceivable to me to be
parochial; I write in the role of "physician to the human gene pool,"
wherever it is represented. In this self-appointed role I feel very humble:
we still know so little about the composition and dynamics of that gene
pool. In this chapter, I will consider genetic technologies and approaches that are directed at populations however defined, wherever
found. In the next chapter we consider measures directed toward
specific individuals. In restricting myself to genetic considerations, I do
not mean to imply that these are paramount. Rather, among the complex
of problems humanity faces, this is the aspect which I am best qualified
to address.

Genetic Medicine for Populations

Humankind's Genetic Dilemma

That our species is departing extensively from the circumstances under which it came into being has been self-evident for many years. It is a reasonable proposition that some of these departures are certainly, and others probably, not in our genetic best interests. More recently, the surge of discovery concerning the details of our genetic material has revealed the full complexity of structure and function that must respond to these changes, a complexity no scientist of conscience can pretend to understand. Even if we understood this genetic structure and how it functions, we currently lack the knowledge to replace the "natural system" under which our DNA evolved (to which "natural system," of course, there will be no *voluntary* return) with one of our own devising. That "natural system" in its own way was undoubtedly no less complex than the structure and functioning of the human genome is now known to be. Fortunately, given the size of the human gene pool and the evidence that as selective factors change the frequencies of the various genes responsible for human attributes change *relatively* slowly (in political terms), we can be confident that, short of cataclysmic happenings, the species will not change in any major way in the next several hundred years.

With a return to the Yanomama-type population dynamics under which we evolved unthinkable, and not yet able to visualize (let alone implement) a substitute dynamic consistent with the future genetic needs of the species, the best we can expect to devise at present is an interim genetic strategy. It is only one of a set of frequently overlapping strategies that society must devise for the future. To set forth such a strategy at this time may seem both premature and presumptuous in view of our vast areas of ignorance concerning human genetics, but I would hope the principles to be enunciated are sufficiently robust that they will not be perceived in a few generations to be as outdated as we now perceive the eugenic approaches of two generations ago. Basically, I am not proposing to attempt to improve our human genetic resources, but rather to attempt to protect the hard-won wisdom the evolutionary process has imposed on the human gene pool. I will introduce the elements of this strategy in a sequence that correlates roughly with both their feasibility/urgency and their effectiveness, the latter judged not only by improvement in the protection and realization of the human genetic potential per unit financial expenditure but by ethical considerations as well.

In the evolutionary panorama, as reflected in the fossil record, the average mammalian species has persisted something like one million years.[3] (For technical reasons, this is a somewhat soft figure but adequate for our purposes.) Depending on just where, for these purposes, the line separating *Homo habilis* from *H. sapiens* is drawn, our "species" (in the broad sense), at roughly 200,000 years of age, is still relatively young. It is, however, by no means clear that the usual yardstick for species age can be applied to humans. Our intellect—which again I suggest does not differ significantly from that which came into being to deal with the problems of a hunting-gathering society—has now enabled us to devise a culture that not only abrogates the population structure under which the species arose, but rides rough-shod over the complex and poorly understood ecosystem of which we are inextricably a part. It may be later in the human era than we think. In Chapter 11 I have gone to some pains to convince the reader we are not about to be rescued from our dilemma by some mysterious burst of evolution.

In the past, human adaptation to a changing world has contained a large element of trial and error. Relatively little thought has been directed at the long-range consequences of human action. Our species has now so perturbed the system—the potential in our trials and errors is so grave—that unlike any other species that ever lived, we hold our destiny in our own hands. Given our current understanding of the worldwide human condition, not to attempt to initiate major steps to meet the various challenges is by default no less a decision than to attempt to formulate the necessary readjustments. My present concern as to whether humankind can meet the array of challenges runs deep. It is not particularly comforting to find myself in this respect in some distinguished company ranging, amongst others, from Max Born and John Platt to Robert Heilbroner and Lewis Thomas.[4]

Primum Non Nocere

This Hippocratic aphorism, enunciated for the guidance of physicians some 2000 years ago, remains as valid today as then: *Above all, do no harm.* Hippocrates wrote for simpler times, when the results of the limited therapies should have been immediately apparent. Even so, as recently as two and three hundred years ago, the most illustrious physicians of the day failed to recognize the disastrous impact on an already ill patient of violent emetics and purgatives, or repeated bleedings. We are today, we believe, much more sophisticated, but so are our procedures. Even in this century, physicians have learned by painful experience that the gratifying response to treatment with X-rays of a painfully engorged breast in a newly lactating woman or an enlarged thymus in newborn infant carries an increased risk 20 to 40 years later of cancer in these organs. Although Hippocrates wrote for the physician, the aphorism is equally relevant for geneticists oriented toward gene pool problems. It is by design that what follows is generally a program of stabilization/prevention rather than active, gene-oriented intervention; preventionism is always more conservative than interventionism.

With this prologue, let us begin to consider genetic measures directed at entire populations, enunciated as a series of objectives.

Objective 1: An Egalitarian Control of Population Growth

In the clash between the needs of an expanding population and the declining resource base to support even the present population, the human geneticist has little to bring to the resource-base issue. Unless, however, the current generation of geneticists is prepared to retreat from the traditional concern of the genetically oriented for future generations, he should have much to say about population control. The relative success of the crash program to control population growth in Japan in the 1950s (cf. Chapter 6, p. 86) and in China in the 1970s illustrates what can be done. Examples of what a more moderate approach to limiting population growth can accomplish are currently provided by Thailand and Indonesia.[5] As can be deduced from a recent annual report of the United Nations (1990), in the 15 years between the period 1965–70 and 1980–85, the total fertility rate (TFR) rate of the developing countries fell by 30%. This is 47% of the way from the initial TFR of 6.0 to the replacement level of 2.1.[6] On balance, then, fertility rates, except in parts of Africa, Latin America, and the Near East, have been moving in the desired direction, although that trend may now be slowing.[7] The challenge is to accelerate that trend. A recent epidemiological analysis suggests that in the worst affected areas, the current AIDS epidemic in Africa will within a few decades change population growth rates from positive to negative values.[8] The cynical will comment that nature is stepping in to reassert her balance, but the physician who understands the suffering involved must insist civilization has failed if there is no better means of population control. There is in fact a large experience to be drawn on in approaching the control of population growth.

The most obvious immediate manifestation of failure to reconcile population with resources will be an increase in the amount of malnutrition in the world, accompanied by localized famines. The tragedy in the Sahel in the late 1960s and the more recent famines in Ethiopia, Mo-

345

zambique, Somalia, and southern Sudan illustrate the vulnerability of some local populations, especially when the drought is compounded by local administrative/political considerations. Famine is of course an ages-old spectre of mankind. The great famines of the past occurred, however, at a very different state in human history; now to accept famine as a control on population growth—anywhere—is no less an abject abdication of the responsibilities of any truly civilized (as opposed to "primitive") society than is the acceptance of disease as the controlling agent.

The goal for humanity is not to achieve the maximum number of people the world can (limpingly) support, but to regulate our numbers in a manner consistent with a quality existence. King[9] and Martin[10] have written with sensitivity and insight of how the current efforts to bring health to all, the guiding principle of the World Health Organization, may have the unanticipated effect of nutritional misery for many more than if the appropriate public-health measures had not been introduced. King phrases the dilemma humankind faces thusly: "Hitherto, living man has been the measure of all things; now many people are beginning to wonder if this measure should not be the integrity of the ecosystem or the welfare of future communities. Unfortunately, these premises conflict" (p. 666). Subsequently, Martin, in a thoughtful extension of this position presented as a World Health Forum roundtable, suggested (among other items) the pragmatism of reducing mortality rates only as rapidly as the economy could absorb the increased populations. The virulent responses to this suggestion by representatives of the developing nations illustrated how profound is the alienation between North and South, but failed to address the issue of how, given the trends we have been discussing, the basic issue is best met.

What policy should guide our efforts at population control? For years I have argued that anything other than a simple quantitative policy, of the same number of children for each couple, is undesirable and unworkable.[11] Any attempt at a qualitative policy, i.e., varying the number of children according to parental attributes, requires massive value judgments that cannot be supported on social or scientific grounds. This policy in effect would attempt—*for the time being*—to stabilize or even reduce the world's population in a way that leaves the relative proportions of the world's peoples essentially the same, i.e., limits the changes in the human gene pool from generation to generation 1) to those inherent in genetic transmission, i.e., the chance deviations of the genetic attributes of the offspring generation from those of the parental generation, and 2) to those that are the inescapable consequence of whatever biological selection is taking place. No other approach will enlist the necessary international cooperation. I would argue for a goal of two children per couple, with strong social disincentives to produce more. Because of failure to marry, infertility, and early death, the ultimate average achieved family size might be about 1.8 if the world subscribed to the policy. Implementing such a policy will appear to many to infringe on civil and religious rights, but in a few more years the consequences of not implementing such a policy will be so clear that population control will assume the imperatives that control of an epidemic now commands. Even were this two-child policy to be implemented by 1996, because of the present age composition of the world's population, population numbers will increase for at least four decades, peaking in the neighborhood of 6.7 billion in 2030.[12]

The ultimate goal is a population number that the planet can comfortably sustain on renewable resources for the indefinite future. Let us assume that the proposed population policy is miraculously in place by 1996. Even four generations (100 years) of strict compliance with the two-children per couple guideline (realized = 1.8) would, after the abovementioned peak at 6.7 billion, only reduce the population to some 4 billion. Since the size of the population that a worldwide sustainable agriculture policy could support is not yet clear,[13] 4 billion people may still be too large a number for the earth to support indefinitely in comfortable circumstances, and special incentives for one-child families may be in order. The critical factors in the final determination of the world's sustainable population size are the control of soil erosion and of the deteriorating agricultural situation discussed in Chapter 16. It must again be stressed that the proposed stabilization of the human gene pool is best viewed as a long-term "temporary" measure, pending acquisition of the wisdom to approach the subject of human reproduction in a more mature fashion.

Even such an apparently even-handed policy as this will not go unchallenged. For instance, one of the major attempts in recent years at framing the magnitude of the population problem was the report entitled "The Limits to Growth"[14] of the so-called Club of Rome, a loosely organized group of concerned individuals from the industrialized nations. In the words of the president of that group, Alexander King, " 'The Limits to Growth' was seen by many people of the South as an attempt by the imperialistic countries of the North to tell the South to stop making so many children so that we, who have fouled up the world and made all the mess of everything, could go on as we are."[15]

Representatives of some ethnic minorities often refer to efforts to restrict their numerical growth as "genocide." What I take to be the historical basis for this concern is shown in the table on page 348,

Estimates of World Population by Regions, 1650–1950 (United Nations, 1953).

Estimated Populations in Millions

Series of Estimates and Date	World Total	Africa	North America[4]	Latin America[5]	Europe and Asia (exc. U.S.S.R.)[6]	Asiatic U.S.S.R.[6]	Oceania	Area of European Settlement[7]
Willcox's estimates[1]								
1650	470	100	1	7	257	103	2	113
1750	694	100	1	10	437	144	2	157
1800	919	100	6	23	595	193	2	224
1850	1,091	100	26	33	656	274	2	335
1900	1,571	141	81	63	857	423	6	573
Carr-Saunders' estimates[2]								
1650	545	100	1	12	327	103	2	118
1750	728	95	1	11	475	144	2	158
1800	906	90	6	19	597	192	2	219
1850	1,171	95	26	33	741	274		335
1900	1,608	120	81	63	915	423	6	573
United Nations estimates[3]								
1920	1,834	136	115	92	997	485	9	701
1930	2,008	155	134	110	1,069	530	10	784
1940	2,216	177	144	132	1,173	579	11	866
1950	2,406	199	166	162	1,242	594	13	935

[1]Willcox, Studies in American Demography (1940), p. 45. Estimates for America have been divided between northern America and Latin America by means of detailed figures presented on pp. 37–44.
[2]Carr-Saunders, World Population (1936), p. 42.
[3]United Nations, Demographic Yearbook 1949–50 (1950), p. 10; and United Nations, "The past and future growth of world populations . . . " (1951), Table II; the 1940 figures are unpublished estimates of the United Nations.
[4]United States, Canada, Alaska, St. Pierre and Miquelon.
[5]Central and South America and Caribbean Islands.
[6]Estimates for Asia and Europe in Willcox's and Carr-Saunders' series have been adjusted so as to include the population of the Asiatic U.S.S.R. with that of Europe, rather than Asia. For this purpose, the following approximate estimates of the population of the Asiatic U.S.S.R. were used: 1650, 3 million; 1750, 4 million; 1800, 5 million; 1850, 8 million; 1900, 22 million.
[7]Includes North America, Latin America, Europe, and the Asiatic U.S.S.R., and Oceania.

derived from a United Nations publication of 1953. It is an effort to reconstruct the growth of the world's population between 1650 and 1950. In 300 years the total population had increased by a factor of 5. However, let us consider the contributions to this growth of each of the major ethnic groups. To do so, I will commit some oversimplifications. I hope those who attack them will suggest improvements. We will equate the Americas and Oceania in 1650 with American Indians, Polynesians, and Australian aborigines. Even with a generous allowance for hybridization, this subgene pool by 1950 had increased by a factor of less than 2. We will equate Africa with Negro; if one-sixth of the gene pool of the Americas were negro in origin—a generous estimate—then the Negro gene pool had increased by a factor of 2.5. Equating Asia with Mongoloids and East Indians, this gene pool had increased by a factor of about 5. Finally, by this same reckoning, the Caucasoid pool (exclusive of East Indians) had increased (last column) from an estimated 113 million persons to an estimated 935 million persons, a factor of 8. It is the "Western World" and its extensions that are most responsible for the population explosion. The role of Caucasoids in the relatively slow rate of increase of the American Indian, Polynesian, and Australian aborigine is a sordid and sickening record most of us prefer to suppress. With respect to the historical role of Caucasoids in inhibiting the rate of growth of the Negro gene pool, there is no scarcity today of eloquent spokesmen, to whom I will leave the question.

Representatives of ethnic minority groups the world over will also claim that, to the extent they suffer higher infant and childhood mortality rates, this uniform policy would be to their disadvantage. Although on the surface this appears to be a legitimate objection, it should be pointed out that because the base of the age pyramid is so much broader in respect to most minority groups than for the national majority groups, i.e., because in general there are relatively more young people in minority groups, these groups should for the next several generations grow in numbers more than the majority Caucasoid populations of the Western World, even with slightly higher mortality rates.

It is clear that the percentage composition in terms of ethnic groups of the world's gene pool today is very different from that of 300 years ago, and any stabilization of population numbers will tend to perpetuate that shift. However, with all due respect for the history involved, I see no way to turn the clock back. Moreover, simply because of the distribution of resources, any failure to limit population growth now is apt to have the most dire consequences for those whose numbers have increased most slowly. I can see no alternative but to live with history and attempt to forestall the even greater inequities that might result from failing resources whose loss will be felt most keenly by groups already the most marginal. I would term this approach the Noah's Ark principle. Let us be sure we protect the diversity of the world's people. And if, in finding a biblical parallel, I can make intellectual contact with some of the religious groups so resistant to population control, so much the better.

The difficulties in the implementation of this policy are formidable, but if the successes of the past mentioned earlier (see also note 16) could be achieved with the relatively crude measures then available, how much better the prospects with the recent, improved low-dosage oral contraceptives and the advent of long-acting means of fertility control, such as levonorgestrel implants (the Norplant System) for women and, still in the experimental stage, pyremethamine and testosterone derivatives for men. It now seems clear that the occasional complications attendant on the use of the new intrauterine devices are less than the complications of pregnancy over the same period of time. Public education as to the use of condoms as part of the effort to control the AIDS epidemic should result in an increased availability and use of this device. Male sterilization through ligation of the vas deferens is a very simple surgical procedure. Furthermore, with respect to the failures that will occur in any birth-control program, the development of synthetic steroids with a high affinity for the progesterone receptor, such as RU-486, renders very early pregnancy termination possible with much less inconvenience than the traditional (later) abortion.[17] It is a shocking commentary on the political acquiescence to opinionated minorities, that antiabortion groups have thus far thwarted the introduction of RU-486 into the U.S. market, despite its endorsement by the American Medical Association, the American Public Health Association, the American Association for the Advancement of Science, and the American Institute of Biological Sciences.[18]

The availability of such a drug as RU-486 should go far to dispel the worldwide disgrace—in which religious leaders and antiabortion lay groups are willing accomplices—of an estimated 10 to 20 million unsafe, illegal, "back alley" abortions that claim roughly 200,000 lives annually, with permanent gynecological damage in as many more.[19] However, the availability of such abortifacients will not obviate the need for continuing access to legal abortion. Since the U.S. Supreme Court's Roe vs. Wade decision in 1972, deaths from abortion have dropped by 90%. For instance, in the U.S. in 1958, there were an estimated 314 deaths from illegal abortions, as compared with 6 deaths in 1985; these 6 occurred on a base of an estimated 1.6 million abortions. This lower mortality derives primarily from ready access to skilled obstetricians and gynecologists.

Were access to abortion again to become severely restricted, we can anticipate a return to the earlier mortality figures.

True, with strong social pressure for birth control there is an implied loss of personal discretion, but I have already argued that the present situation is for countries the social equivalent of a major military emergency, in which the very existence of a country is perceived as threatened, with some loss of personal liberties to be borne for the sake of the future. There will undoubtedly be meaningful differences in the speed with which nations can implement such a program, either because of problems of communications in remote areas or because of religious conditioning. How is compliance assisted and ultimately assured? If noncompliant nations develop food shortages, what is the responsibility of nations in compliance to come to their aid? Should compliant nations prohibit immigration from noncompliant nations?

Some years ago Garret Hardin, alarmed by the population forecasts we have been discussing, suggested that each country needs to think of itself as a group of survivors in a lifeboat, afloat on a stormy sea after a maritime disaster, with no sure hope of salvation save through its own efforts.[20] The less-developed nations interpreted this as a policy of abandonment, that having exploited and impoverished them, the more developed nations should now simply walk away. When Hardin spoke at the University of Michigan, he was picketed, and, as the person delegated to introduce him, I had a few problems with audience control (not to mention, guilt by association). The lifeboat analogy is dramatic, and makes its point. With my experience at the outbreak of World War II (p. 8), it was especially poignant. Certainly the less-developed countries need to make the maximum effort at reconciling population with resources, and very soon. Clearly the U.S. and the other developed nations do not have the resources—nor if they did, should they expend them—to attempt to mitigate all the suffering ahead if there is no fundamental change in world population policy. But a literal lifeboat policy is neither ethical nor practical. Not ethical, because the developed countries do bear worldwide responsibilities. Not practical, because hungry and desperate people can be stopped at national borders only by a carnage that would signal the end of human compassion. Like it or not, we're all in this together.

It will be convenient for the industrialized nations to view the issue of population control as primarily one of the Third World. If, however, the real problem be conceived as one of bringing the world into a better energy balance, then the need for the industrialized nations to curb their consumption of energy is no less pressing than the need for the Third World to exercise restraint in population growth. Indeed, it will be necessary for the U.S., in particular, to control its wasteful use of energy if it is to assume a real leadership role in the adjustment ahead. If in this presentation I emphasize the population side of the question, it is because others can speak to the energy issue so much more authoritatively.

The variety of social readjustments required by the twin concepts of a "steady-state" population characterized by zero population growth and an ecosystem in reasonable balance are enormous, but beyond the scope of this book. I would not like to appear to underestimate them. To mention only one, such a program can temporarily (several generations) enormously increase the gerontocracy issue mentioned in Chapter 17. Unfortunately, the issues created by a continuation of present trends appear to be even more recalcitrant. To me the most hopeful recent sign of a political recognition of the ecological dilemma and the actions it demands is the appearance of Senator (now Vice President) Al Gore's superb book, *Earth in the Balance*.[21] It is a chilling commentary on the myopic White House "leadership" of the past decade with respect to this issue that when Gore became the nominee for the Vice Presidency, the spokeswoman for the Bush–Quayle reelection effort stated that the Republican campaign saw the book's pointed recommendations for new taxes on coal and oil, more foreign aid, and more governmental action to curb global environmental deterioration as weapons to be turned on Gore![22] The substantive issues raised by the book were not mentioned by the White House during the presidential campaign, other than through the oblique and pejorative identification of Gore as "Mr. Ozone." The readjustments to come imply "downsizing" a wide variety of materialistic human activities. I think of one of the sparsely elegant *haiku* of the 17th century Japanese monk–poet Basho:

> My house burned down
> Now I can better see
> The rising moon.

Objective 2: Euphenics

A proposition with which human geneticists have not established sufficient identity is the optimization of genotype expression in contemporary and future society. In my 1961 Harvey Lecture I referred to this as "culture engineering," as contrasted with the "genetic engineering" of the eugenicist (and now potentially of the molecular geneticist). A few years later Lederberg[23] suggested the term "euphenics," which I find more felicitous.[24] Both terms embody the concept of consciously shaping the milieu in which humans function, to ameliorate the expression of all our varied genotypes (as contrasted with the permissive attitude toward a pell-mell rush to change that characterizes most modern cultures). The challenge of euphenics is to ensure that each individual maximizes his genetic potentialities. In a sense, we are revisiting the heredity–environment issue of past generations, but now, in the light of knowledge of the extent of genetic variation, viewing the question as an interaction rather than a dichotomy.

This effort at optimization extends to the mind as well as the body. We consider first euphenics directed at physical traits, which may be practiced at either the individual or the population level. With respect to the latter we need constantly to remind ourselves that human beings evolved under very different circumstances than prevail over much of the world today, and that there has been inadequate time for whatever genetic adaptations to "civilization" are appropriate to take place. Our long evolutionary history as a gatherer and scavenger, who may have begun to profit significantly from hunting only 100,000 or 200,000 years ago, leaves us remarkably omnivorous, ready to eat at any time, but not well adapted to highly refined (concentrated), low-fiber foods. We also acquired food by virtue of a rather active lifestyle. Salt in large quantities was not a component of our primate-like diet. Chapter 9 documents in part just how greatly diet and lifestyle have changed with civilization.

The prevalence of obesity, diabetes mellitus, atherosclerosis, certain cancers, and hypertension in the more affluent societies—rare diseases in tribal cultures, even when allowance is made for the age differential— is a manifestation of indulgence in a diet and lifestyle for which evolution has not prepared us. Some years ago, studies in collaboration with Stefan Fajans and Jerome Conn revealed that long before the kind of diabetes that overtakes older persons usually comes to the attention of the physician, there are already subtle changes in how those at high risk for diabetes tolerate a sugar load (i.e., a glucose tolerance test).[25] This, plus the apparent absence of diabetes in unacculturated tribal groups versus its high frequency in reservation-dwelling American Indians (see Chapter 9) led me to what is now known as the "thrifty genotypes hypothesis."[26] I suggested that in hunting–gathering times, when meals were irregular and sometimes infrequent, it was important to have a quick insulin trigger, i.e., to release insulin promptly after a meal, to get the full benefit of the glucose that resulted from that meal. Now, however, that trigger in the genetically predisposed who overindulge in high sugar content foods and become obese is being so excessively stimulated that eventually the body responds in a variety of unfavorable ways that together result in diabetes. What was an adaptive mechanism has in affluent societies become a liability. If the hypothesis is correct, it provides one of the better examples of the biological price of progress.

The appropriate genetic studies to date have suggested that there is a broad range in genetic susceptibility to the diseases mentioned in the preceding paragraph. This finding invites debate, as to whether it would be more effective in attempts at reducing these diseases to promulgate measures to be adopted by total populations, or to identify especially susceptible individuals who are then individually encouraged to adopt a particular lifestyle. I will take the position that the most effective course at present is to adopt recommendations applicable to the entire population but to reinforce these recommendations with regard to individuals with particular predispositions as these individuals come to attention through various screening procedures. The identification of these individuals, heretofore largely dependent upon family history and a medical evaluation, may in the future be greatly focused by the DNA technologies which, guided by the family history, will then contribute to establishing which young members of a pedigree have inherited the "susceptibility" genes now being identified by geneticists. Unfortunately, from the standpoint of public acceptance, implementation of the indicated measures to reduce these diseases would often require a personal (and social) discipline that many would find onerous. It is convenient to think of these diseases of civilization as an inevitable price of the twentieth century affluence of some societies, to be miraculously managed by one's physician should they develop.

The literature on the specific details of the relationship between diet, exercise, and health, and how to modify the first two to human advantage, is enormous.[27] Although controversy abounds, there is consensus on a number of points. For instance, there is clear evidence that a high fat intake, especially of the so-called saturated fatty acids composing so much of animal fat, increases the cholesterol level and thus the risk of atherosclerotic cardiovascular disease of the brain, heart, and kidneys. This same high fat diet probably increases the risk of cancer of the colon, prostate, and breast. On the other hand, diets high in "bulky" plant foods—fruits, vegetables, legumes, and whole-grain cereals—are associated with a lower occurrence of coronary heart disease and cancers of the lung, colon, esophagus, and stomach. Diabetes mellitus is also less common in populations consuming such diets. High salt intake is almost certainly a factor in the development of hypertension, at least in some individuals. Regular physical exercise is associated with a decreased risk of coronary heart disease. The obesity (defined as 20% above "ideal" weight) affecting at least 20% of North Americans is accompanied by diabetes mellitus, hypertension, cancer of the uterus, gall bladder disease, and degenerative arthritis. Over a 30-year period, nonsmoking men whose weight was 20% or more above normal, as determined by the Metropolitan Life Insurance Company tables, had a relative risk of dying three to four times greater than that of men of desirable weight (100–109%). Call it euphenics or call it common sense, there is now little room for argument with the proposal that health (i.e., genotype expression) would be substantially improved by a diet and exercise schedule more like that under which we humans evolved.

It is important to recognize that these disease associations are statistical. No one suggests that an improved diet will eliminate these diseases, but it would lessen their impact. This lessening might come through what has been termed a "compression of morbidity," resulting from the fact that with onset of illness delayed but average life expectancy (i.e., onset of terminal senescence) unaltered, the time available for *chronic* disease is foreshortened.[28] An additional fraction of people would enjoy good physical health longer. The reverse side of this coin is that more people will live into the age when senescence phenomena take over. Thus, in meeting one socially desirable goal, a second social problem, an excessive number of senior citizens, is exacerbated.

In addition to these common genetically conditioned susceptibilities whose expression is so environment-dependent, there are also rare individual susceptibilities, which suggest that certain individuals should observe particular idiosyncratic constraints not shared by the general population. Again the first steps have been taken in the observance of such constraints. A suitable example is provided by the measures now in place to accommodate genetic defects in the sex-linked gene encoding for the enzyme glucose-6-phosphate dehydrogenase. This is the most common enzymopathy known, estimated to affect 400 million people worldwide. When males who carry the gene on their single X-chromosomes, or the less common females homozygous for the same gene, are exposed to a number of drugs, contract certain infections, or eat fava beans, red blood cell destruction and an anemia result. The drugs involved include some antimalarials, some antibiotics, and some antipyretic–analgesics. The defect is especially common in Black males and males of Eastern Mediterranean origin (Italy, Greece, Iraq, Iran, etc.). Like the sickle-cell trait, the condition is thought to have achieved its high frequency by conferring protection against malaria, but the evidence is less clear than for sickling. It is now sound medical practice to test for the occurrence of this defect in males of these ethnic origins, prior to the administration of these drugs, and employ alternative medication for those who test positive. This is a kind of euphenics.

The euphenic potentialities with respect to the functioning of the mind may be even greater than with respect to the body. (I really do not subscribe to a mind–body dichotomy, but it is at this juncture convenient.) In a society confronted with the array of problems on which I have commented so briefly, and where the intelligent application of technology is so vital, each mind will need to function at its maximum capacity. The potential of a society to produce thinking, problem-solving minds is unknown. The example of Ancient Greece is both provocative and haunting. Galton (1892) in *Hereditary Genius*[29] was one of the many who have drawn attention to the population base underlying the flowering of Greece. Between 530 and 430 B.C. the district of Attica produced 14 persons whom he would classify as illustrious (genius or near genius), namely:

> Statesmen and Commanders: Themistocles (mother an alien), Miltiades, Aristeides, Cimon (son of Miltiades), Pericles (son of Xanthippus, the victor at Mycale)
> Literary and Scientific Men: Thucydides, Socrates, Xenophon, Plato
> Poets: Aeschylus, Sophocles, Euripides, Aristophanes
> Sculptor: Phidias

During that period Attica's population amounted to about 90,000 native free-born persons, 40,000 resident aliens, and a laboring and artisan population of 40,000 slaves. The abovementioned 14 persons were all drawn from the native free-born, who over a period of a century should amount to about 270,000 persons, or about 135,000 males, of whom only one-half would survive to the age of 26, and one-third to the age of 50. No matter how we criticize the sampling or the scoring process involved, or whether the estimate of population base is too low by a factor of 2, clearly here was an extraordinary flowering of ability. To Galton this was a genetic phenomenon, which somehow quickly ran its course. Aware as we now are of gene frequencies and slow rates of genetic change, surely we must see this extraordinary manifestation of human capabilities as in whole or part the result of the creation of an unusual intellectual environment for the human mind. It is sobering how little we really know about the hereditary–environment interaction in the realm of the mind.

The espousal of euphenics at a time when such profound readjustments are in the offing, will strike some as indulgently idealistic. In fact, this is perhaps a prime time for euphenic considerations. In the U.S., with the cost of medical care annually mounting more rapidly than the increase in the GNP, the public and the political establishment are receptive to "cultural" modifications that will reduce the frequency of atherosclerosis, diabetes mellitus, stroke, etc. Economic developments alone may reduce the consumption of (fatty) meat, and further pressure toward euphenic practices will also come from the diminishing agricultural base and the secondary consequences of an affluent diet on the environment. At present almost half the energy used by U.S. Agriculture eventually is devoted to the livestock sector. Amazingly, in the U.S., over one-third of all raw materials consumed (including fossil fuels) are directly or indirectly devoted to the production of livestock. In the feedlots of the world, it requires about 6.0 kilograms of grain (including soybean meal) to produce 1 kilogram of pork, and 4.8 kilograms to produce 1 kilogram of beef.[30] These feedlots and cattle ranges contribute to environmental degradation, since the animal wastes of the feedlots have become a major source of water pollution, and the methane released by livestock contributes importantly to the Greenhouse Effect (cf. Chapter 16, note 5). As the number of people has doubled since midcentury, so has the number of four-legged livestock. To move the world's current population of 5.4 billion people to the American-style diet to which they aspire would require $2\frac{1}{2}$ times as much grain as the world produces for all purposes. It simply isn't going to happen. Unfortunately, poor countries in which malnutrition is rampant export grain as a "cash crop." Were the industrialized nations to cut back in their consumption of animal protein, either acting on principal or stimulated by a faltering economy, one immediate result would be serious economic dislocation for these countries. This is but one small example of the complexity of the readjustment implied in moving back to a greater consumption of grain products. But, if the world does evolve to a relatively greater consumption of grain, in what form are these grain products best consumed? Specifically, is it not sound euphenics to encourage the use of grain products that have been minimally refined, thereby retaining both the bulk and the total mineral and vitamin content of the grain? (Now we buy the bulk back, as bran, and add it to the diet, and get our vitamins in pills.)

The greatest challenge to euphenics is with respect to intellectual functioning. Not only is unequal access to education basically unfair (and not encountered in tribal cultures), but it results in underutilization of the most important asset society has as it faces ahead: brain power. This aspect of euphenics is well served by the current efforts, the world over, to broaden and improve the educational systems. Up to this point in time, cultural evolution has unfolded with much more thought to material than biological well-being; now the data are coming in that for the first time permit proper consideration of the biological factors.

The application of euphenic principles is best begun early—the earlier the better.[31] The primary need is during childhood and adolescence, when health and mental habits that will persist for life are formed. Indeed, the practice of euphenics begins prenatally. Given the sometimes lasting impact on physical and mental development of prematurity/low birth weight, a strong effort directed at ensuring that each fetus enjoys a full 40 weeks *in utero,* in a normal (drug-free) environment, might have more impact on the general level of functioning of the gene pool than much of the gene therapy we will be discussing later, and be much more cost effective. The allocation of euphenic resources to youth in the face of the demands of the gerontocracy promises to be a very troubling issue in the years just ahead.

Many of my molecularly oriented colleagues in genetics would not recognize much of what I term euphenics as a legitimate concern of genetics. In fact, the genotype–phenotype interaction has been a prime focus of interest in classical genetics, now temporarily shunted to one side by recent genetic enthusiasm for precisely defining and then manipulating the genotype. Molecular geneticists need to be reminded that they have a vital role to play in euphenics, in bringing critical definition to the tremendous variation of genotype in the genotype side of the genotype–environment interaction.

Objective 3: Keeping Mutation Rates as Low as Possible

The most obvious way to hold mutation rates down in through control of exposure to mutagens. In the preceding chapter, I have described in some detail how in the late 1950s and early 1960s the current guidelines for exposure to ionizing radiation were reached. The subsequent recommendations of the U.S. Federal Radiation Council and the International Commission on Radiological Protection were that members of the general population should not be subjected over a 30-year period to more than 5 *r* units of radiation (0.05 Gy) to the gonads from nonmedical sources, with the further recommendation that actual exposures be kept as low as reasonably achievable. These 5 *r* are, of course, over and above natural, background exposures. In point of fact, the *average* of the various nonmedical exposures to which the individual members of the population of the United States are subject is currently estimated to be more like the equivalent of 0.3 *r* (0.003 Gy) over a 30-year period.[32] This does not include radon exposures. The treatment of the subject by the National Academy of Sciences Committee of 1955–1956, and by the various governmental agencies mentioned above that followed, all accepted the hypothesis that there was no threshold to the genetic effects of radiation, so that even these low additional exposures should entail some genetic damage.

Since those recommendations were reached, there have been the two important scientific developments, already discussed in Chapter 13. One is that the genetic doubling dose for humans of chronic, ionizing radiation is, on the basis of the studies on the children of A-bomb survivors, very likely some three to five times higher than was assumed when these recommendations were promulgated, with a corresponding decrease in risk. The other is a growing body of knowledge on the complex biochemical mechanisms by which the cell repairs genetic damage. These defenses explain why the mutagenic effects of radiation are less when a given dose is delivered chronically: the repair mechanisms can better keep up with the damage. Whether there is some (very low) rate of damage for which repair is complete (or almost complete) is a moot point: the conservative course is to assume not.

I regard it as a piece of relatively good news for society that our genetic material does not appear to be as susceptible to the mutagenic effects of ionizing radiation as was at one time feared. In view of these developments, an argument could be made to relax the current recommendation concerning the genetically permissible exposures to radiation. However, although the genetic risks of radiation exposures appear to be less than surmised some 40 years ago, the carcinogenic risks of radiation appear somewhat higher.[33] These risks now dominate thinking about the protection of the public; as argued in the preceding chapter, observance of standards of exposure based on cancer risks should be quite adequate as regards genetic risks.

Also, as argued in the preceding chapter, the situation is much more ambiguous with respect to the results of current exposures of humans to chemical mutagens. The subject is so much on the public mind that studies as described in the previous chapter would seem to be a prudent aspect of any comprehensive genetic program for the future. As noted earlier, my best guess is that at currently regulated levels of chemical exposures, there is no more of a genetic problem than there is with respect to the regulated levels of radiation exposure. However, one aspect of a physician's activities, even, and perhaps especially if he is gene-pool oriented, is reassurance. We need the studies of the effects of chemical mutagens on humans described in the last chapter.

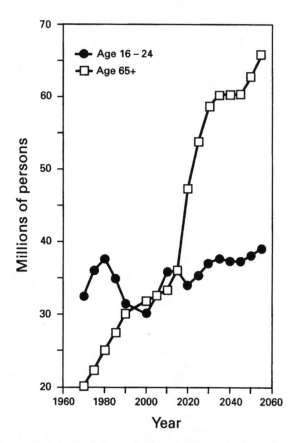

The changing age structure of the United States, more or less characteristic of all industrialized nations. For the foreseeable future, the age group 16–24 will remain rather constant in numbers, whereas the number of senior citizens (greater than age 64) will increase dramatically.

20

Genetic Medicine for Individuals

The previous chapter dealt with genetic measures for populations. This chapter will consider the counterpart genetic measures for individuals. In fact, the dichotomy is not as clean as I make it out to be, but the distinction is convenient.

Objective 4: The Provision of Genetic Counseling and Prenatal Diagnosis

Assuming that major effects to limit population growth will be initiated worldwide within the next decade, it follows that the ever-present desire of all potential parents for normal, healthy children can only be increased in a society where parents are attempting to limit themselves to two offspring. Genetic counseling coupled with prenatal diagnosis and elective abortion where indicated is for the foreseeable future the major approach to meeting these parental concerns and reducing the burden of chronic genetic disease on parents and society (which in most countries is assuming the costs of these genetic diseases).

As noted in Chapter 2, until some 30 years ago, the primary role of genetic counseling was to provide risk figures to prospective parents with respect to the likelihood of a defective child, on the basis of which the parents could plot a reproductive course. This is still the primary function of counseling, but now the geneticist has at his disposal an armamentarium that renders counseling much more dynamic, aggressive, and satisfactory to the patient. At the same time, the medical, ethical, legal, and social problems of counseling have become more complex. In a variety of situations, the birth of a child afflicted with a specific severe genetic disease can be anticipated at an early stage of pregnancy, when a therapeutic abortion is feasible. Were a two-child family to become a social objective, prenatal diagnosis coupled with abortion when indicated can offer parents of certain genotypes, who wish to be monitored, the near-certainty of two children free of the particular disease in question. Let us briefly consider six illustrative (but by no means exhaustive) examples of the types of situations in which "counseling" takes on new meaning, in ascending order of complexity and descending order of present feasibility.[1]

1. *Down syndrome,* the chief manifestation of which is mental deficiency, is usually due to a supernumerary chromosome 21, which may have been contributed by either the father or mother (but usually the mother). This supernumerary chromosome results from a chromosomal accident (nondisjunction) during meiosis. *In utero* diagnosis is accomplished by either placental (chorionic villi) biopsy or aspiration of amniotic fluid by means of a procedure called amniocentesis, followed by culture of the biopsy tissue or of the cells floating free in the fluid, as the case may be, and cytologic study of the resulting cells some two weeks later. The former procedure becomes practical as early as a gestational age of nine weeks, and the latter at 12 weeks, so that, allowing time for the appropriate studies, the diagnosis can become available as early as 12–15 weeks of gestation, when abortion carries a minimum of risk. Except under unusual medical circumstances, abortion for genetic indications is not performed after the 20th week of pregnancy.

The frequency of the syndrome is between 1.0 and 1.5 cases per 1000 newborn infants, but the probability of occurrence is strikingly related to the age of the mother: at maternal age 35, the frequency is 1 in 300, at age 45, 1 in 100, and over maternal age 45 it is 1 in 50. In most populations about 11% of all pregnancies occur at a maternal age greater than 34; monitoring this 11% would detect about 27% of all the fetuses with this syndrome, and such monitoring is now considered good obstetrical practice. Indeed, in the current litigious atmosphere, the physician who fails to recommend amniocentesis to an "older" pregnant woman who subsequently delivers a child with Down syndrome will almost surely find himself subject to a medical malpractice action. The practical issue is what age constitutes "older," and the tendency is to extend the procedure to ever younger mothers. This raises issues of cost-effectiveness in the practice of genetic medicine, issues to which I come later.

2. *Hexosaminidase A deficiency (of which Tay-Sachs disease is the best known manifestation)* is a recessively inherited degenerative disease of the central nervous system with, in the most severe form, onset

about age six months and death usually between ages two and four, after a prolonged and distressing course. Most children with Tay-Sachs disease are the offspring of Ashkenazi Jews of East European origin, in which group it has a frequency of about 1 in 4000 births. The carrier (heterozygous) state can be detected by the demonstration of half-normal levels of the enzyme hexosaminidase A in blood serum or leukocytes. More recently, it has become clear that only 3 different mutations in the hexosaminidase gene account for 95% of the mutant alleles in the Jewish population. Carriers of some one of the alleles responsible for Tay-Sachs disease have a frequency of about 1 in 30 in Ashkenazi Jews. The cause of the high frequency of this disease in this ethnic group is unknown.

Since, for this disease as well as the next three diseases to be discussed, a carrier state can be defined, one strategy of disease control would be population screening for carriers, at least in high risk groups, with the identification of marriages that might result in affected children. Such couples then have the option of refraining from reproducing or creating a family through adoption. This strategy, which in theory would avoid the troublesome issue of abortion, is at present simply not economically, emotionally, or sociologically valid. A much better alternative strategy is early prenatal diagnosis of the disease followed by abortion. The commonly pursued preventative approach to hexosaminidase A deficiency in high-risk populations consists in testing pregnant women for the occurrence of the carrier state on the occasion of their first prenatal visit. If a women tests positive, her husband is then tested. If both are positive, there is a 1 in 4 probability that the fetus will have the disease, and amniocentesis with culture of the cells is indicated. The prenatal diagnosis of Tay-Sachs disease requires the demonstration of the critical enzyme deficiency in the cultured cells or the DNA defect. Between 1969 and 1992, here in the United States, 1633 fetuses were screened for the disease, among whom 307 were found to be affected (M. Kaback, personal communication). All of these latter pregnancies were terminated except for a few in whom the diagnosis was not reached prior to the twentieth week of gestation. If the first visit to a physician occurs shortly after cessation of menses is certain, the diagnosis can be made sufficiently early for a therapeutic abortion to be performed by 16 weeks of gestation. The program has resulted in a 95% decrease in the incidence of the disease in the United States. A sometimes overlooked aspect of a program like this, directed at a recessively inherited disease, is that the fetus will usually be found to be normal, with the resulting parental reassurance, and if, on the other hand, the fetus is abnormal and the pregnancy terminated, the parents can if they desire quickly proceed with another pregnancy.

3. *Sickle-cell anemia,* discussed at some length in Chapter 3, is a severe, chronic, recessively inherited anemia largely confined to Negroes. In African-Americans, the disease affects about 1 child in 1000; carriers of the defective allele have a frequency of about 1 in 14. In Africa below the Sahara, the frequency varies according to the region but if all areas are averaged, then the frequencies of affected and carriers are similar to those in the U.S. The carrier condition for the disease (sickle-cell trait) can be detected with a simple sickling test or by the electrophoresis of hemoglobin. The detection of the disease in the early fetus is complicated by the fact that, as discussed in Chapter 14, the fetus does not elaborate the type of hemoglobin ($\alpha_2\beta_2$) in which the defect is manifest until about the sixth month of gestation; before that, the fetus makes a type of hemoglobin in which the defect is not manifest. However, a relatively simple, very specific test on the DNA of cultured cells derived from the fetus of carrier parents will identify those fetuses destined to develop the disease, at a stage in development when abortion is still readily feasible. In fact, a recent development in molecular genetics is a direct test of cells from a placental biopsy or obtained by amniocentesis for the predisposition to the disease (without the intermediate step of cell culture), thus enabling the diagnosis of disease to be made some two weeks earlier than when cell culture is employed.

4. *Thalassemia major* due to defects in the β-globin chain (the type we studied in Rochester) is especially prevalent in the eastern Mediterranean Basin, where it affects about 1 child per 2000 births, with a carrier frequency of about 1 in 20. The disease is even more common in Thailand. Its high frequency in these countries appears to result, as in the case of the sickle-cell trait, from the protection the trait confers against malaria; it is a balanced polymorphism. As for the sickle-cell trait, the detection of carriers is relatively "low-tech," requiring a simple hematological examination. Unfortunately, since, as just noted, the β-hemoglobin chain is not made early in fetal life, the potential diagnosis must depend on studies of the fetus' DNA, and, unlike the sickling phenomenon, specific to a single DNA defect, no less than 90 different defects in the β-globin gene capable of causing thalassemia have been identified. There is thus at the DNA level no single test for the disease. Fortunately, from the diagnostic standpoint, only two of these 90 different known thalassemia alleles identified on the basis of worldwide studies account for 50–75% of all the affected genes in the Mediterranean region. Therefore, testing for homozygosity for just these two genes would identify between 25% and 56% of affected fetuses, and the efficiency can be increased by adding to the screening procedure (at increasing cost) tests for additional known thalassemia genes in de-

creasing order of their frequency. What can be accomplished with a determined effort at prenatal diagnosis and abortion in a high frequency area is illustrated by the experience of Sardinia and Cyprus. In Sardinia, where one in seven individuals is a carrier of one of the defective β-thalassemia alleles and one in 250 births was a child with thalassemia major, a determined prenatal screening program has resulted by now in a decrease of 95% in the birth of children with the disease. The situation there is particularly favorable for prenatal diagnosis because 95% of the defective alleles result from the same mutation. A similar program in Cyprus, where the frequency of β-thalassemia is essentially the same as in Sardinia, has already resulted in a 70% decrease in the birth of children with thalassemia major. Both programs are overwhelmingly cost effective.[2] Genetic counseling and prenatal diagnosis have led to similar but less striking decreases of the disease in Greece and Italy.

5. *Cystic fibrosis* is a recessively inherited disorder resulting in a severe deficiency in the secretions of the pancreas and the tracheal cells of the lung. With a frequency of 1 in 2500 American Caucasian newborns, it is the most common, serious, recessively inherited disease of Caucasian populations. The disease is much less common in African- and Asian-Americans. Unlike the three diseases just mentioned, there is no simple biochemical or hematological test for the carrier state. The diagnosis of the carrier state must be made at the DNA level; thus far well over 200 different disease-causing gene defects have been recognized. One particular lesion is found in some 70% of the defective genes in a North American Caucasian population, and there are presumably other predominant genetic lesions in other populations. In contrast to the preceding two diseases, where, as we have discussed, the carrier state is thought to confer some protection against malaria, the cause of the relatively high frequency of the cystic fibrosis allele in Caucasians remains unresolved.

The multiple possible strategies for disease prevention in this situation are currently under active discussion.[3] To begin with, there is a considerable spectrum in the severity of the disease, more so than in the diseases just discussed; and some patients, with current treatment, live near-normal lives, although the average life expectancy of a newborn infant with the disease still does not exceed about 40 years. The case for abortion is not so compelling as for the four diseases just discussed. Given the decision to limit the birth of children with the disease, however, the least ambitious strategy to lower the frequency of the disease is to intervene after the birth of an affected child. Assume society has accepted the two-children-per-family concept. Among first born-children, 1 in 2500 will exhibit the disease. If, following the birth of such a child, the genetic defect in each of the carrier parents was characterized, it would be possible to monitor the next (and presumably last) pregnancy for a recurrence of the disease, with the expectation that 1 in 4 fetuses would be found afflicted and the pregnancy could be terminated if so desired by the parents. Unfortunately, a little arithmetic reveals that only 1 in 8 of the affected infants to be born in the population will be detected by this approach. This is a *de minimus* strategy, applicable to all recessively inherited diseases that can be diagnosed *in utero*.

The care of a child with cystic fibrosis is prolonged, expensive, and emotionally demanding. Many geneticists and informed parents are ready to consider more aggressive measures to anticipate the birth of an affected child. Carriers for any one of the many disease-producing defects in the gene have a frequency of about 1 in 25 persons. The least demanding use of the ability to detect carriers is to survey as many of the genetic relatives of a child with cystic fibrosis as possible for the carrier state for the specific allele(s) causing the disease in the child, and then in due time screen the spouses of those found to carry the gene, to identify marriages prone to produce affected children. Most important to the strategy are the unaffected siblings of affected children, 2 among 3 of whom will be carriers; each of the latter has a 1 in 25 chance of marrying a carrier. Thus, on the average, 2 in every 75 marriages involving the normal siblings of patients should have the genetic background to produce affected children. Again, unfortunately, a little more arithmetic of the type in the last paragraph reveals that this strategy, even if pushed very hard (to more remote relatives of the patient), will not detect more than about 10% of the potential cases of the disease.

At the other extreme in the application of genetic screening to this disease, some geneticists visualize screening all pregnant Caucasian women at the time of their first prenatal visit, for heterozygosity for the most common defective allele (the 70-percenter). If a woman tests positive, then, just as for Tay-Sachs disease or sickle-cell anemia, her partner should be tested for the carrier state for the same allele. If he is positive, prenatal monitoring could be initiated; because diagnosis can be made directly from sampling chorionic villi, the disease can be diagnosed at some 12 weeks of gestation. If testing is limited to the most common allele, then about 49% of the couples capable of producing an affected child will be identified. If the testing of both parents were extended to, say, the 16 mutations most commonly associated with cystic fibrosis, then about 88% of carriers and 77% of the couples capable of producing an affected child could be detected, but this greater efficiency is at a considerable increase in screening costs.

The debate concerning the optimum strategy for the prevention of this disease will undoubtedly continue for the next decade. Even without a genetic policy, a concerned individual can request screening. Among the facilities offering screening, the number of cystic fibrosis alleles included in the test varies widely, but as an example, one commercial genetic service in the U.S. now offers screening for the 16 most common mutations, at a 1993 cost of less than $100. Furthermore, in the latter case testing is based on cells derived from the lining of the cheeks, i.e., does not involve a venapuncture.

6. Still a different strategy is indicated for certain dominantly inherited disorders, such as *Huntington chorea,* with a frequency of 1 to 1.5 cases per 10,000 persons (see Chapter 12). A particularly troublesome aspect of this disorder from the standpoint of genetic management is that most persons who develop the disease have already completed reproduction before the first symptoms of the disease appear. At this writing, the gene for Huntington chorea has just been identified. Short of a cure based on the new understanding of the structure of the defective allele, control of the disease can follow from identifying the precise genetic lesion in each family in which the disease occurs and then testing all of the young, apparently normal children of an individual with the disease for the carrier state, and, when the carriers are identified, monitoring their unborn children for the presence of the Huntington chorea gene, with, if so desired, abortion of the 50% of fetuses who would be carriers. Because the disease is dominantly inherited and the onset of the disease is usually after reproduction is complete, so that selection against the disease is negligible, the great majority of cases occur in rather extensive pedigrees. If it develops that many of the pedigrees of Huntington chorea share the same genetic lesion, then a DNA "probe" developed for one specific family might have broad applicability to the diagnosis of carriers in other families. The ability to detect carrier fetuses at a very early stage carries the potentiality for virtually eradicating this disorder.

Some Practical Considerations

Four of the six examples of prenatal diagnosis that have just been discussed draw upon the techniques of molecular genetics, i.e., are examples of the rapidly emerging field of "DNA diagnostics." There is no doubt that this area will constitute one of the principal applications of molecular genetics to humankind's medical problems. DNA diagnostics will in the future permit screening for a wide variety of genetic carrier states and disease predispositions. These developments are, however, creating a number of issues currently under active debate, issues in addition to those already touched upon. Much of the debate involves the question of cost-effectiveness. Cost-effectiveness is not a simple concept, since it must somehow take into consideration not only the direct medical costs but also the emotional trauma and disruption of normal family life that a chronic disease in a child entails. Ignoring these considerations for now, in a world of limited medical resources, to what financial lengths should prospective parents and society go to prevent the birth of a child destined to present at birth, or develop later, a chronic genetic disease that will require specialized care for the child as long as she or he lives?

At the moment, the cost of institutional care for a child with Down syndrome, or the cost of terminal care for a child with Tay-Sachs disease, are such that the present screening programs resulting in prenatal diagnosis are considered cost-effective for these two diseases. On the other hand, it has been estimated that if all pregnant women were screened for the carrier state for the alleles most commonly associated with cystic fibrosis, with appropriate follow-up studies where indicated, in the United States the cost of detecting each fetus with cystic fibrosis might be between 1.0 and 2.2 million dollars, the cost depending on the completeness of the screening and follow-up.[3] The direct medical cost of lifetime care of an affected individual is difficult to estimate but is probably between $200,000 and $300,000. At this point in time the majority of geneticists would question whether screening entire populations for the carrier state could be considered cost-effective in the strict financial sense, but the improving molecular technologies for carrier detection will keep this question under close scrutiny in the future. In fact, since the above estimate of the cost of prenatal diagnosis appeared, new and more efficient screening techniques have already been suggested.[4] As a general principle, the more uncommon the recessive disorder, the more expensive the detection of the affected fetus; cystic fibrosis represents a relatively favorable situation.

There is, however, another set of considerations in connection with large-scale screening programs, such as the programs mentioned in connection with the discussion of cystic fibrosis, sickle-cell anemia, and thalassemia. First, there is the need for the utmost diagnostic accuracy. Given the requisite level of accuracy, there is the further problem of transmitting this type of information to the patient in a comprehensible fashion. Then, legitimate concern has been expressed that if the results of screening programs became generally disseminated, they might be prejudicial with respect both to life and health insurance, marriage, and employment. The efforts in the U.S. directed at the prenatal diagnosis of

sickle-cell anemia, followed where desired by abortion of affected fetuses, already provides a tragic example of good intentions gone astray.[5]
The National Sickle Cell Anemia Control Act of 1972 confused the
sickle-cell trait with sickle-cell anemia, an issue I thought I had deci-
sively settled 23 years earlier! This confusion immediately transferred to
the workplace, with unfortunate results for individuals with the sickle-
cell trait who were perceived as suffering serious disease. The requirements for a satisfactory and acceptable screening program for genetic
disease and carrier states, a program which neither stigmatizes nor
arouses undue anxiety, have been well presented;[6] they are rigorous but
can in principle be met. Unfortunately, there are already examples—in
addition to those resulting from sickle-cell testing—where diagnoses of
presymptomatic genetic disease (or carrier states) or even family histories of genetic disease have compromised access to health insurance.[7] If
genetic-screening programs intensify, and as private health insurance
programs strive to meet the growing cost of medical care, this issue
could loom ever larger.

Second, prenatal diagnosis is not without risk to the fetus; approximately 1 in 200 fetuses are "spontaneously" aborted following the
procedure, and the majority of these will be normal. However, some
fraction of these aborted fetuses would undoubtedly have been lost even
without the procedure, so that the precise risk figure is undoubtedly less
than 1 in 200.

Finally, then, the question will surely arise of the level of disease
severity that justifies an abortion. This will be a hard-fought issue that
cannot be turned aside. There is a continuum in the severity of genetic
defect. Few physicians or parents who have cared for a child with Tay-
Sachs disease or thalassemia major would question the wisdom of terminating a pregnancy when this diagnosis was reached during the first
trimester. At the other extreme, one can visualize attempts at almost
frivolous genetic justifications for pregnancy termination, such as polydactyly, that would certainly be inappropriate. A potentially very troublesome issue is termination on grounds of sex of child. Given the desire
of most parents for at least one son, termination on grounds of sex could
lead to an imbalance of the sex ratio that would have serious sociological
implications. Population geneticists would not regard sex of the fetus an
indication for abortion. A lesson to be drawn from the experience to date
is the desirability of an "omnibus" screening program, combining the
high-frequency genetic diseases of various ethnic groups, so that no
single group feels especially singled out for action.

The total frequency of clinically severe genetic disease toward
which prenatal diagnosis might ultimately be directed is difficult to

estimate at this time. Virtually all the chromosomal abnormalities and a wide range of congenital defects can be diagnosed early in pregnancy. Well over 100 different biochemical defects have been diagnosed prenatally.[8] The diagnostic problem with all three of these disease categories is that monitoring is so often initiated only after the birth of the first affected child, which, as already discussed for cystic fibrosis, is not very efficient. On the other hand, it is certainly not practical (and perhaps not desirable) that each fetus have a "prenatal physical." I suggest that with the effort now in train to render the provision of medical services as efficient as possible, the present is the time for a careful analysis of what would constitute an ethical, socially acceptable, and cost-effective screening program for a total population, a program of course requiring periodic revision. Although the diseases and abnormalities toward which such a program might be directed afflict under one percent of all births,[8] the chronicity and severity of these diseases results in a disproportionate impact on families and medical services.

The efficacy of measures such as I have just described in the reduction of genetic disease depends, of course, on the vigor with which they are pursued and the extent to which early diagnosis is followed by pregnancy termination. Moreover, experience to date has amply illustrated the role of error (clerical or diagnostic), confusion, nonavailability of the service, and parental decisions not to abort in these programs. Furthermore, there are, of course severe, recessively inherited genetic diseases which for various reasons it is not yet feasible to diagnose prenatally. In short, even with vigorous implementation of a program of prenatal diagnosis and abortion, there will still for the foreseeable future be substantial numbers of genetically impaired children requiring therapy, either of the conventional type (see the next section) or, now, the much discussed gene therapy.

The established strategies for control of genetic disease through prenatal diagnosis all involve recourse to abortion not later than 20 weeks of gestation and preferably 3 or 4 weeks earlier. This is currently ethically unacceptable to some. To me, the overriding ethical responsibility in such a crowded and complex world is to bring into it wanted children who can fully participate. That is the desire of most parents. Those who feel otherwise are free to decline abortion, but it is now a medical responsibility to inform such parents of the risk to prospective children. Although I have written in a rather impersonal way on therapeutic abortion, I do not underestimate its emotional impact. I know even better the emotional impact on parents of a severely and chronically handicapped child. Abortion to prevent a serious, fatal disease cannot possibly be as traumatic as watching a loved child eke out a

370

miserable existence en route to an early death. Note that I treat abortion as an individual decision; I cannot imagine that as the facts become understood, any civilized society would prohibit abortion on the indication of serious genetic disease.

A technical issue intrudes at this point. Assume a society committed to two-child families. Parents who choose not to abort a fetus destined to be afflicted as a child with some recessively inherited disease, a child who will usually not live to produce progeny, will in fact transmit fewer of the disease genes in question to the next generation than parents who abort an affected fetus and replace it with a normal child. This is because in this circumstance two thirds of the normal-appearing children of these latter parents will be carriers of the gene in question. The parents with the child who dies early, on the other hand, may not replace it with another child, and so will be less likely to transmit the abnormal allele to the next generation. Thus, abortion under these circumstances may be considered dysgenic, but demonstrably less so than the application of successful gene therapy to the same situation.

The extreme in prenatal diagnosis results from the current ability, when parents are known to be carriers of such a genetic disease as cystic fibrosis or hemophilia, to effect *in vitro* fertilization using parental gametes, followed by genetic screening of the resulting developing egg at the 8-cell stage (using one of the 8 cells), and if the embryo is found to be normal, implanting it into the biological mother. To date, of 660 preembryos that have been reimplanted, 33% gave rise to successful pregnancies. The estimated total cost of the procedure is $13,000 (1992 dollars).[9] This is a technically dazzling feat that neatly sidesteps the abortion issue, while at the same time raising in an acute form the question of cost-effectiveness in genetic medicine, since, assuming the carrier parents have been identified, the cost of early prenatal diagnosis would be approximately $2000 in a hospital setting, and the cost of an abortion if indicated, about $1000. In a situation where the government funds universal health care, should the more expensive procedure be supported? This is but one example of the many difficult issues to be worked through in the wake of the new genetic medicine.

Objective 5: The Treatment of Genetic Disease
(Exclusive of Gene Therapy)

Children and adults with genetic disease require at one time or another all the supportive measures known to modern medicine, and because their disease is chronic, often require these measures over extended periods of time. Treatment includes surgical measures, ranging from such simple procedures as amputation of a supernumerary digit in polydactyly to the complexity of a kidney transplant for an individual whose own kidneys no longer function adequately because of inherited polycystic kidney disease or hereditary cystinosis. In this section, however, we are not concerned with those general measures, but with therapy specifically directed toward the basic genetic defect.

As noted earlier, simply inherited genetic diseases affect about 1.0% of all children, and, in addition, a substantial proportion of adults will ultimately develop diseases resulting from a combination of genetic predisposition and environmental precipitants. The diverse treatments of these diseases (where treatment exists) is primarily by dietary management or replacement therapy,[1] and now occupies a significant fraction of physician time, and is, of course, a very obvious component of the total genetic program we are considering. The importance of the early diagnosis and treatment of some of these diseases is such that in the State of Michigan (as in most other states), a system that screens all newborn infants for five (usually inherited) metabolic disorders is in place. The two most common of these disorders are, respectively, due to lack of the thyroid hormone and to the inability to metabolize a particular amino-acid, phenylalanine; the other three are very rare. The lack of thyroid hormone results in a disease known as cretinism, and the amino acid disorder, in phenylketonuria, neither of which is life-threatening. Without early therapy, both of these diseases are accompanied by mental defect, and most affected children would require institutionalization or the equivalent in domiciliary care. In 1989, in the State of Michigan, 146,018 newborns were screened and 57 children affected with one of these five diseases detected. Each mother was charged $20 for the screening, and testing was mandatory, so that the total budget for the program was some $2,920,360. Each year of institutional care for a child with either of these diseases would cost about $40,000. Thus, if each of the children diagnosed annually averaged ten years of institutional care, the total cost to society would be $22,800,000. The program is clearly cost-efficient for society, and certainly for the State.

The situation with respect to phenylketonuria requires some further discussion. The mental retardation in this disease is often accompanied by seizures and psychotic behavior. By virtue of a very special diet low in phenylalanine, the levels of phenylalanine and related compounds in the blood and body tissues can be held at near-normal levels. If the condition is detected shortly after birth and this special diet initiated immediately, the affected child will have normal or near-normal intelligence. For best results, the diet should be continued throughout life.

However, the diet is unpalatable, and most affected children discontinue it when they reach their teens. Even though very high levels of phenylalanine and related compounds generally reappear in the blood and other tissues when the special diet is discontinued, the brain seems to have lost much of its early sensitivity to these compounds, and the affected children can function within normal limits.

Unfortunately, as women with the disease who had been treated as children and then discontinued the diet (and were within the limits of normal mentality) began to reproduce, it developed that a high proportion of their children were mentally defective. This was not for genetic reasons (the children would almost always only be heterozygotes for the responsible gene), but because the high maternal blood levels of phenylalanine reaching the fetus via the placenta were very toxic to the developing fetal brain. To prevent this, it would appear to be necessary that these women initiate the rather unpalatable phenylalaninerestricted diet as soon as possible after pregnancy is recognized or, in these days of planned parenthood, as soon as a pregnancy is contemplated. Compliance with this constraint has already been shown not to come easily. Unless compliance can be assured, the treatment of children with this disease has the potential for increasing the frequency of mental retardation! This disease also illustrates an additional problem created by medical advances: an individual with a genetic disease who would not previously have reproduced will now do so, transmitting a gene for phenylketonuria to all his or her children.

The techniques of molecular genetics may lead to substantial improvements in the treatment of genetic disease. For many inherited diseases, the defective gene product remains unidentified. Until recently, cystic fibrosis was in this category. The chromosomal localization and the identification of the DNA sequence of the gene's exons in 1989 permitted the prediction of the nature of the defective protein in this disease. This protein exhibited a striking homology to a large family of proteins known to be involved in the transport of various substances across cell membranes; a variety of lines of evidence indicate this particular protein is involved in chloride ion transport (although it may have other functions as well). This finding immediately suggested new pharmacological approaches to the disease, approaches under active pursuit at the present time.[10] Other similar examples will certainly follow.

Objective 6: Gene Therapy

Given the many problems in the treatment of genetic disease, it is not surprising that with the advances in molecular genetics, within the past several decades an entirely new therapeutic approach, gene therapy, has caught the imagination of geneticists (not to mention the public). A salient discovery of molecular genetics has been the almost ridiculous ease with which genes can be moved from the nucleus of one cell to the nucleus of another, both within and between species, and, under appropriate conditions, then maintain function in their new location. The techniques employed to move genes about are as diverse as the use of viral vectors (i.e., carriers), direct injection of the gene of interest into the cell of choice, encapsulating the gene in a fat globule (liposome) which is taken up by the cell, or even electrophoration ("treating," with a DNA cocktail, cells whose permeability has been altered in various ways). These technologies create the obvious possibility of introducing into the nuclei of the cells of selected tissues of a person homozygous for a defective gene associated with a severe disease, one or more genes normal with respect to the trait in question, i.e., *human gene therapy*. Gene therapy is not limited to the substitution of good genes for bad. In principle, it extends to the manipulation of genetic control mechanisms. For instance, the child with sickle-cell anemia (or β-thalassemia) has as a fetus made perfectly normal fetal hemoglobin ($\alpha_2\gamma_2$), adequate to its needs. The severe anemia of this child developed after birth, when abnormal β-globin chain production took over (and γ-chain production virtually ceased), and the anemia could be substantially alleviated if the production of fetal hemoglobin could be reactivated; such reactivation would be a form of gene therapy.

Molecular geneticists with a particular interest in the human condition, in their quest for a relevance for all their new insights, are turning to gene therapy just as inevitably as the surgeon, with current life support and immunosuppressive techniques, turned to organ transplantation. Gene therapy is the response of some geneticists to the ancient commitment of medicine to healing, and efforts to explore the potential of gene therapy are inevitable. Furthermore, it seems to avoid the contentious issue of the abortion of seriously defective children raised by genetic counseling and prenatal diagnosis. Unfortunately, the publicity currently accorded to gene therapy far exceeds its immediate prospects. This early in these developments, given the absolutely fantastic developments in molecular genetics during the past several decades, it would be foolish of me to attempt to set limits on what might ultimately be possible. Some conjectures, however, are not out of order. (I exclude from this discussion the use of gene-transfer techniques to engineer the elaboration of human gene products by nonhuman organisms, such as insulin by appropriately altered microorganisms. I also exclude from the discussion the possibility of treating cancer by genetically engineered cells that have the potential of destroying the cancer.)

There are three principal approaches to human gene therapy at present. The first approach would remove a genetically defective tissue, such as bone marrow, introduce into the nuclei of the critical cells of this explant normal genes with respect to the trait in question, and reimplant the treated bone marrow into the same individual, where it presumably would now function normally. For various technical reasons, three genes among the many associated with genetic disease have emerged as prime candidates for the bone-marrow approach. The first is the gene encoding for the enzyme hypoxanthine-guanine phosphoribosyl transferase (HPRT), the absence of which results in Lesch-Nyhan disease, a devastating sex-linked disorder characterized by the accumulation of uric acid, with resultant gouty arthritis and kidney disease, and by neurological symptoms, especially an impaired intellect and severe self-mutilation. The second and third prime candidates are the genes encoding for adenosine deaminase (ADA) and purine nucleoside phosphorylase (PNP), the absence of either of which enzymes results in a severe (and rapidly fatal) disturbance of the immune system. The frequency of the first disease is about 1 in 50,000 to 100,000 births whereas the others are extremely rare: no more than 40 to 50 children with ADA deficiency and perhaps 10 with PNP deficiency are thought to be alive at present. It is important for perspective to bear in mind how very rare these diseases are. They are considered "ideal" candidates for gene therapy because the tissue toward which the therapy is directed, bone marrow, is so accessible and because restoration of as little as 5–10% of normal enzyme activity may be sufficient to offset the genetic defect. Furthermore, since DNA synthesis (and cell growth) is defective in all three of these diseases, correction of the defect might in time allow a relatively few treated marrow cells when introduced into the body to outgrow the defective cells and eventually become the dominant cell element.

A second approach to gene therapy would take advantage of the fact that the replacement gene may not always have to be introduced into the tissue in which it normally functions in order to alleviate disease symptoms. For instance, there is a type of dwarfism due to a genetically determined failure of certain cells in the pituitary gland to produce growth hormone. With current techniques, it may be possible to introduce into muscle tissue genes that will produce growth hormone in affected children. A much more important example would be the direct introduction of genes that would produce insulin into the muscle tissue of a diabetic child. In both instances, this therapy would work because the gene product would be carried by the blood stream to the tissues that need it. Now the problem, assuming successful gene transfer, becomes the duplication of the body's extremely complex mechanisms for regulating the amount of the gene product in circulation, since too much of either growth hormone or insulin results in disease as severe as too little of the gene product. This form of gene therapy might be most successful in situations such as are represented by the bleeding disorders, where only a fraction of the amount of the gene product normally present seems adequate to control the symptoms of the disease.

The third principal approach to gene therapy would, for specific inherited diseases, introduce into a tissue-specific retrovirus engineered to be nondisease producing, the normal gene for the trait in question and then infect the patient with the altered retrovirus. The expectation is that the virus would invade the nuclei of a sufficient number of cells, and the normal gene thereby introduced would function sufficiently well, so that the symptoms of the disease would be relieved. The prime candidate diseases would, at present, be the three mentioned earlier. With respect to the Lesch-Nyhan disorder, while treatment of the bone marrow as discussed in an earlier paragraph might relieve the accumulation of uric acid in the blood that characterizes this disease, it would probably not completely alleviate the neurological symptoms. These latter would be treated through infecting the patient with an appropriately engineered virus with a predilection for the central nervous system. In this specific case, the most appropriate carrier virus would be the herpes simplex virus, which is normally responsible for the common "cold sore," the virus now modified by insertion of a functional normal allele of the Lesch-Nyhan gene and also modified so as not to produce disease.

Issues Created by Gene Therapy

The prospect of gene therapy has initiated a great deal of discussion on the part of scientists, ethicists, legislators, and the general public.[11] Prospects for the first two types of therapy just enumerated are much further along than for the third, and it is the first two which have attracted most of the discussion. The prime candidates for treatment are devastating diseases, usually fatal in the first two decades of life, where unusual therapeutic risks might be justified.[12] Even so, there is consensus that: "Application of gene therapy for a human genetic disease should require evidence that it is safe, might prove beneficial, is technically possible, and is ethically acceptable."[13] There are, unfortunately, several unknowns in gene therapy that render the application of these criteria difficult.

First, at this moment, there is no way to control the exact position of the newly introduced genes when they insert into chromosomes. In gene therapy, thousands of the patient's abnormal cells will be treated one way or another; each treated cell may experience gene insertions in its chromosomes at several (different) locations. Should even one of these introduced genes insert itself into a chromosome in such a way as to activate a gene associated with cancer, the result could be a cancer some years later. The lag period of 20 to 30 years between exposure to the atomic bombs and development of a radiogenic cancer indicates that the assessment of the carcinogenic potential of gene therapy will not come quickly. Should gene therapy be associated with the subsequent development of cancer in a significant number of patients, that would not be an automatic contraindication of its use. Society accepts chemotherapy for childhood leukemia despite the treatment-induced increased frequency of secondary malignancies in these children. However, such a development would certainly stimulate an even closer look at gene therapy. Research on controlling the sites of gene insertion currently has a high priority.

The second issue is the efficacy and duration of the treatment: it is no great victory simply to prolong a devastating existence. In most specialized tissues, there are cells committed to the production of a particular gene product, with finite life spans, and then there are the much less common stem cells, which throughout life feed new cells into the production line. For a lasting therapeutic result with gene therapy, it is these rare stem cells that must be altered, but in a way consistent with their continuing division over a span of many years. Whether without some selective advantage the altered stem cells would over a long period divide more slowly than the unaltered cells, and ultimately be lost, is simply unknown.

A third issue that would be especially relevant in the event of therapy using a retroviral vector is whether the vector would remain non-disease producing indefinitely and also not escape from the target tissue. Specifically, given the prevalence of retroviral "footprints" in the germinal DNA, how safely can it be assumed that the genes introduced in the course of gene therapy directed at somatic cells will remain confined to somatic cells? As brought out in Chapter 15, there are some one million retroviral footprints *established* in human DNA; my surmise is that few of the retroviruses responsible for these footprints were specifically targeted for the germ line: the original intrusions occurred by accident, with what consequences can only be surmised. Before a final judgment on the ethical acceptability of gene therapy can be reached, data must be at hand, presumably derived from an experimental animal such as the mouse, on the frequency with which random gene insertions into chromosomes result in a cancer or a germ-line mutation.

In Chapters 14 and 15, we considered the complexity of the gene pool as it is currently understood. At the time, these considerations may have seemed like a somewhat academic exercise. In fact, whatever the measures society would adopt directed toward this gene pool—even one as conservative as stabilizing its size—it is imperative that those who discuss them have this complexity clearly in mind. This need is especially acute when we consider gene therapy. What studies of the human genome have revealed thus far is an incredibly rich tapestry of improvisations, the threads running every which way, the fabric somehow eventuating in harmonious and integrated phenotypes. But, because of this variability from gene to gene in structure and control, there is no simple "formula" approach to gene therapy—the field must develop on a case by case basis, and there will be many surprises along the way.

Society has adopted a system of accreditation for both the trades and professions, to ensure a background of knowledge and competence on the part of licensees. Society has not yet really dealt with the issue of deciding when the knowledge base is sufficient to "license" intervention into DNA, but has made significant progress in that direction. Because most of the research on the applications of the emerging DNA technologies is financed by government agencies, it was deemed in the public interest to establish in 1974 an oversight committee for research of this nature, known as the Recombinant DNA Advisory Committee, of which there was a subcommittee devoted to Human Gene Therapy. By an informal agreement, investigators all over the world are submitting their research plans to this subcommittee. On July 30, 1990, the subcommittee, after prolonged discussion, approved the first human gene therapy protocol, a plan to insert normal ADA genes into cultured peripheral lymphocytes from patients with the severe immunodeficiency disease resulting from ADA deficiency, and then to inject these cells back into the patient. Presumably, as long as they persist in the body the altered cells will produce sufficient adenosine deaminase to alleviate the symptoms of the disease. This is still not gene therapy *sensu strictu*; for a variety of technical reasons the proposed procedure is unlikely to be a definitive protocol for the treatment of the disease but did, after 16 years of activity on the part of the Recombinant DNA Advisory Committee, represent the first step toward human gene therapy.[14] As of mid-1992, there were 19 additional protocols that had received approval by the appropriate committee, the majority of the protocols for studies in the U.S. However, most of these protocols involve the treatment of cancer with genetically engineered lymphocytes; only seven protocols involve attempts at true gene therapy.[15]

A variety of other disorders, such as sickle-cell anemia, thalassemia, inherited muscular dystrophy, and cystic fibrosis, have been mentioned as candidates for such therapy. These, numerically, are much more significant targets for gene therapy. The issues raised by gene therapy are very different from the currently much more tangible issues raised by prenatal diagnosis and elective abortion. Although the two approaches are not antithetical—even with a well-implemented program of prenatal diagnosis there will still for several reasons be children with the targeted diseases in question—the next several decades should witness a vigorous discussion at many levels as to where the primary emphasis should be placed. Should the necessary preliminary studies indicate some possibility for successful gene therapy, then geneticists will be forced to confront head on the question to which the next chapter is devoted: In a world of diminishing resources and containment of medical expenditures, where in the names of cost efficiency and humanity should the primary genetic efforts be directed—at population control, at prenatal screening and abortion, at euphenics, or at gene therapy? With respect to the latter, I suggest that the case for immediate gene therapy for the four diseases mentioned in opening this paragraph is not as compelling as the case for the three rare diseases mentioned earlier as prime candidates for therapy, and surely, before acquiescing in gene therapy for the latter four, society must insist on a prolonged experience with gene therapy for these rare and truly desperate diseases, and on the animal experimentation that will determine the oncogenic and other risks of the shotgun introduction of foreign genes into the nuclei of a variety of somatic cells (see below).

At this juncture, the population geneticist in me is compelled to reiterate a technical point made in the earlier discussion of phenylketonuria, a point made repeatedly in the histories of medicine and genetics. All successful therapy of genetic disease that results in extended survival and reproduction that would not otherwise have occurred has implications for the gene pool. Genes for deleterious traits that would formerly have been lost through the early death of their possessor will now be transmitted to the next generation. Since the mutation pressure that contributed to the present frequency of these genes will presumably continue, then successful therapy implies that the frequency of these genes will increase in future populations.

This increase in the frequency of such genes, however, will be very,

very slow (because mutation rates for specific genes are so relatively low). For instance, consider a fatal, recessively inherited disease of childhood, with a frequency in a population of 1 in 100,000 births. In theory, the mutation rate necessary to sustain this disease at this frequency is 1×10^{-5}. The frequency of the deleterious gene in the population is $\sqrt{0.00001}$ or 0.003; the frequency of the normal gene is 0.997. If completely successful gene therapy were suddenly to become available for this disease, so that there was no negative selection, each generation a proportion of 0.00001 of the remaining normal genes would be converted to the abnormal allele. But even after 100 generations of such therapy, the frequency of the normal allele will only have decreased to 0.996 [i.e., $0.997 \ (1 - 1 \times 10^{-5})^{100}$]. The concerns over the implications of gene therapy need not include the rapidity with which it will alter the gene pool. The early eugenicists viewed the medically mediated, improved reproductive prospects of those with genetic diseases as an imminent cause of etiolation of the gene pool. While, in principle, these concerns are based in fact, the changes in gene frequencies will be so gradual that this threat is negligible by comparison with the pressing issue of control of population growth.

Germ-Line Gene Therapy

A principle reason for the sometimes shrill debate over the prospect of gene therapy for selected diseases is the obvious fact that if it is feasible to introduce foreign genes into the somatic cells of the body, it is also feasible to introduce genes into germ cells, and such experiments have in fact already been successfully performed with mice. No responsible individual or organization is currently advocating human germ line therapy. The haphazard (i.e., untargeted) introduction of selected genes into the human germ line—which is what the current technology involves—is clearly undesirable; already some 10% of the germ-line transgenes produced in mice have been found to be associated with deleterious insertional mutations, revealed by inbreeding.[16] The inclusion of Chapters 14 and 15 in this book was intended to make clear the tremendous complexity the human genome has assumed, and the relative poverty of our current understanding of how it holds together and functions so successfully. The ability of humankind to wreak damage on complex biological systems is so sufficiently documented that yet another example seems unnecessary. It will require decades of careful observation of the results of human somatic cell gene therapy and experimentation with the mouse and other animal models to identify the collateral consequences of the present technologies for somatic gene therapy and to develop better methods of gene delivery than currently exist, before a discussion of germ-line therapy can be productive. Germ-line therapy, to be acceptable, requires the precise removal of the defective gene and the equally precise insertion of its normal allele in its place—plus the demonstration that the normal gene is not inserting elsewhere in the genome. Were, some day, the precise correction of defective genes in germ cells to become possible, then society would face a whole series of wrenching decisions concerning the advisability of, and the administrative framework for, applying the new technology to the human species.

These societal decisions may be closer than many, including myself, would like to think. On September 14, 1992, I responded to an invitation to brief the aforementioned Recombinant DNA Advisory Committee on the subject of spontaneous and induced human mutation rates. In extending this invitation, the Executive Secretary to the Committee, Dr. Nelson Wivel, carefully explained to me that in order to prepare for possible future considerations of germ-line therapy, the Committee felt the time had come to begin to consider germ-line dynamics. One insight into these dynamics is provided by the study of spontaneous and induced mutation: spontaneous, because it reveals the frequency of unrepaired mishaps in our DNA in the absence of known external perturbing factors; induced, because it reveals the magnitude of the response to known mutagenic insults, i.e., informs us concerning the homeostatic properties of the genome. After I had completed the straightforward presentation on mutation the Committee had requested, I allowed myself the license of a few words on the wisdom of their "germ-line" exercise. Germ-line therapy would by any name be the new eugenics. I spoke as one of the very few persons—perhaps the only such—currently active in human genetics who had personally reviewed the contents of the Eugenic Records Office (see Chapter 1) and knew at first hand the pitifully inadequate knowledge base upon which, only two generations ago, an action-oriented eugenics program was based. I told the Committee that, although it was inevitable that some individuals think along these lines, it was to me inconceivable that a Committee of this prestige began the kind of formal deliberations that inevitably would suggest the imminence of germ-line therapy, deliberations based on a knowledge base that in another generation will be seen to be as unsatisfactory as that of the old eugenics movement. Perhaps, with 30 to 40 years experience with a somatic cell gene therapy whose implications are still virtually untouched, it will become appropriate to consider germ-line therapy, but to do so now would be another example of the shortsighted intellectual arrogance that has already so profoundly disturbed the natural order of which we are a part.[17]

The Special Case of Aging

We come now to a subject many geneticists do not at first thought associate with their discipline: aging. Mammalian species vary greatly in their average life span. The house mouse lives about three years, but its same-sized cousin, the deer mouse of the forest (*Peromyscus*), lives twice that long. Humans have a maximum life span of 110 to 120 years, but chimpanzees about half of that. The latter difference has evolved in some five to seven million years, i.e., since the two evolutionary lines separated. These species differences are manifested by cells in tissue culture: normal cells derived from long-lived species will undergo more cell divisions and maintain function longer than cells derived from shorter-lived species. Interestingly, the female chimpanzee continues to reproduce until near the end of her life span; the prolongation of human survival chiefly involves the addition of postreproductive years. Students of human evolution seem to concentrate on the emergence of upright posture and a large cranial capacity as key events, but this delay of senescence is clearly seminal to the ability of humans to develop a culture. Within each species, there is marked variation in the timing of "natural death" among the individuals comprising the species. The term "natural" is most imprecise—let us agree only that some individuals seem to "age" more rapidly than others.

Earlier (Chapter 17) I mentioned four years of service on the Council of the National Institute of Medicine. These four, very interesting years were an exercise in controlled schizophrenia. The Institute had two faces. The face addressing the medical and social problems of America's aging population was drawn and anxious, for reasons already discussed. The figure on page 360 illustrates the projected dramatic increase in the elderly component of industrialized societies, even as the youthful component remains relatively constant. The face addressing the scientific basis of aging was smiling, beaming with the optimistic belief that soon we would understand the basis for aging—and with understanding comes the opportunity for manipulation. Clearly, however, any significant prolongation of life could only exacerbate a problem with which American society was already dealing badly, the clash between the needs of senior citizens and of children discussed in Chapter 17.

In any discussion of the biology associated with aging, we must make a clear distinction between the diseases associated with old age—

cancer, stroke and heart disease—and the process of aging, or senescence, characterized by wide-spread cell loss and impairment of function of the remaining cells. Although senescence may set the stage for the aforementioned diseases, there is an aging process apart from these specific events. In order to consider the possibility of intervention into this process, we must consider briefly theories of the nature of the determinants in this aging process. In passing, however, let us note that the diseases associated with aging may be subject to control mechanisms similar to those governing senescence. Humans and mice have about the same risk of dying of cancer. Miller[18] calculates, however, that because of the smaller size and shorter life span of the mouse, the likelihood that a comparable cell in the two species will develop into a cancer is about 90,000-fold *lower* in a human than in a mouse. The basis for this evolutionary adjustment remains obscure.

Theories of Aging

Among the many theories of aging, three somewhat overlapping theories are currently receiving particular attention.[19]

1. *The theory that aging is genetically programmed* postulates that aging is basically due to deactivation of "longevity-assurance" genes and/or the activation of "senescence" genes. A precedent for such genetic programming is to be found in the already-discussed hemoglobin system, where as development proceeds different hemoglobin genes are expressed. Of particular pertinence to this theory is the demonstration that cultured normal human diploid cells, no matter how well cared for, will undergo only a finite number of cell divisions, this number depending on the age of the individual supplying the cells; those cells nearing the end of their ability to divide contain relatively large amounts of an RNA that in experiments inhibits the division of younger cells. Experiments with cultured cell lines indicate that there are four different genes or groups of genes involved in this phenomenon; one of these four functional units has been located on chromosome 4.[20] A very active chase is underway to identify the precise nature of the genes involved. While these important experiments on cultured cells substantially further understanding of the genetic control of aging, they leave unanswered the question of how the "senescence switch" is normally controlled in the intact organism.

2. The *free-radical theory* of aging holds that the reaction of so-called free radicals with proteins and DNA results in a progressive accumulation of damage to essential body components that ultimately

results in the complex phenomenon termed aging. These free radicals are chemicals released in the course of the normal biochemical reactions of the body. The best known of the free radicals is a highly reactive form of oxygen designated as superoxide $(-O_2^{\cdot})$. The body produces or ingests many compounds, such as uric acid, superoxide dismutase, vitamin E, or carotenes, capable of quickly reacting with these free radicals, but some free radicals temporarily escape inactivation, and have been shown to damage many proteins and DNA. The greater the food intake, the greater the production of free radicals. The mean duration of life in rats has been increased by 10–25% following caloric restriction (and the onset of cancer delayed); this has been attributed to a decreased production of free radicals.[21] Whatever the cause, there is evidence that the functions of specific genes respond differentially to caloric restriction in laboratory rodents.[22]

3. The *DNA damage theory* of aging postulates that, efficient though the various DNA repair mechanisms mentioned in Chapter 15 are, in time uncorrected or unrepaired damage to DNA accumulates as somatic mutations, to the point that the cell can no longer function, and cell death or inability to divide ensues. Normal "healthy" DNA interacts not only with the free radicals just mentioned but also with many intracellular molecules and in particular with small methyl groups $(-CH_3)$, which are thought to play a significant role in regulating gene activity. The hypothesis has been advanced that with aging these protective methyl groups are slowly stripped away from the DNA, with the activation of senescence genes when demethylation reaches a certain point. A variant of this theory postulates that repeated somatic cell divisions result in progressive damage to the telomeres that cap the ends of chromosomes, the final result being chromosomes that are no longer functional.

Aging most likely results from a mixture of the processes envisioned by these theories (which of course are not mutually exclusive), but I would place special emphasis on the ultimate activation of senescence genes. This is because the exposure of humans to free radicals and DNA damage should be about the same as that of the omnivorous chimpanzee, whose metabolism is in all respects so similar to our own, but we are functional twice as long as a chimpanzee. Our greater life expectancy is most readily explained by a difference in aging-switch mechanisms. In the animal kingdom, some of the best evidence that aging is programmed comes from the "immortality" of the germ line, which seems to have escaped the programmed death of the somatic cells. However, this evidence must be viewed with reservations, since periodically at meiosis there is the opportunity to cull out (and perhaps repair) some types of genetically deficient cells, and a further culling occurs in early development, where zygotes without an adequate complement of normal genes are eliminated.

The addition of free-radical reaction inhibitors to the diet early in life increased the average survival of a variety of experimental animals, some by as much as 20% (cf note 23). [The chemicals employed in these experimental studies are not without side effects; their routine use in humans is not (yet) advocated.] Since this treatment decreases the frequency of malignancies and other diseases of old age in experimental animals, it would have to be considered a form of euphenics, and, as I have developed the topic, a desirable end point. It must be admitted, however, that the dividing line between this (euphenic) approach to aging and a chemical treatment designed to influence the activation of the aging switch (gene therapy) could in practice become tenuous, especially given the evidence cited earlier that in laboratory rodents, caloric restriction alone may alter the functioning of certain genes.

It is important in considering the statistic that life expectancy for a newborn infant has increased dramatically in the past century, from 47 years in 1900 to 75 years in 1990, to understand that this increase in life expectancy is due primarily to improved nutrition and control of infectious diseases, and to only a minor degree, in recent years, to improved therapy for the diseases of old age. Furthermore, there is no evidence that the process of senescence has been significantly altered. Now, however, no matter which of these DNA-oriented theories provides the major explanation of aging, all of them imply the possibility of human intervention into the process. Earlier, in a discussion of gene therapy, I suggested that, in a child with sickle-cell anemia (or thalassemia), any intervention with the "hemoglobin switch," as a result of which the genes for fetal hemoglobin continued to function into adult life (and the genes for adult-type hemoglobin were not activated), had to be regarded as a form of gene therapy. Precisely the same argument holds with reference to intervention into the aging process. To the extent that measures to forestall aging involve the administration to individuals of chemicals designed to influence the functioning of DNA (or more direct approaches to DNA), these measures logically fall under the category of gene therapy.

It follows that efforts to influence the aging process deserve no less ethical scrutiny than somatic cell gene therapy. The possibility of interfering with the aging process almost surely has far more social implications than all the "gene therapy" I have just discussed. Gene therapy would involve a relatively few unfortunate children and adults, but aging therapy would ultimately involve literally everyone. It is surprising, considering to what extent the potential social impact of influencing the aging process exceeds that of gene therapy, how little publicity research that carries even the remote possibility of thwarting aging has received. The possibility of gene therapy for aging, remote as it may seem at first thought, is approaching the point at which national-level, serious discussions of its advisability must take place. As the chapter frontispiece illustrated, the recent and projected augmentation in the number of senior citizens is already growing at such a pace that it is difficult to argue that the pace needs to be accelerated.

The involvement of so many highly differentiated tissues in aging suggests the phenomenon has a complex genetic basis that draws upon the function of many genes at many loci. This statement is backed by the experiments already mentioned on control of senescence in cells in cell culture, as well as by experiments with house mice demonstrating that strain differences in longevity can be associated with the presence of specific chromosomes.[24] The more complex the phenomenon, the less the probability is of a socially acceptable, successful intervention. Thwarting a "switch mechanism" in, for example, brain tissue could create the asymmetry of a very active mind in an increasingly feeble body (or, even worse, the reverse possibility must be considered).

It was the custom that at one's last meeting as a member of the Council for the National Institute of Aging, one made a little speech. I could not resist commenting on the inherently antithetical nature of the major objectives of the Institute, arising from the drive to increase the pool of senior citizens when accommodating even the present number was proving so taxing. The dissolution of the extended family system— the chief source of social reinforcement in early societies—will continue, and this, coupled with the increasingly straitened times upon us, can only intensify the plight of our senescent citizens. In the meantime, if funding is available, research on managing DNA in ways to prolong life will continue to be actively pursued, with no foreknowledge of whether the prolongation will result in added years of vigor or years that are better described as an extension of senescence. Given conformity with ethical standards, there can be no forbidden research in a democratic society, but I would assign a low priority to research directed at altering the functioning of the "aging switches," as contrasted to research on the diseases of old age (which if successful of course implies some prolongation of average life span).

One thing is certain: the problems inherent in a gerontocracy that were mentioned earlier will not diminish. Increasingly, as resources become more constrained, the issue will become how to provide adequately for the old and still insure the realization of the genetic potential of the young, on whom, ultimately, so many of the benefits for the old depend. In the context of the argument I am developing, it would be a serious mistake to neglect the euphenics of the young for palliative measures directed toward their grandparents and great grandparents. The tensions inherent in the conflicting needs of these two ends of the age scale will not be easily resolved.

Calligraphy by P.B. Neel (1983). A literal translation is "vision" or "dream," particularly appropriate for this chapter.

21

Physician to the Gene Pool

I open this final chapter with a reaffirmation of the belief that humankind, by its unbridled proliferation and wanton exploitation and degradation of Planet Earth, has over the course of the past several centuries created a crisis for itself and for countless other species. Other species before us have faced crisis—Norman Myers estimates that "of all the species that have ever existed, possibly half a billion or more, there now remain only a few million"[1]—but we are the first species in crisis to be able to articulate its problem and consider counter measures. We are the first species to hold its future in its own hands. We may not meet the responsibility that statement implies, but there is no way we can escape it.

Genetics, I emphasize again, is only one of the many disciplines that must participate in the effort to meet this crisis. In fact, other disciplines, through national and international efforts, are already devoting much greater effort to evaluating the future and attempting to develop programs than are geneticists. The relative inactivity of geneticists in the discussion of these population problems—geneticists who above all others should by training be ombudsmen for the future—is peculiar. It is a legitimate speculation that the amazing developments in molecular genetics of the past 30 years have been so enthralling, and understandably so, as to divert professional (and public) attention away from gene pool problems. During recent years, this preoccupation of geneticists with molecular genetics has undoubtedly been reinforced by the so-called Human Genome Project, a Department of Energy/National Institutes of Health-funded effort to establish the precise nucleotide sequence of all the human chromosomes (as well as the sequences of a number of other species). In addition, any remaining attention an informed public might have directed to gene pool problems has, focused

389

to a considerable extent by geneticists, been predominantly concerned with exposure to mutagens, a concern now to be seen, as laid out in Chapter 13, as somewhat out of proportion to the risks. At this point, however, it is clear that the continuing definition of the content of the human genome, down to its last minutiae, will not contribute materially to the solution of humankind's major genetic problems, as I have defined them. If geneticists do not increasingly concern themselves with the issues I have been discussing, they will simply not contribute substantially to the decisions shaping the gene pool of future generations that are being forced upon society.

It is important at this juncture to emphasize once again that, despite the incredible advances of the past several decades, as yet we really know relatively little about the nature and functioning of the human gene pool and the forces that shaped it. It should be obvious by now that I have a respect for the intricacy of the human genome and of the population structure that brought it into being approaching awe. The evolution from the concept in my student days of genes as proteinaceous beads on a DNA string to the concepts of today is one of the great leaps of the human mind, but that leap provides a sobering view of the enormous and, to me, still frightening complexity that can now be manipulated. The occurrence of rogue cells (cf. Chapter 9) is a humbling reminder of how much we have yet to learn about the operation of the gene pool and its management of perturbing factors.

In these last chapters, I have found it convenient, in considering the contribution of genetic knowledge to the upcoming crisis, to distinguish between the long-range and the short-range genetic problems of the species. The long-range problems arise primarily from abandoning the population structure and the selective pressures under which humankind evolved. Fortunately, the gene pool, in the course of four billion years, has evolved a resourcefulness, a toughness, we still do not completely understand. The evolutionary wisdom encoded in the human gene pool will not degrade quickly. Although our species may in time acquire the understanding to deal with these long-range issues, in the meantime there are much more pressing problems, arising from the demands a rapidly growing population is placing upon an equally rapidly shrinking set of natural resources.

Perspectives and Priorities

Let us now move to an examination of the precise priorities to be assigned to the six genetic objectives discussed in the preceding two chapters. From the standpoint of one who believes the most important resource of the species is its collective gene pool, the *first genetic priority,* in the very stressful times about to unfold, is to decrease the gene pool to a sustainable size in a way that will preserve insofar as possible all of its poorly understood diversity. This requires an immediate worldwide program of population control by means at hand, whose objective for the foreseeable future is no more than two children per family, wherever that family be. No flashy genetic technologies are involved. As I discussed in Chapter 10, many, if not most, of the so-called primitive peoples seem to have tried to reconcile the rate of entry of new life into their societies with the ability of their cultures to accommodate it. That commitment, made by minds no less sentient than our own, was lost with the advent of agriculture and the great religions, a loss that resulted in runaway population growth as modern medicine and sanitation evolved. The balance cannot be redressed until all secular and religious authorities recognize the urgent need for population control. Concurrently with this population control, the gene pool must be protected against "unnecessary" damage. This requires minimizing exposures to man-made mutagens to the point where society would judge that the trade-off between the benefits to which these exposures were incidental more than matched the estimated cost in morbidity and mortality of the mutagenic exposures.

How does the geneticist enter into the issue of population control? My friends sometimes ask me whether I really expect them to pass out condoms on street corners or man Norplant booths at gathering places for young people. No, but what we can do through our various professional societies, the world over, is develop resolutions concerning the gravity of the situation, endorse the concept of evenhanded restraint on family size, and volunteer our voice in the impending debate. But having entered into this arena, the geneticist must be prepared to fight the necessary battles for the acceptance of the concepts he endorses. In the current genetic culture, it will be much more convenient to stay in the laboratory, teasing new genes out of those three billion nucleotides, than to become involved in the coming battles over population control. A redirection of research-funding priorities at the national level, from the current emphasis on "hard, molecular" genetic science to more attention to various aspects of the issue of population control (plus the other priorities listed below), would help nudge genetic research toward a balance more appropriate to the times.

The *second genetic priority* is to attempt, within the resources available, to optimize the expression of *all* genotypes. This effort quickly dichotomizes. There is, on the one hand, the need for the early recognition and treatment of a variety of relatively uncommon genetic predispositions and serious genetic disorders, as briefly characterized earlier. There is, on the other hand, the need to recognize that humankind, by virtue of its evolutionary history, is maladapted to the excessive caloric intake/relatively high fat/low physical activity/high salt regime that characterizes so many persons in the affluent societies and manifests itself in obesity and an increase in diabetes mellitus, hypertension, high tissue uric acid levels (gout), and even cancer. These predispositions are sufficiently prevalent that a strong educational program directed toward the lifestyle of *total* populations is the most effective genetic approach.

In addition, the techniques of molecular genetics offer such promise in the detection of susceptible individuals that when individuals with a positive family history for a variety of inherited diseases present themselves in a medical setting, it will become part of medical practice to screen this enriched sample for the especially predisposed, who can then elect (or not) to engage in particular preventive measures. It may also be possible in the future to give nature a helping hand with respect to dietary manipulation: as noted earlier, in experimental animals (rats and mice) the addition to the diet of antioxidants prolonged average survival—in one experiment by 20%—but did not increase the maximum life span. In humans, given the current average life expectancy of 75 years, and a maximum life span of 110 years, such a dietary prolongation would correspond to an additional 20 years of life expectancy!

There is, however, a "catch 22" aspect of euphenics: to the extent that euphenics thwarts the development of the diseases of civilization and results in a more productive midlife, it can only result in an increase in the number of senior citizens and exacerbate the problems already manifest in an emerging gerontocracy. Thus, there is an inescapable paradox inherent in my espousal of euphenics, on the one hand, and my concern with the emergence of a gerontocracy, on the other hand. Successful euphenics will undoubtedly result in even more old people than the current projections, thus increasing the social tensions inherent in adjusting to this development. The working years must be prolonged. Under these circumstances, research on thwarting the "aging switch(es)" would in a rational society emerge with a low priority. Let us adjust to a present problem before pursuing research that can only intensity that problem.

The euphenics of the mind deserve no less attention than the euphenics of the body. Any amelioration of the difficult circumstances in which it appears the next several generations will function requires that every intellect realize its fullest capacity. No society can afford the sad luxury of a wasted mind. In recent years, the *per capita* public expenditure on the physically and intellectually handicapped has far exceeded the expenditure on the intellectually gifted. The compassion elicited by the handicapped must not unduly divert resources from the education of those minds with the problem-solving capacity essential to meeting the societal problems now upon us. At issue is the enormous problem of the appropriate allocation of funds for education; in the present context, let me note only that, although not usually seen as such, this is a major area for the practice of euphenics.

It is anticipated that with the limitations on family size, the usual parental concerns that their offspring be normal, productive members of society can only be heightened. The *number three priority* in the genetic program I am advancing is an increased provision of genetic counseling and prenatal diagnostic services, with the objective of providing the option of abortion when *severe* genetic disease is encountered in a fetus. It is presumed that those parents who elect abortion in this circumstance will then choose to have a child free of the particular disease in question. These prenatal diagnostic services will draw upon all of the recent advances in cytogenetics and biochemical and molecular genetics. A major research effort designed to render prenatal diagnosis of serious genetic defect more cost effective is warranted.

It is not by accident that in this discussion of a genetic program that considers both the probability of favorable impact on gene pool problems and cost-effectiveness, I assign *somatic cell gene therapy*, including therapy directed toward the aging process, should this become possible, *the lowest priority*. Human gene therapy for those diseases where it is appropriate occupies this low-priority position because of its probable minimal impact on "gene pool problems" as I have defined them, but such therapy directed toward aging is assigned this position because of the major societal issues that would at this time be created by the ability to manipulate the "aging switch(es)," and especially the potential negative impact on the euphenics of the young. It is imperative that the subject of genetic manipulation of the aging process begin to attract the same level of ethical consideration that has characterized gene therapy. I would never argue against research in these two areas but suggest that from the societal standpoint there are much higher genetic priorities. Furthermore, for the foreseeable future the "high-tech" nature of gene therapy will limit its application to the advanced nations. Even in the advanced nations, given current efforts at medical costs containment, the probable relatively high cost of gene therapy, as contrasted with prenatal diagnosis and abortion where feasible, should lead to increasing emphasis on the latter alternative.

There is also an image problem associated with gene therapy. There is some danger that as a result of the current attention devoted to this topic, the decision makers attempting to plan for the future will relegate human geneticists to the role of "gene mechanics" and "DNA jockeys." It should be apparent by now that my view of what genetic concepts have to offer the future extends well beyond the confines of medical genetics. What I see as the larger responsibilities of the human geneticist have received relatively little attention in recent years. It is a great paradox that the human geneticists (read: eugenicists) of 70 years ago, short on specific knowledge concerning the basis of human inheritance, were long on concern for the future, whereas the human geneticists of today, increasingly long on specific knowledge, fearing the opprobrium of an eugenic label, appear to have retreated from that concern for the future. In a world where some difficult decision must soon be made, if only by default, it is incumbent upon the genetic-minded to come forward with a more holistic approach to the genetic aspects of the present human dilemma than is now evident.

Unfortunately, without some reordering of genetic research priorities at the national level, a continuing emphasis on the prospects for gene therapy will undoubtedly dominate research on the "service" aspects of human genetics for the next decade. The subject has momentum. In the hands of those who try to interpret scientific developments for the public, discussions of gene therapy readily lend themselves to hyperbole.[2] To lay audiences it has an almost magic quality. Unlike population control and euphenics, it does not challenge religious beliefs nor call for self discipline. As the quintessence of high biomedical technology, it is a magnet for research support. But is it, in truth, *the* issue with which geneticists should be identified in the public mind, let alone in their own minds? There is a further very sobering fact underlying this discussion: unless population is reconciled with resources shortly, the general economic deterioration will mean such a decline in medical services that gene therapy can exist only in a few affluent pockets of society. Society's practical return from its by now large investment in molecular genetics will be much more in the areas of prenatal diagnosis and the improved early detection and understanding of disease that leads to better prevention and treatment than in gene therapy.

The preceding chapter set forth the very stringent requirements that must be imposed on research on germ-line gene therapy. The precise replacement of a defective gene in the nucleus of a germ cell presents such technical challenges that there is no reason at present to include this subject in a genetic program directed at the *immediate* future. More than that, I fear that to hold out promise that this development will make a significant contribution to the genetic problems facing the world's populations would only distract from the job at hand.

It is likely that a program that so espouses prenatal diagnosis followed by abortion as that I have presented will encounter ethical/religious concern and even opposition. To those who take the position (which I do not) that early abortion is morally equivalent to the killing of an innocent person, I would point out, with Muller,[3] that ours is a society that already, under prescribed conditions, condones the killing of innocent persons, most recently on a large scale in the civilian casualties of the Persian Gulf War. He writes:

> . . . appropriateness of killing is a function of that which is perceived to be socially efficacious. Where killing is socially detrimental, it is forbidden; where it is beneficial, it is condoned. This includes the killing of *innocent* persons. . . . Abortion is perceived as sinister by many, when, in fact, it may be likened to our other methods of institutionalized killing, all of which currently serve the purpose of social pragmatism. (pp. 319–321)

The issue, as Potter has stressed,[4] is increasingly whether "sanctity of life" takes precedent over "meaningful life." To those who argue that the continuing presence of the seriously genetically defective among us would be a humbling reminder of the need to offer thanksgiving and compassion, I suggest that, despite all we can do in the way of eliminating genetic disease, there will still be no lack of human tragedies to test that compassion. I remain no stranger to that gut surge of emotion that wells up at the sight of a child, overwhelmingly handicapped by a genetic disorder, struggling gamely to make it on an impossibly tilted playing field, and parents distraught at the disaster with which they are confronted. As ethical standards evolve to recognize the new realities, isn't it a worthy goal to seek to minimize such situations? It seems clear that the right-to-life advocates will embrace the promise of gene therapy as the alternative to the currently practical, early abortion of genetically defective fetuses. This endorsement of an uncertain future development should not be permitted to obfuscate the issue. In an era of social pressure for two-child families, just as there should be no coercion of an individual to abort a genetically defective child when abortion is personally abhorrent, so no individual should be forced by law to carry a defective child to term despite her convictions, and thus compromise her prospects for two normal children.

In this troubled world, the genetic program I have suggested will appear to many to be both hopelessly idealistic and incredibly naive. It involves profound behavior modification, and this rather rapidly. In fact, it is only one of a number of apparently idealistic and naive programs which must be adopted to ensure that—worldwide—the grandchildren of the current younger generation inherit a world in reasonable balance between resources and population, a world permitting the realization of the human genetic potential. Implementing these programs requires both a political vision and a public awareness of which there are currently only faint glimmerings. It is possible that our species is not up to the task of extricating itself from the situation it has created. I prefer to think that as the current situation becomes better understood, the reaction to it will result—worldwide—in the political creativity which characterized the beginnings of the U.S. and other similar leaps of the human mind, with, at all levels, leadership emerging worthy of the times. The alternative is a continuation and accentuation of the nationalistic trends now so evident in many parts of the world, each nationalistic group striving in the name of security for a reproductive advantage over its neighbor, with the inevitable consequence an acceleration of current resource depletion.

Within recent years, two events with enormous portent for the future have transpired. On the one hand, with respect to the Gulf War, the United Nations has shown the ability to act in concert when confronted with a development perceived to threaten world stability. This engenders hope for similar resolute action in the face of the no less threatening issues described earlier, including action regarding the genetic component of the issues. On the other hand, the collapse of communism in Eastern Europe, with the emergence of governments much more preoccupied with internal problems than with balance-of-power issues, not only the better permits the entire world to address common issues, but, for all countries, should substantially reduce the burden of armaments. There is the opportunity for an epochal redirection of the human ingenuity, which previously concentrated on weapons of mass destruction, to an equal effort devoted to the enhancement of the human condition everywhere. Do not see this as some kind of new idealism but as the only realism that can sustain a faltering society. I expect some to dismiss this program as "impractical." My answer is: look where practical men have gotten us.

Human history presents no more dramatic time of decision than the present. My premise is that although the stresses and strains of a transition to a sustainable society will be enormous, reason will obtain, the necessary discipline will emerge, and law and order prevail, both nationally and internationally. It is too early to think the unthinkable: that humans have evolved into societies with so many strong, conflicting interests that the capacity or resolve to formulate/implement the necessary long-range planning on a worldwide basis no longer exists. Should this optimistic premise not prove correct, then the populations of the world will crash in various unpleasant ways. A measure of the human gene pool will survive, but humankind will have squandered much of the genetic legacy from the past five million years of human evolution.

Notes

Chapter 1

1. Neel, J.V. 1937. Phenotypic variability in mutant characters of *Drosophila funebris*. *J. Exp. Zool.* **75**:131–142.
2. Osler, W. 1904. Aequanimitas, with Other Addresses to Medical Students, Nurses, and Practitioners of Medicine. Philadelphia: P. Blakiston's Son and Co.
3. Hadorn, E., and Neel, J. V. 1938. Der hormonale Einfluß der Ringdrüse (Corpus Allatum) auf die Pupariumbildung bei Fliegen. *Roux Arch f. Entwicklungsmechanik der Organismen* **138**:281–304.
4. Neel, J.V. 1940. The interrelations of temperature, body size, and character expression in *Drosophila melanogaster*. *Genetics* **25**:225–250; Neel, J. V. 1940. The pattern of supernumerary macrochaetae in certain *Drosophila* mutants. *Genetics* **25**:251–277.
5. This seminar, in 1939, was undoubtedly the seed that resulted, ten years later, in his textbook, *Principles of Human Genetics*, which over the next 25 years, and in two subsequent editions, was more widely used than any other on this subject.
6. Neel, J. V. 1942. A study of a case of high mutation rate in *Drosophila melanogaster*. *Genetics* **27**:519–536.
7. Heron, D. 1913. Mendelism and the Problem of Mental Defect—A Criticism of Recent American Work. London: University of London, Publication of the Galton Laboratory, No. 7.
8. Haller, M. H. 1962. Eugenics: Hereditarian Attitudes in American Thought. New Brunswick, N.J.: Rutgers University Press.
9. Ludmerer, K. M. 1972. Genetics and American Society. Baltimore: Johns Hopkins University Press.
10. Kevles, D. J. 1985. In the Name of Eugenics. New York: Alfred A. Knopf.

11. Allen, G. E. 1986. The Eugenics Record Office at Cold Spring Harbor, 1910–1940: An essay in institutional history. *OSIRIS*, 2nd series, **2**:225–264.
12. Neel, J. V. 1943. Concerning the inheritance of red hair. *J. Hered.* **34**:93–96.
13. Reed, T. E. 1952. Red hair colour as a genetical character. *Ann. Eugenics* **17**:115–139.

Chapter 2

1. Neel, J. V. 1958. Editorial. Medicine's genetic horizons. *Ann. Int. Med.* **49**:472–476.
2. Neel, J. V. 1961. A geneticist looks at modern medicine. *Harvey Lectures* **56**:127–150.

Chapter 3

1. Wintrobe, M. M. 1942. Clinical Hematology. Philadelphia: Lea and Febiger, p. 517.
2. Dameshek, W. 1943. Familial Mediterranean target-oval cell syndromes. *Amer. J. Med. Sci.* **205**:643–660.
3. Neel, J. V., and Valentine, W. N. 1944. Hematologic and genetic study of the transmission of thalassemia (Cooley's anemia, Mediterranean anemia). *Arch. Int. Med.* **74**:185–196.
4. Valentine, W. N., and Neel, J. V. 1945. The frequency of thalassemia. *Am. J. Med. Sci.* **209**:568–572.
5. Neel, J. V. 1947. The clinical detection of the genetic carriers of inherited disease. *Medicine* **26**:115–153. [I note with pleasure that in 1992, this article was reprinted in part in a section entitled, "Classics in Medicine" in the Journal *Medicine,* with commentary by Barton Childs.]
6. Neel, J. V. 1949. The inheritance of sickle cell anemia. *Science* **110**:64–66; Neel, J. V. 1951. The inheritance of the sickling phenomenon, with particular reference to sickle cell disease. *Blood* **6**:389–412.
7. Pauling, L., Itano, H. A., Singer, S. J., and Wells, I. C. 1949. Sickle cell anemia, a molecular disease. *Science* **110**:543–548.
8. Garrod, A. E. 1908. The Croonian Lectures on inborn errors of metabolism. *Lancet* **2**:1–7, 73–79, 142–148, 214–220. [The republication of these lectures with a commentary by Harris is today the most available source of these unusual lectures. (Harris, H. 1963. Garrod's Inborn Errors of Metabolism. London: Oxford University Press.)]
9. Ingram, V. M. 1957. Gene mutations in human haemoglobin: The chemical difference between normal and sickle cell haemoglobin. *Nature* **180**:326–328.
10. Powell, W. N., Rodarte, J. G., and Neel, J. V. 1950. The occurrence in a

family of Sicilian ancestry of the traits for both sickling and thalassemia. *Blood* 5:887–897.

11. Itano, H. A., and Neel, J. V. 1950. A new inherited abnormality of human hemoglobin. *Proc. Natl. Acad.Sci. USA* 36:613–617.

12. Neel, J. V. 1952. Perspectives in the genetics of sickle cell disease. *Blood* 7:467–471.

13. Raper, A. B. 1950. Sickle-cell disease in Africa and America—a comparison. *J. Trop. Med. Hyg.* 53:49–53; Lehmann, H. 1951. Sickle-cell anaemia and sickle-cell trait as homo- and heterozygous gene-combinations. *Nature* 167:931–933.

14. Lambotte-Legrand, J., and Lambotte-Legrand, C. 1951. L'anémie à hématies falciformes chez l'infant indigène du Bas Congo. *Mémoirs Royal Colonial Belge* 19:93.

15. Neel, J. V. 1953. Data pertaining to the population dynamics of sickle cell disease. *Am. J. Hum. Genet.* 5:154–167.

16. Allison, A. C. 1954. The distribution of the sickle-cell trait in East Africa and elsewhere, and its apparent relationship to the incidence of subtertian malaria. *Trans. Roy. Soc. Trop. Med. & Hyg.* 48:312–318.

17. Lambotte-Legrand, J., and Lambotte-Legrand, C. 1955. Le pronostic de l'anémie drépanocytaire au Congo Belge (à propos de 300 cas et de 150 décès). *Ann. Soc. Belge Med. Trop.* 35:53–37.

18. Vandepitte, J. M., Zuelzer, W. W., Neel, J. V., and Colaert, J. 1955. Evidence concerning the inadequacy of mutation as an explanation of the frequency of the sickle cell gene in the Belgian Congo. *Blood* 20:341–350.

19. Livingstone, F. B. 1960. The wave of advance of an advantageous gene: The sickle cell gene in Liberia. *Human Biol.* 32:197–202.

20. There is now evidence from DNA studies for three (some say five) independent origins of the sickle-cell mutation. The three alleles have been designated the Senegal type (observed in Atlantic West Africa), the Benin type (observed in central West Africa), and the Bantu type (observed in Bantu-speakers) (reviewed in Nagel, R. L., Fabry, M. E., Pagnier, J., Zohoun, I., Wajcman, H., Baudin, V., and Labie, D. 1985. Hematologically and genetically distinct forms of sickle cell anemia in Africa. *New Engl. J. Med.* 312:880–884). The type or types of allele(s) present in Liberia had not been established at this writing. However, the principle of a spreading gene is not altered by these findings.

21. Neel, J. V., Hiernaux, J., Linhard, J., Robinson, A. R., Zuelzer, W. W., and Livingstone, F. B. 1956. Data on the occurrence of hemoglobin C and other abnormal hemoglobins in some African populations. *Am. J. Hum. Genet.* 8:138–150.

22. Olesen, E. B., Olesen, K., Livingstone, F. B., Cohen, F., Zuelzer, W. W.,Robinson, A. R., and Neel, J. V. 1959. Thalassemia in Liberia. *Brit. Med. J.* 1:1385–1387.

23. Neel, J. V. 1956. The genetics of human haemoglobin differences: Problems and perspectives. (The Galton Lecture for 1955.) *Ann. Hum. Genet.* 21:1–

30; Neel, J. V. 1957. Human hemoglobin types: Their epidemiological implications. (The Cutter Lecture for 1956.) *New Engl. J. Med.* **256**:161–171.
23. Willcox, M. C. 1975. Thalassemia in Northern Liberia. A survey in the Mount Nimba area. *J. Med. Genet.* **12**:55–63: Bienzle, U. J., Kamp, H., Feldheim, M., Reimer, A., Steffen, E., and Guggenmoos-Holzmann, I. 1983. The distribution and interaction of haemoglobin variants and the β-thalassemia gene in Liberia. *Hum. Genet.* **63**:400–403.
24. Miller, L. H., Mason, S. J., Clyde, D. F., and McGinnis, M. H. 1976. The resistance factor to *Plasmodium vivax* in Blacks: The Duffy-blood-group genotype FyFy. *New Engl. J. Med.* **295**:302–304.
25. Rucknagel, D. L., and Neel, J. V. 1961. The hemoglobinopathies. In: *Progress in Medical Genetics,* Vol. I, A. Steinberg (ed.). New York: Grune & Stratton.

Chapter 4

1. Oughterson, A. W., and Warren, S. 1956. Medical Effects of the Atomic Bomb in Japan. New York: McGraw-Hill.
2. This is a point on which it is obviously difficult to obtain systematic data, but for anecdotal verification of the relevance of Tsuzuki's comment, see pp. 203 and 210 in: Saga, J. 1987. Memories of Silk and Straw. Tokyo: Kodansha International.
3. Snell, F. M., Neel, J. V., and Ishibashi, K. 1949. Hematologic studies in Hiroshima and a control city two years after the atomic bombing. *Arch. Int. Med.* **84**:569–604.
4. In 1947, the Manhattan Engineering District was superseded by the Atomic Energy Commission. This in turn became the Energy Research and Development Administration in 1975, and then the Department of Energy in 1977.
5. Los Alamos Scientific Laboratory. 1950 (revised). The effects of atomic weapons. Washington: U.S. Gov't. Printing Office.
6. The terminology for radiation exposure has evolved over the past 50 years. Here the term r is an abbreviation for "roentgen," arbitrarily defined as the quantity of ionizing radiation that liberates 2.083×10^9 ion pairs in a cubic centimeter of air (at $0 \geq C$ and at a pressure of 760 mm of mercury). Roentgen is also defined, in terms of electrical charge, as 2.5×10^{-4} coulomb per kilogram of air. We will in this treatment express radiation exposure in r when this was the original presentation but use the newer terminology of gray and sievert units (see Chapter 13) when this becomes appropriate. This rather austere terminology needs to be translated into more familiar language. In 1980, the average annual individual dose from diagnostic radiology to the gonads of the parents of the next generation in the U.S. (the genetically significant dose) was about .03 r. With respect to the specific procedures of diagnostic radiology, in males without gonadal shielding, an X-ray examina-

tion of the lower spine delivers about 2 *r* to the gonads and an X-ray of the abdomen, about 1 *r*. In females, the corresponding doses are 7 *r* and 2 *r*. However, most of these exposures are of no genetic significance because the patients have completed reproduction. (Exposure of the U.S. population from diagnostic medical radiation. Washington: National Council on Radiation Protection and Measurements Report No. 100, pp. 38–41.)

7. Genetics Conference. 1947. Genetic effects of the atomic bombs in Hiroshima and Nagasaki. *Science* **106:**331–333.
8. Kevles, D. J. 1978. The Physicists. New York: A. A. Knopf.

Chapter 5

1. Craven, W. F., and Cate, J. L. 1953. The Army Air Forces in World War II, Vol. 5, The Pacific: Matterhorn to Nagasaki, June 1944 to August 1945. Chicago: University of Chicago Press.
2. Beser, J. 1988. Hiroshima and Nagasaki Revisited. Memphis (Tennessee): Global Press.
3. Chinnock, F. W. 1969. Nagasaki: The Forgotten Bomb. Cleveland: World Publishing Co.
4. Marx, J. L. 1971. Nagasaki: The Necessary Bomb? New York: MacMillan Co.
5. Nagai, T. 1984. The Bells of Nagasaki. Translated by W. Johnson. Tokyo: Kodansha International.
6. Committee for the Compilation of Materials on Damage Caused by the Atomic Bombs in Hiroshima and Nagasaki. 1981. Hiroshima and Nagasaki: The Physical, Medical, and Social Effects of the Atomic Bombings. New York: Basic Books.
7. Wyden, P. 1984. Day One: Before Hiroshima and After. New York: Simon and Schuster.
8. In Wyden's "Day One: Before Hiroshima and After," the following paragraphs appear:

> Shinzo Hamai, known among his people as "The A-Bomb Mayor," understood that the survivors' psychological bind would be unending. "They know there is no effective treatment," he said. "They feel doomed, I do not know how long this mental suffering will continue." His political opponents accused Hamai of "selling the bomb." Eventually he had to leave his party and run for reelection as an independent, but his loyalty to his fellow survivors never faltered. Since he was one of the lucky *hibakusha* (survivors) not visibly stigmatized by the bomb, he could face the past as well as the future.
>
> Hamai was pleased, therefore, when a young American medical officer called on him in 1947 and announced that the United States Government

would start a clinic to determine the health consequences of the bombing with scientific precision. The mayor offered a site conveniently near downtown. The American objected; the area might be subject to flooding. They wanted their Atomic Bomb Casualty Commission atop Hijiyama Hill, where it would tower 500 feet over the city. Hamai advised against this choice and offered another hill site. He pointed out that a military cemetery and a memorial to an ancient emperor made Hijiyama sacred ground. The citizens would resent having an American institution desecrate this choice spot. The occupiers were adamant. So was Hamai. Eventually he was summoned to the welfare ministry in Tokyo and surrendered after being threatened with "disagreeable consequences." (pp. 336–337).

This account is not quite correct. There was indeed a young American medical officer (myself) who called on Mayor Hamai to discuss possible sites for a research facility. The rejection of the Mayor's suggestion was, however, a later action on the part of the ABCC, in which I did not participate.

9. There has been much speculation as to whether there was significant residual radioactivity following the bombings, such that persons who came close to the hypocenter following the bombings, in the course of relief operations or otherwise, or even American servicemen whose duties brought them to the two cities up to 12 months following the bombings, received significant amounts of radiation. From an examination of the evidence then available, I felt the contribution of residual radiation to the radiation exposure of the average survivor was relatively minor and could be ignored. We accordingly made no effort to obtain information on our registrants' whereabouts on the first few days following the bombings. The later detailed treatment of this subject by K. Takeshita (*J. Radiat. Res.* **16,** Supplement, 24–31, 1975) confirms this position; although there was significant radioactivity in the region of the hypocenter for several days after the bombings, human activity there was minimal.

10. Oya, S. (ed.) 1968. Japan's Longest Day, compiled by the Pacific War Research Society. London: Souvenir Press.

11. For a charming account of the postwar Japan with which we became familiar in the course of these studies, see Schull, W. J. 1990. Song Among the Ruins. Cambridge (Mass.): Harvard University Press.

12. Among the American pediatricians who did so much to ensure the quality of the data, by conducting diagnostic clinics on newborn infants and the examinations at age nine months were John Wood, James Yamazaki, Wayne and Jane Borges, Robert Kurata, Stanley and Phyllis Wright, and George Plummer. A major responsibility of this group became the training of the young Japanese physicians who would join the organization for a year or two shortly after graduation from medical school, to improve their pediatric background and their command of English.

13. Eisenbud, M. 1990. An Environmental Odyssey: People, Pollution, and

Politics in the Life of a Practical Scientist. Seattle: University of Washington Press.
14. Neel, J. V., and Schull, W. J. 1956. The Effect of Exposure to the Atomic Bombs on Pregnancy Termination in Hiroshima and Nagasaki. Washington, D.C.: National Academy of Sciences–National Research Council Publ. 461.
15. Neel, J. V., and Schull, W. J. 1954. Human Heredity. Chicago: University of Chicago Press.

An entirely unanticipated offshoot of this activity materialized in 1959, when we learned in a roundabout way that a pirated translation of our book had been published in the U.S.S.R. in 1958. Since at this time the suppression of the classic genetic viewpoint that had characterized the Lysenko era still lingered, we were amazed, our book being of course uncompromisingly the antithesis of Lysenkoism. We have never understood quite how this translation slipped through the net, but it was a real pleasure when, during my first visit to the U.S.S.R., in 1984, well-worn copies of our text emerged wherever I went, and it became apparent how useful it had been to the "genetic underground" that surfaced with the downfall of Lysenko in the 1950s.

16. See Chapter 4, note 4.
17. Neel, J. V. 1958. A study of major congenital defects in Japanese infants. *Am. J. Hum. Genet.* **10**:398–445.

Chapter 6

1. Neel, J. V., Kodani, M., Brewer, R., and Anderson, R. C. 1949. The incidence of consanguineous matings in Japan. *Am. J. Hum. Genet.* **1**:156–178.
2. McIntosh, R., Merritt, K. K., Richards, M. R., Samuels, M. H., and Bellows, M. T. 1954. The incidence of congenital malformations: A study of 5,694 pregnancies. *Pediatrics* **14**:505–522.
3. Baird, P. A., Anderson, T. W., Newcombe, H. B., and Lowry, R. B. 1988. Genetic disorders in children and young adults: A population study. *Am. J. Hum. Genet.* **42**:677–693.
4. The many efforts to develop a figure for the frequency of simply inherited disease and chromosomal abnormalities were most recently brought together in the 1990 report of the Committee on the Biological Effects of Ionizing Radiation entitled, "Health Effects of Exposure to Low Levels of Ionizing Radiation" (BEIR V), Washington: National Academy Press. Our figure is comfortably within the range of other estimates.
5. Kishimoto, K. 1962. Preliminary report of activities of the consanguinity study group of the Science Council of Japan. *Eugen. Quart.* **9**:5–13; Tanaka, K. 1962. Consanguinity study on Japanese populations. In: *The Genetics of Migrant and Isolate Populations* (E. Goldschmidt, ed.) Baltimore: Williams and Wilkins.

6. Schull, W. J., and Neel, J. V. 1965. The Effects of Inbreeding on Japanese Children. New York: Harper and Row.
7. In the former Soviet Union, some of the exposures resulting from the Chernobyl accident of 1986, plus other inadvertent exposures at the Semipalatinsk test site and Chelyabinsk nuclear reprocessing plant, rival those in Japan, but it is not yet clear that significant studies of the potential genetic effects of these exposures can be mounted.
8. Ishikuni, N., Nemoto, H., Neel, J. V., Drew, A. L., Yanase, T., and Matsumoto, Y. S. 1960. Hosojima. *Am. J. Hum. Genet.* **12**:67–75.
9. Schull, W. J., and Neel, J. V. 1972. The effects of parental consanguinity and inbreeding in Hirado, Japan. V. Summary and interpretation. *Am. J. Hum. Genet.* **24**:425–453.

Chapter 7

1. Muller, H. J. 1950. Our load of mutations. *Am. J. Hum. Genet.* **2**:111–176.
2. Neel, J. V. 1958. The study of natural selection in primitive and civilized human populations. *Hum. Biol.* **3**:43–72; see also Neel, J. V., and Schull, W. J. 1972. Differential fertility and human evolution. In: Evolutionary Biology, Vol. 6 (eds. Th. Dobzhansky, M. K. Hecht, and W. C. Steere). New York: Appleton-Century-Crofts, pp. 363–378; Neel, J. V. 1975. The study of "natural" selection in man: Last chance. In: *The Role of Natural Selection in Human Evolution,* F. M. Salzano (ed.) Amsterdam: North-Holland.
3. The precise date of entry of the first Amerindians into the Americas continues to be the subject of brisk controversy, current estimates ranging from approximately 40,000 to 15,000 years ago. Three interesting treatments are:

   Dillehay, T. D., and Collins, M. B. 1988. Early cultural evidence from Monte Verde in Chile. *Nature* **332**:150–152; Wolkomir, R. 1991. New finds could rewrite the start of American history. *Smithsonian* **21**:130–144; Dillehay, T. D., and Meltzer, D. J. 1991. The First Americans: Search and Research. Boca Raton (FL): CRC Press.

4. For a hilarious account of how the relationships between very civilized men can deteriorate under jungle conditions, I recommend:

   MacCreagh, G. 1926. White Waters and Black. Chicago: University of Chicago Press.

5. Maybury-Lewis, D. 1967. Akwē-Shavante Society. Oxford: Clarendon Press.
6. Neel, J. V., Salzano, F. M., Junqueira, P. C., Keiter, F., and Maybury-Lewis, D. 1964. Studies on the Xavante Indians of the Brazilian Mato Grosso. *Am. J. Hum. Genet.* **16**:52–140.
7. Neel, J. V., and Salzano, F. M. 1967. Further studies on the Xavante Indians. X. Some hypotheses–generalizations resulting from these studies. *Am. J. Hum. Genet.* **19**:554–574.

Chapter 8

1. The most graphic and comprehensive accounts are by Chagnon, Lizot, and
Good (Chagnon, N. 1968. Yanomama, the Fierce People. New York: Holt,
Rinehart and Winston; Chagnon, N. 1974. Studying the Yanomama. New
York: Holt, Rinehart and Winston; Lizot, J. 1984. Les Yanomamai Centraux.
Paris: École des Haut Études en Sciences Sociales. Cahiers de l'Homme,
Nouvelle Série XXII; Good, K., with Chanoff, D. 1991. Into the Heart. New
York: Simon & Schuster). Otto Zeries supplies an early account of their
material culture (Zeries, O. 1964. Waika: Die Kulturgeschichte Stellung der
Waika-Indianer im Rahmen der Völkerkundes Südamerikas. Band I. Mu-
nich: Klaus Renner Verlag).

   "Yanomama" is the remarkable account of a white woman who, captured
as a child of 11, lived with the Indians for 20 years, and, escaping to a
mission, told her story to Ettore Biocca (Biocca, E. 1970, Yanóama, New
York: Dutton. Translated from the Italian by Dennis Rhodes). At first the
apparently total recall of this illiterate woman for these 20 years, accurate
wherever it could be checked against Chagnon's taperecorded material,
stretched my imagination. Slowly I realized that the events she recounts
had been discussed so many times in the constricted world of the villages
in which she lived that they had become much more ingrained in her
memory than the events of the varied, somewhat helter-skelter lives we live
are for us.

   Those who enjoy superb photography should see the books by Re and Re
(Re, G., and Re, F. 1984. Gli Ultimi Yanomami. Torino: Point Couleur Edi-
zione) and Goetz (Goetz, I. S. 1969. Uriji jami. Caracas: Asociacion Cultural
Humbolt. Translated into English by P. T. Furst). Note the variable spellings
of the tribal name; I take no position as to which is more correct.

2. Esmeralda had been the highest outpost on the Orinoco of the Spanish
Colonial Empire, with a sizable garrison and mission. It has gone downhill
ever since. Von Humbolt, of his visit there in 1801, wrote (p. 511): "If the *villa*
of Esmeralda, with a population of twelve or fifteen families, be at present
considered as a frightful abode, this must be attributed to the want of cultivation, the distance from every other inhabited country, and the excessive
quantity of moschettos" (von Humbolt, A., and Bonpland, A. 1914. Personal
narrative of travels to the equinoctial regions of the New Continent during
the years 1799–1804. Translated into English by H. M. Williams. London:
Longman, Hurst, Rees, Orme, and Brown.). We were to get to know Esmeralda well. The "families" in von Humbolt's time were headed by retired
soldiers or persons banished there for various crimes. By our time, it was
reduced to a few families charged with keeping termite mounds off the
runway and a nondescript storehouse where one could also sling a hammock
when waiting for a plane. Von Humbolt mentions the largely unfriendly
contacts of the Spanish explorers with Indians higher on the Orinoco who
must have been the Yanomama, and later explorers of the region had similar
experiences. To my knowledge, the first semipermanent contact with them in

modern times, at least in Venezuela, was established by James Barker of the New Tribes Mission in 1952.

3. This aspect of tribal disruption is difficult to document in detail, but see: Posey, D. A. 1987. Contact before contact: Typology of post-Colombian interaction with Northern Kayapó of the Amazon Basin. *Bol. Mus. Par. Emilio Goeldi, Sér Antropol.* **3**:135–154; Aquiar, G. F. 1991. Ethnohistory, intertribal relationships, and genetic diversity among Amazonian Indians. *Hum. Biol.* **63**:743–762.

4. Distributed by National Audiovisual Center, Washington, D.C. 20409.

5. Richards, P. W. 1964. The Tropical Rain Forest. Cambridge: Cambridge University Press.

6. Forsyth, A., and Miyata, K. 1984. Tropical Nature. New York: Charles Scribner & Sons.

7. For an engaging but somewhat dramatic account of the dangers in Yanomama-land, written in the best tradition of the adventurer–traveler, I recommend "In Trouble Again" by Redmond O'Hanlon (New York: Random House. 1988).

8. Charlie occupies a special place in my heart. I first met him during the Venezuela stopover of 1964, a young Venezuelan dentist who was passionately attracted to the Amerindians. He wished to join forces with us, so I brought him to Ann Arbor for a year of training. One month into his stay there, he discovered sky-diving. The discovery was not good for his scientific training. When he returned to Venezuela, he soon organized a sky-diving rescue team, which shortly distinguished itself with several spectacular exploits. He joined us on a number of our expeditions to the Yanomama, and then moved on to help organize a series of government-backed expeditions to other remote parts of Venezuela. About this time, he invented a complex explorer survival knife, which may well become *the* standard for military personnel in the jungle. His well-published exploits and engaging personality led to an appointment as Minister of Youth, a position rather quickly terminated when his disregard for Government protocol became apparent. Charlie was the best amateur photographer I have ever met—nothing we every engaged in in the jungle was so important that it couldn't be interrupted for a picture. He has written several beautifully illustrated books on his experiences:

> Brewer-Carias, C. 1975. Venezuela. Caracas: Oficina Central de Informacion.
>
> ——1978. Roraima, the Crystal Mountain. Caracas: Oficina Central de Informacion.
>
> ——1978. The Lost World of Venezuela and Its Vegetation. Caracas: Oficina Central de Informacion.
>
> ——1983. Sarisariñama. Caracas: Oficina Central de Informacion.

Reconciling in the field the sometimes disparate interests of Charlie, Napoleon, and the genetics team served to remind me, back in Ann Arbor, of how easy it really was to chair a Medical School Department.

Chapter 9

1. Neel, J. V., Salzano, F. M., Junqueira, P. C., Keiter, F., and Maybury-Lewis, D. 1964. Studies on the Xavante Indians of the Brazilian Mato Grosso. *Am. J. Hum. Genet.* **16**:52–140; Weinstein, E. D., Neel, J. V., and Salzano, F. M. 1967. Further studies on the Xavante Indians. VI. The physical status of the Xavantes of Simões Lopes. *Am. J. Hum. Genet.* **19**:532–542.
2. Salzano, F. M., and Neel, J. V. 1976. New data on the vision of South American Indians. *Bull. Pan Am. Hlth. Organ.* **10**:1–8.
3. Donnelly, C. J., Thomson, L. A., Stiles, H. M., Brewer, C., Neel, J. V., and Brunelle, J. A. 1977. Plaque, caries, periodontal disease, and acculturation among Yanomama Indians, Venezuela. *Community Dent. Oral Epidemiol.* **5**:30–39.
4. The best quantitative nutritional data have been published by Lizot (Lizot, J. 1978. Population, resources and warfare among the Yanomami. *Man* **12**:497–517), who argues convincingly that there is no lack of protein in the Yanomama diet, an important point in view of the well-publicized speculations to the contrary (cf. esp. Harris, M. 1972. I don't see how you can write anything of value if you don't offend someone. *Psychol. Today* **8**:64–69; 1974. Cows, pigs, wars and witches; The riddle of culture. New York: Random House.
5. Neel, J. V., Mikkelsen, W. M., Rucknagel, D. L., Weinstein, E. D., Goyer, R. A., and Abadie, S. H. 1968. Further studies of the Xavante Indians. VIII. Some observations on blood, urine, and stool specimens. *Am. J. Trop. Med. Hyg.* **17**:474–485.
6. Neel, J. V., Andrade, A. H. P., Brown, G. E., Eveland, W. E., Goobar, J., Sodeman, W.A., Stollerman, G. H., Weinstein, E. D., and Wheeler, A. H. 1968. Further studies of the Xavante Indians. IX. Immunologic status with respect to various diseases and organisms. *Am. J. Trop. Med. Hyg.* **17**:486–498.
7. See, for example, Knowler, W. C., Pettitt, D. J., Bennett, P. H., Williams, R. L. 1983. Diabetes mellitus in the Pima Indians: Genetic and evolutionary considerations. *Am. J. Phys. Anth.* **62**:107–114; and Weiss, K. M., Ferrell, R. E., and Harris, C. L. 1984. A New World syndrome of metabolic diseases with a genetic and evolutionary basis. *Yrbk. Phys. Anthro.* **27**:153–178.
8. Spielman, R. S., Fajans, S. S., Neel, J. V., Pek, S., Floyd, J. C., and Oliver, W. J. 1982. Glucose tolerance in two unacculturated Indian tribes of Brazil. *Diabetologia* **23**:90–93.
9. Oliver, W. J., Cohen, E. L., and Neel, J. V. 1975. Blood pressure, sodium intake, and sodium related hormones in the Yanomama Indians, a "no salt" culture. *Circulation* **52**:146–151; Oliver, W. J., Neel, J. V., Grekin, R. J., and Cohen, E. L. 1981. Hormonal adaptation to the stresses imposed upon sodium balance by pregnancy and lactation in the Yanomama Indians, a culture without salt. *Circulation* **63**:110–116.
10. Hecker, L. H., Allen, H. E., Dinman, B. D., and Neel, J. V. 1974. Heavy metal

levels in acculturated and unacculturated populations. *Arch. Envir. Hlth.* **29**:181–185.

11. Rivière, R., Comar, D., Colonomos, M., Desenne, J., and Roche, M. 1968. Iodine deficiency without goiter in isolated Yanomama Indians: Preliminary note. In: Biomedical Challenges Presented by the American Indian. Pan American Health Org.: Washington, D.C., pp. 120–123.

12. Lawrence, D. N., Facklam, R. R., Sottnek, F. O., Hancock, G. A., Neel, J. V., and Salzano, F. M. 1979. Epidemiologic studies among Amerindian populations of Amazônia. I. Pyoderma: Prevalence and associated pathogens. *Am. J. Trop. Med. Hyg.* **28**:548–558.

13. Lawrence, D. N., Neel, J. V., Abadie, S. H., Moore, L. L., Adams, L. J., Healy, G. R., and Kagan, I. G. 1980. Epidemiological studies among Amerindian populations of Amazônia. III. Intestinal parasitoses in newly contacted and acculturating villages. *Am. J. Trop. Med. Hyg.* **29**:530–537.

14. Beaver, P. C., Neel, J. V., and Orihel, T. C. 1976. *Dipetalonema Perstans* and *Mansonella Ozzardi* in Indians of southern Venezuela. *Am. J. Trop. Med. Hyg.* **25**:263–265.

15. Eveland, W. C., Oliver, W. J., and Neel, J. V. 1971. Characteristics of *Escherichia coli* serotypes in the Yanomama, a primitive Indian tribe of South America. *Infect. Immun.* **4**:753–756.

16. There is a large literature on the various ways in which the bacterial flora of the gut may influence health status. See, for example: Gustafsson, B. E. 1982. The physiological importance of the colonic microflora. *Scand. J. Gastroenterol.* **77**:117–131; Bocci, V. 1992. The neglected organ: Bacterial flora has a crucial immunostimulatory role. *Persp. Biol. Med.* **35**:251–260.

17. Neel, J. V. 1971. Genetic aspects of the ecology of disease in the American Indian. In: *The Ongoing Evolution of Latin American Populations,* F. A. Salzano (ed). Springfield: C. C. Thomas, pp. 561–590; Neel, J. V. 1977. Health and disease in unacculturated Amerindian Populations. In: *Health and Disease in Tribal Societies,* Ciba Foundation Symposium 49. Amsterdam: Elsevier–North Holland, pp. 155–177.

18. Neel, J. V., Centerwall, W. R., Chagnon, N. A., and Casey, H. L. 1970. Notes on the effect of measles and measles vaccine in a virgin-soil population of South American Indians. *Am. J. Epidemiol.* **91**:418–429.

19. Denevan, W. M. (ed.) 1976. *The Native Population of the Americas in 1492,* Madison: The University of Wisconsin.

20. Squire, W. 1882. On measles in Fiji. *Trans. Ep. Soc. London* **4**:72–74.

21. Nutels, N. 1968. Medical problems of newly contacted Indian groups. In: Biomedical Challenges Presented by the American Indian. Washington: Pan American Health Organization, pp. 68–76.

22. Bloom, A. D., Neel, J. V., Choi, K. W., Iida, S., and Chagnon, N. A. 1970. Chromosome aberrations among the Yanomama Indiana. *Proc. Natl. Acad. Sci. USA* **66**:920–927.

23. Bloom, A. D., Neel, J. V., Tsuchimoto, T., and Meilinger, K. 1973. Chromosomal breakage in leukocytes of South American Indians. *Cytogenet. Cell Genet.* **12**:175–186.

24. Awa, A. A., and Neel, J. V. 1986. Cytogenetic "rogue" cells: What is their frequency, origin, and evolutionary significance? *Proc. Natl. Acad. Sci. USA* **83**:1021–1025.
25. Neel, J. V., Awa, A. A., Kodama, Y., Nakano, M., and Mabuchi, K. 1992. "Rogue" lymphocytes among Ukrainians not exposed to radioactive fall-out from the Chernobyl accident: The possible role of this phenomenon in oncogenesis, teratogenesis, and mutagenesis. *Proc. Natl. Acad. Sci. USA* **89**:6973–6977.
26. Two books which serve as an excellent entry to this complex field are: Shapiro, J. A. (ed.) 1983. Mobile Genetic Elements. New York: Academic Press; Lambert, M. E., McDonald, J. F., and Weinstein, I. B. (eds.) 1988. Eukaryotic transposable elements as mutagenic agents. Banbury Report 30. Cold Spring Harbor, N.Y.: Cold Spring Harbor Laboratory Press.
27. Neel, J. V. 1971. Genetic aspects of the ecology of disease in the American Indian. In: The Ongoing Evolution of Latin American Populations (F. A. Salzano, ed.) Springfield: C. C. Thomas, pp. 561–590; Neel, J. V. 1974. Control of disease among Amerindians in cultural transition. *Bull. Pan Am. Health Org.* **VIII**:205–211; Neel, J. V. 1977. Health and disease in unacculturated Amerindian populations. In: Health and Disease in Tribal Societies (Ciba Foundation Symp. 49 n.s.) Amsterdam: Elsevier, pp. 155–177.
27. Neel, J. V. 1982. Infectious disease among Amerindians. *Med. Anthropol.* **6**:47–54.
28. Pan American Health Organization. 1968. Biomedical Challenges Presented by the American Indian. Pan American Health Organization Publication 165. Washington: Pan American Health Organization.

Chapter 10

1. Krzywicki, L. 1934. Primitive Society and Its Vital Statistics. London: MacMillan.
2. The most recent year for which U.S. age data by sex are available is 1980 (Bureau of the Census. 1983. 1980 Census of Population, Vol. 1., Chap. 8, General Population Characteristics, Part 1, Table 41).
3. Firth, R. 1936. We, the Tikopia. Boston: Beacon Press, p. 376. "Faces turned down" is a euphemism for smothering at birth.
4. MacCluer, J. W., Neel, J. V., and Chagnon, N. A. 1971. Demographic structure of a primitive population: A simulation. *Am. J. Phys. Anthropol.* **35**:193–207.
5. Neel, J. V. 1980. On being headman. *Perspect. Biol. Med.* **23**:277–294.
6. Neel, J. V., and Weiss, K. M. 1975. The genetic structure of a tribal population, the Yanomama Indians. XII. Biodemographic studies. *Am. J. Phys. Anthropol.* **42**:25–51.
7. Cohen, M. N. 1989. Health and the Rise of Civilization. New Haven: Yale University Press.

8. Chagnon, N. A. 1974. Studying the Yąnomamö. New York: Holt, Rinehart, and Winston, esp. p. 160.
9. Deevey, E. S. 1960. The human population. *Scientific American* **203**(3):195–204.
10. Spielman, R. S., Neel, J. V., and Li, F. H. F. 1977. Inbreeding estimation from population data: Models, procedures, and implications. *Genetics* **85**:355–371.
11. Thompson, E. A., and Neel, J. V. 1970. Probability of founder effect in a tribal population. *Proc. Natl. Acad. Sci. USA* **75**:1442–1445.
12. Neel, J. V. 1970. Lessons from a primitive people. *Science* **170**:815–822; Neel, J. V. 1978. The population structure of an Amerindian tribe, the Yanomama. *Ann. Rev. Genet.* **12**:365–413; Neel, J. V. 1982. The wonder of our presence here: A commentary on the evolution and maintenance of human diversity. *Persp. Biol. Med.* **25**:518–558.
13. Eisely, L. 1957. The Immense Journey. New York: Random House.

Chapter 11

1. Ward, R. H., and Neel, J. V. 1970. Gene frequencies and microdifferentiation among the Makiritare Indians. IV. A comparison of a genetic network with ethnohistory and migration matrices: A new index of genetic isolation. *Am. J. Hum. Genet.* **22**:538–561; Ward, R. H. 1972. The genetic structure of a tribal population, the Yanomama Indians. *Ann. Hum. Genet., Lond.* **36**:21–43.
2. Cavalli-Sforza, L. L., and Edwards, A. W. F. 1964. Analysis of human evolution. In: Genetics Today, S. J. Geerts, ed. Oxford: Pergamon Press, pp. 923–933.
3. Spielman, R. S., da Rocha, F. J., Weitkamp, L. R., Ward, R. H., Neel, J. V., and Chagnon, N. A. 1972. The genetic structure of a tribal population, the Yanomama Indians. VII. Anthropometric differences among Yanomama villages. *Am. J. Phys. Anthropol.* **37**:345–356; da Rocha, F. J., Spielman, R. S., and Neel, J. V. 1974. A comparison of gene frequency and anthropometric distance matrices in seven villages of four Indian tribes. *Hum. Biol.* **46**:295–310; Spielman, R. S., Migliazza, E. C., and Neel, J.V. 1974. Regional linguistic and genetic differences among Yanomama Indians. *Science* **184**:637–644; Neel, J. V., Rothhammer, F., and Lingoes, J. C. 1974. The genetic structure of a tribal population, the Yanomama Indians. X. Agreement between representations of village distances based on different sets of characteristics. *Am. J. Hum. Genet.* **26**:281–303.
4. Tanis, R. J., Ferrell, R. E., Neel, J. V., and Morrow, M. 1974. Albumin Yanomama-2 a "private" polymorphism of serum albumin. *Ann. Hum. Genet., Lond.* **38**:179–190; Ward, R. H., and Neel, J. V. 1976. The genetic structure of a tribal population, the Yanomama Indians. XIV. Clines and their interpretation. *Genetics* **82**:103–121; Tanis, R. J., Neel, J. V., and de Araúz, R. T. 1977. Two more "private" polymorphisms of Amerindian tribes. LDHB

Notes

Gua-1 and ACP-1 B Gua-1 in the Guayamí in Panama. *Am. J. Hum. Genet.*
29:419–430; Neel, J. V. 1980. Isolates and private polymorphisms. In: Population Structure and Disorders, A. Eriksson, ed. London: Academic Press, pp. 173–193.
5. Neel, J. V. 1978. Rare variants, private polymorphisms, and locus heterozygosity in Amerindian populations. *Am. J. Hum. Genet.* **30:**465–490.
6. Smouse, P. E., Vitzhum, V. J., and Neel, J. V. 1981. The impact of random and lineal fission on the genetic divergence of small human groups: A case study among the Yanomama. *Genetics* **98:**179–197.
7. For a condensed discussion of both the remarkable progres and continuing areas of disagreement regarding human evolution see: Tuttle, R. H. 1988. What's new in African paleoanthropology. *Ann. Rev. Anthropol.* **17:**391–426; Simon, E. L. 1989. Human origins. *Science* 245:1343–1350.
8. Smouse, P. E., and Neel, J. V. 1977. Multivariate analysis of gametic disequilibrium in the Yanomama. *Genetics* **85:**733–752; Smouse, P. E., Neel, J. V., and Liu, W. 1983. Multiple-locus departures from panmictic equilibrium within and between village gene pools of Amerindian tribes at different stages of agglomeration. *Genetics* **104:**133–153.
9. Neel, J. V., and Ward, R. H. 1970. Village and tribal genetic distances among American Indians, and the possible implications for human evolution. *Proc. Natl. Acad. Sci. USA* **65:**323–330.
10. Neel, J. V., Layrisse, M., and Salzano, F. 1977. Man in the tropics: The Yanomama Indians. In: Population Structure and Human Variation, G. A. Harrison (ed), International Biological Programme, Vol. II. Cambridge: Cambridge University Press, pp. 109–142; Neel, J. V. 1978. The population structure of an Amerindian tribe, the Yanomama. *Ann. Rev. Genet.* **12:**365–413; Neel, J. V. 1982. The wonder of our presence here: A commentary on the evolution and maintenance of human diversity. *Perspect. Biol. Med.* **25:**518–558; Neel, J. V. 1983. Some base lines for human evolution and the genetic implications of recent cultural developments. In: How Humans Adapt: A Biocultural Odyssey, D. J. Ortner (ed). Washington, D.C.: Smithsonian Press, pp. 67–102.
11. The reference most commonly cited as initiating Wright's emergence as a commanding figure in evolutionary theory is "Evolution in Mendelian Populations," published in 1931 in *Genetics* **16:**97–159. For a masterful discussion of Wright's thinking and impact on evolutionary genetics, see "Sewall Wright and Evolutionary Theory," by William B. Provine (Chicago: University of Chicago Press, 1986).
12. Among their numerous papers on this subject, the two most relevant are: Eldredge, N., and Gould, S. J. 1972. Punctuated equilibria: An alternative to phyletic gradualism. In: Models in Paleobiology, T. J. M. Schopf (ed). San Francisco: Freeman, pp. 82–115; Gould, S. J., and Eldredge, N. 1977. Punctuated equilibria: The tempo and mode of evolution reconsidered. *Paleobiology* **3:**115–151.
13. Neel, J. V. 1984. Human evolution: Many small steps, but not punctuated equilibria. *Perspect. Biol. Med.* **28:**75–103.

413

14. Wilson, P. J. 1980. Man, the Promising Primate. New Haven: Yale University Press.
15. For a recent and much more extended assessment of the status of "punctuated equilibria," the reader might consult: Hoffman, A. 1989. Arguments on Evolution. Oxford: Oxford University Press.
16. I hesitate to open the Pandora's box of dissent over the timing of the Amerindians arrival in the New World. For two recent objective presentations of the elements in the puzzle, I recommend: Dillehay, T. D., and Meltzer, D. J. (eds.) 1991. The First Americans: Search and Research. Boca Raton, FL: CRC Press; Hoffecker, J. F., Powers, W. R., and Goebel, T. 1993. The colonization of Beringia and the peopling of the New World. *Science* **259**:46–53.
17. Arvelo-Jiménez, N., and Cousins, A. L. 1991. False promises. *Cultural Survival* **16**:10–13.
18. Darwin, C. 1862. The Various Contrivances by which Orchids are Fertilized by Insects. London: J. Murray.

Chapter 12

1. Neel, J. V., and Falls, H. F. 1951. The rate of mutation of the gene responsible for retinoblastoma in man. *Science* **114**:419–422; Falls, H. F., and Neel, J. V. 1951. Genetics of retinoblastoma. *Arch. Ophthalmol.* **46**:367–389.
2. Knudson, A. G. 1971. Mutation and cancer: Statistical study of retinoblastoma. *Proc. Natl. Acad. Sci. USA* **68**:820–823.
3. Reed, T. E., and Neel, J. V. 1955. A genetic study of multiple polyposis of the colon (with an appendix deriving a method of estimating relative fitness). *Am. J. Hum. Genet.* **7**:236–263; Crowe, F. W., Schull, W. J., and Neel, J. V. 1956. A Clinical, Pathological, and Genetic Study of Multiple Neurofibromatosis. Springfield, IL: C. C. Thomas; Reed, T. E., and Neel, J. V. 1959. Huntington's chorea in Michigan. II. Selection and mutation. *Am. J. Hum. Genet.* **11**:107–136; Shaw, M. W., Falls, H. F., and Neel, J. V. 1960. Congenital aniridia. *Am. J. Hum. Genet.* **12**:389–415.
4. Vogel, F., and Rathenberg, R. 1975. Spontaneous mutation in man. In: Advances in Human Genetics, H. Harris and K. Hirschhorn, eds. New York: Plenum Press, pp. 223–318; Neel, J. V. 1983. Frequency of spontaneous and induced "point" mutations in higher eukaryotes. *J. Hered.* **74**:2–15.
5. Neel, J. V., Satoh, C., Goriki, K., Fujita, M., Takahashi, N., Asakawa, J., and Hazama, R. 1986. The rate with which spontaneous mutation alters the electrophoretic mobility of polypeptides. *Proc. Natl. Acad. Sci. USA* **83**:389–393.
6. Harris, H., Hopkinson, D. A., and Robson, E. B. 1974. The incidence of rare alleles determining electrophoretic variants: Data on 43 enzyme loci in man. *Ann. Hum. Genet., London* **37**:237–253.
7. Atland, K., Kaempfer, M., Forrsbohm, M., and Werner, W. 1982. Monitoring

for changing mutation rates using blood samples submitted for PKU screening. In: Human Genetics, Part A: The Unfolding Genome, B. Bonné-Tamir (ed.) New York: A. R. Liss, pp. 277–287.

8. Kimura, M., and Ohta, T. 1969. The average number of generations until extinction of an individual mutant gene in a finite population. *Genetics* **63**:701–709.

9. Li, F. H. F., Neel, J. V., and Rothman, E. D. 1978. A second study of the survival of a neutral mutant in a stimulated Amerindian population. *Am. Naturalist* **112**:83–96.

10. Thompson, E. A., and Neel, J. V. 1978. Probability of founder effect in a tribal population. *Proc. Natl. Acad. Sci. USA* **75**:1442–1445.

11. Chakraborty, R., and Neel, J. V. 1989. Description and validation of a method for simultaneous estimation of effective population size and mutation rate from human population data. *Proc. Natl. Acad. Sci. USA* **86**:9407–9411.

12. Neel, J. V., and Rothman, E. D. 1978. Indirect estimates of mutation rates in tribal Amerindians. *Proc. Natl. Acad. Sci. USA* **75**:5585–5588; Neel, J. V., Mohrenweiser, H. W., Rothman, E. D., and Naidu, J. M. 1986. A revised indirect estimate of mutation rates in Amerindians. *Am. J. Hum. Genet.* **38**:649–666.

13. Jacobs, P. A. 1975. The load due to chromosomal abnormalities in man. In: The Role of Natural Selection in Human Evolution, F. M. Salzano, ed. Amsterdam: North-Holland, pp. 337–352.

14. For those who might at this stage wish to pursue further the molecular basis of mutation, three superb texts are: Alberts, B., Bray, D., Lewis, J., Raff, M., Roberts, K., and Watson, J. D. 1989. Molecular Biology of the Cell. New York: Garland. Second edition; Lewin, B. 1990. Genes IV. New York: Oxford University Press; Singer, M., and Berg, P. 1991. Genes and Genomes. Mill Valley, CA: University Science Books.

15. Haldane, J. B. S. 1957. The cost of natural selection. *Jour. Genetics* **55**:511–524.

16. Green's estimate is to be found in the volume edited by Lambert et al; see ref. 26 of Chapter 9.

17. Muller, H. J. 1950. Our load of mutations. *Am. J. Hum. Genet.* **2**:111–176.

Chapter 13

1. Schull, W. J., Neel, J. V., and Hashizume, A. 1966. Some further observations on the sex ratio among infants born to the survivors of the atomic bombings of Hiroshima and Nagasaki. *Am. J. Hum. Genet.* **18**:328–338.

2. Kato, H., Schull, W. J., and Neel, J. V. 1966. A cohort-type study of survival in the children of parents exposed to atomic bombings. *Am. J. Hum. Genet.* **16**:214–230; Neel, J. V., Kato, H., and Schull, W. J. 1974. Mortality in the children of atomic bomb survivors and controls. *Genetics* **76**:311–326.

3. Awa, A. A., Honda, T., Neriishi, S., Sufuni, T., Shimba, H., Ohtaki, K.,

Nakano, M., Kodama, Y., Itoh, M., and Hamilton, H. B. 1987. Cytogenetic study of the offspring of atomic bomb survivors, Hiroshima and Nagasaki. In: Cytogenetics: Basic and Applied Aspects, G. Obe and A. Basler, eds. Berlin: Spring-Verlag, pp. 166–183.

4. Furusho, T., and Otake, M. 1985. A search for genetic effects of atomic bomb radiation on the growth and development of the F_1 generation. V. Stature of 6- to 11-year-old elementary school pupils in Hiroshima. RERF Tech. Rep. 9-85. Radiation Effects Research Foundation, Hiroshima.
5. Neel, J. V., Satoh, C., Goriki, K., Asakawa, J., Fujita, M., Takahashi, N., Kageoka, T., and Hazama, R. 1988. Search for mutations altering protein charge and/or function in children of atomic bomb survivors: Final report. *Am. J. Hum. Genet.* **42**:663–676.
6. The current "final word" on the estimation of the amounts of ionizing radiation received by the survivors of the atomic bombings will be found in: Roesch, W. C. (ed.) 1987. U.S.–Japan Joint Reassessment of Atomic Bomb Radiation Dosimetry in Hiroshima and Nagasaki: Final Report. Hiroshima: Radiation Effects Research Foundation.
7. Yoshimoto, Y., Schull, W. J., Kato, H., and Neel, J. V. 1991. Mortality among the offspring (F_1) of atomic bomb survivors, 1946–1985. *J. Radiat. Res.* **32**:327–351.
8. Yoshimoto, Y., Neel, J. V., Schull, W. J., Kato, H., Soda, M., Eto, R., and Mabuchi, K. 1990. Malignant tumors during the first 2 decades of life in the offspring of atomic bomb survivors. *Amer. J. Hum. Genet.* **46**:1041–1052.
9. Neel, J. V., Schull, W. J., Awa, A. A., Satoh, C., Kato, H., Otake, M., and Yoshimoto, Y. 1990. The children of parents exposed to atomic bombs: Estimates of the genetic doubling dose of radiation for humans. *Am. J. Hum. Genet.* **46**:1053–1072.
10. Russell, W. L., Russell, L. B., and Kelly, E. M. 1958. Radiation dose rate and mutation frequency. *Science* **128**:1546–1550.
11. Abrahamson, S., and Wolf, S. 1976. Reanalysis of radiation-induced specific locus mutations in the mouse. *Nature* **264**:715–719.
12. As this book goes to press, there is a further potentially truly amazing development with respect to the estimation of radiation doses in Japan. The neutron exposures in Hiroshima were in the past estimated primarily by the "activation" of certain cobalt atoms by the neutrons released by that atomic bomb. More recently, new and very sensitive techniques permit similar studies of such other atoms as europium, sulfur, and chlorine. These measurements are pointing to a significantly greater neutron component in the radiation spectrum of the Hiroshima bomb than currently assumed (*Science* **258**:394, 1992). It is too early to be specific, but it is possible that as the result of this upward revision in the dose sustained by survivors, the estimation of the genetic doubling dose for humans of acute ionizing radiation could increase to about 2.5 Sv equivalents. This possibility should be kept in mind when in Chapter 21 we attempt to place the genetic risks of radiation in perspective. At the very least, this development renders the present estimate of the genetic risks of radiation conservative.

13. Neel, J. V., and Lewis, S. E. 1990. The comparative radiation and genetics of humans and mice. *Annu. Rev. Genet.* **24:**327–362.
14. Neel, J. V., and Schull, W. J. (eds.) 1991. The Children of Atomic Bomb Survivors: A Genetic Study. Washington, DC: National Academy Press.
15. Committee on the Biological Effects of Ionizing Radiations, National Research Council 1990. Health Effects of Exposure to Low Levels of Ionizing Radiation (BEIR V). Washington: National Academy Press.

Chapter 14

1. For those who might wish to pursue the hemoglobin paradigm further, it has been excellently reviewed in great detail in the following references:

   Bunn, H.F., and Forget, B. G. 1986. Hemoglobin: Molecular, Genetic, and Clinical Aspects. Philadelphia: W. B. Saunders.
   Collins, F. S., and Weissman, S. M. 1984. The molecular biology of human hemoglobin. *Prog. Nucleic Acid Res. Mol. Biol.* **31:**315–458.
   Dickerson, R. E., and Geis, I. 1983. Hemoglobin: Structure, Function, Evolution, and Pathology. Menlo Park, CA: Benjamin/Cummings.
   Orkin, S. H., and Kazazian, H. H. 1984. The mutation and polymorphism of the human β-globin gene and its surrounding DNA. *Annu. Rev. Genet.* **18:**131–171.
   Stamatoyannopoulos, G., Nienhuis, A. W., Leder, P., and Majerus, P. W. 1987. The Molecular Basis of Blood Disease. Philadelphia: W. B. Saunders.
   The complexity of the system controlling the expression of the hemoglobin genes is well illustrated in:
   Gumucio, D. L., Shelton, D. A., Bailey, W. J., Slightom, J. L., and Goodman, M. L. 1993. Phylogenetic footprinting reveals unexpected complexity in trans factor binding upstream from the ε-globin gene. *Proc. Natl. Acad. Sci. USA* **90:**6018–6022.

2. The data discussed in the latter (posthemoglobin) half of this chapter are very current and scattered through many papers. It has seemed beyond the specific scope of this book to provide detailed references. General summaries touching upon some of this material will be found in note 14 of Chapter 12, plus Scriver, C. R., Beaudet, A. L., Sly, W. S., and Valle, O. (eds.) 1989. The Metabolic Basis of Inherited Disease, 6th ed. New York: McGraw-Hill.

Chapter 15

1. Most of the material in this chapter can be extracted from the three references provided in note 14 of Chapter 12, but for some of the more recent

developments referred to in this chapter I will provide references as we proceed.

2. This general subject is covered in the references given in note 25 of Chapter 9. Specific references to human mutations caused by mobile genetic elements include: Dombroski, B. A., Mathias, S. L., Nanthakumar, E., Scott, A. F., and Kazazian, H. H. 1991. Isolation of an active human transposable element. *Science* **254**:1805–1808; Miki, Y., Nishisho, I., Horii, A., Miyoshi, Y., Utsunomiya, J., Kinzler, K., Vogelstein, B., and Nakamura, Y. 1992. Disruption of the APC gene by a retrotransposal insertion of L₁ sequence in a colon cancer. *Cancer Research* **52**:643–645; Wallace, M. R., Andersen, L. B., Saulino, A. M., Gregory, P. E., Glover, T. W., and Collins, F. S. 1991. A *de novo Alu* insertion results in neurofibromatosis type 1. *Nature* **353**:864–866.

3. For an exposition of this viewpoint, see: Wills, C. 1991.The Wisdom of the Genes. Oxford: Oxford University Press.

4. For a clear review of this confusing topic, see: Deininger, P. L. 1989. SINEs: Short interspersed repeated DNA elements in higher eucaryotes. In: Mobile DNA (eds. D. E. Berg and M. M. Howe). Washington: American Society for Microbiology, pp. 619–636; Hutchinson, C. A., III, Hardies, S. C., Loeb, D. D., Shehee, W. R., and Edgell, M. H. 1989. LINEs and related retroposons: Long interspersed repeated sequences in the eucaryotic genome. In: Mobile DNA (eds. D.E. Berg and M. M. Howe). Washington: American Society for Microbiology, pp. 593–617.

5. Jeffreys, A. J., Royle, N. J., Wilson, V., and Wong, Z. 1988. Spontaneous mutation rates to new length alleles at tandem-repetitive hypervariable loci in human DNA. *Nature* **332**:278–281.

6. Lindahl, T. 1977. DNA repair enzymes acting in spontaneous lesions in DNA. In: DNA Repair Processes (eds. W. W. Nichols and D. G. Murphy). Miami: Symposia Specialists, pp. 225–240.

7. For a more sophisticated discussion of the potential instability of the gene pool and some conjectures as to how it is controlled, see: Kricker, M. C., Drake, J. W., and Radman, M. 1992. Duplication-targeted DNA methylation and mutagenesis in the evolution of eukaryotic chromosomes. *Proc. Natl. Acad. Sci. USA* **89**:1075–1079.

8. National Research Council, Committee on the Biological Effects of Ionizing Radiations (BEIR V). 1990. Health Effects of Exposures to Low Levels of Ionizing Radiations. Washington, DC: National Academy Press.

9. Ames, B. N. 1979. Identifying environmental chemicals causing mutations and cancer. *Science* **204**:587–593; Ames, B. N. 1989. Mutagenesis and carcinogenesis: Endogenous and exogenous factors. *Env. Mol. Mutagenesis* (Supplement 16) **14**:66–77.

10. A useful orientation to the interest in, and complexity of, these repair enzymes will be found in Collins, A., Johnson, R. T., and Boyle, J. M. (eds.) 1987. Molecular biology of DNA repair. *J. Cell Sci.,* Supplement 6:353; Sutherland, B. M., and Woodhead, A. D. (eds.) 1990. DNA Damage and Repair in Human Tissues. New York: Plenum Press.

11. Mohrenweiser, H. W., Larsen, R. D., and Neel, J. V. 1989. Development of

molecular approaches to estimating germinal mutation rates. I. Detection of insertion/deletion/rearrangement variants in the human genome. *Mut. Res.* **212**:241–252.

12. There is a truly burgeoning literature on this subject. Complexity theory deals with the organizational properties of highly intricate systems whereas chaos theory emphasizes nonlinear system dynamics, leading to unstable, aperiodic behavior, heretofore applied to such phenomena as the mechanics of turbulent fluids and meteorology. Three very readable introductions are: Glass, L., and Mackey, M. C. 1988. From Clocks to Chaos. Princeton, NJ: Princeton University Press; Krasner, S. (ed.) 1990. The Ubiquity of Chaos. Washington, DC: American Association for the Advancement of Science; Levin, R. 1993. Complexity: Life at the Edge of Chaos. New York: MacMillan Co. This espousal of complexity theory does not supersede my endorsement of the Wright model but is an extension rendered necessary by the accumulation of genetic data not available to Wright.

Chapter 16

1. Out of plethora of possible references, I have listed here a number of recent, mostly book-length treatments that will lead the reader to many other publications on the subject.

Ehrlich, P. R., and Ehrlich, A. 1990. The Population Explosion. New York: Simon & Schuster.

Goldsmith, E., Allen, R., Allaby, M., Davoll, J., and Lawrence, S. 1972. A blueprint for survival. *The Ecologist* **2**:2–43.

Lamm, R. D. 1985. Megatraumas: America at the Year 2000. Boston: Houghton Mifflin.

McKibben, B. 1989. The End of Nature. New York: Random House.

Potter, V. R. 1988. Global Bioethics: Building on the Leopold Legacy. Lansing: Michigan State University Press.

Scientific American, September 1989, "Managing Planet Earth," **261**, 3:46–175.

Weiner, J. 1990. The Next One Hundred Years. New York: Bantam Books.

World Commission on Environment and Development. 1987. Our Common Future. Oxford: Oxford University Press.

The annual reports of the Worldwatch Institute on the State of the World are especially valuable.

Earlier references that bear rereading are:

Heilbroner, R. L. 1974. An inquiry into the human prospect. New York: W. W. Norton.

Platt, J.R. 1970. Perception and Change: Projections for Survival. Ann Arbor: University of Michigan Press.

Ehrlich, P. 1968. The Population Bomb. New York: Ballantine.

2. U.S. Geological Survey, Department of the Interior. 1983. The High Plains regional aquifer. Water Fact Sheet published by U.S. Geological Survey.
3. Two excellent articles that provide access to the vast literature on this subject are: Pimental, D., and Levitan, L. 1986. Pesticides: amounts applied and amounts reaching pests. *Bioscience* **36**:86–91; Pimental, D., McLaughlin, L., and Zepp, A. 1991. Environmental and economic effects of reducing pesticide use. *Bioscience* **41**:402–409.
4. Anderson, C. 1992. Researchers ask for help to save key biopesticide. *Nature* **355**:661.
5. A rounded and encyclopedic treatment of this subject is the four-volume report of the Panel on Policy Implications of Greenhouse Warming, Committee on Science, Engineering, and Public Policy, National Research Council (Washington, National Academy Press, 1991). A concise statement of the uncertainties in the projections will be found in: Broecker, W. S. 1992. Global warming on trial. *Natural History*, April. pp. 6–14.
6. Hofmann, D. J., Oltmans, S. J., Harris, J. M., Solomon, S., and Deshler, T. 1992. Observation and possible causes of new ozone depletion in Antarctica in 1991. *Nature* **359**:283–287.
7. Meadows, D. H., Meadows, D. L., Randers, J., and Behrens, W. W., III. 1972. The Limits to Growth. New York: Universe Books.
8. U.S. Department of State and Council on Environmental Quality. 1980. The Global 2000 Report to the President, Vols. 1–3. Washington, D.C.: U.S. Government Printing Office.
9. See *Nucleus* **14**:1–3 (1992–1993).
10. See also: MacKellar, F. L., and Vining, D. R. 1987. National Resources Scarcity: A Global Survey. In: Population Growth and Economic Development: Issues and Evidence, D. G. Johnson and R. D. Lee, eds. Madison, WI: The University of Wisconsin Press, pp. 259–329; Meadows, D. H., Meadows, D. L., and Randers, J. 1992. Beyond the Limits. Post Mills, VT: Chelsea Green Publ. Co.; World Bank 1992 World Development Report. 1992. New York: Oxford University Press.
11. United States Senate, Committee on Government Operations, Subcommittee on Intergovernmental Relations. 1969. Hearings on S. Res. 78, to establish a select Senate committee on technology and the human environment. Washington: Government Printing Office.
12. Keyfitz, N. 1985. World population growth: Demographic and economic realities. *Issues in Science & Technology*, Winter, 1985, 60–70.
13. I get some delayed satisfaction from the fact that in 1992 the U.S. National Academy of Sciences and the U.K. Royal Society issued a joint statement that finally calls attention to the gravity of the current threats to the earth's capacity to sustain life (Population Growth, Resource Consumption, and a Sustainable World: A Joint Statement by the Officers of the Royal Society of London and the U.S. National Academy of Sciences. Washington: National Academy Press).
14. Ortner, D. J. (ed.) 1983. How Humans Adapt: A Biocultural Odyssey. Washington: Smithsonian Institution Press.

15. Keyfitz, N., op. cit., p. 87.
16. Of the many treatments of the impact of the epidemic diseases, three that I have found especially valuable are:

    Hecker, F. C. 1859. The Epidemics of the Middle Ages. London: Trübner and Co.

    Marks, G., and Beatty, W. K. 1976. Epidemics. New York: Charles Scribner's Sons.

    McNeill, W. H. 1976. Plagues and Peoples. Garden City, NY: Anchor Press.

17. Committee for the Study on Malaria Prevention and Control, Institute of Medicine. 1991. Malaria: Opportunities and Obstacles. Washington: National Academy Press. See also: *World Health,* September–October, 1991, pp. 2–31. The entire number is devoted to the return of malaria.
18. The August 21, 1992, issue of the journal *Science* is largely devoted to documenting the emergence of drug-resistant strains of pathogens.
19. Lederberg, J., Shope, R. E., and Oaks, S. C. (eds.) 1992. Emerging Infections: Microbial Threats to Health in the United States. Washington: National Academy Press. pp. 308.
20. St. John, R. R. 1991. The worldwide AIDS epidemic: A crisis in public health. In: AIDS: The Modern Plague. The President's Symposium for 1991, P. White (ed.) Blacksburg (VA): Virginia Polytechnic Institute and State University Press, pp. 13–27.
21. Keyfitz, N., op. cit., p. 84.
22. Wilson, E. O. (ed.) 1988. Biodiversity. Washington: National Academy Press.
23. Midgley, M. 1983. Animals and Why They Matter. Harmondsworth, England: Penguin Books.
24. Regan, T. 1983. The Case for Animal Rights. Berkeley: University of California Press.
25. Taylor, P. W. 1986. Respect for Nature. Princeton, NJ: Princeton University Press.
26. In the context of how little separates us from the chimpanzee, see the highly readable book by Jared Diamond (The Third Chimpanzee. New York: Harper Collins, 1992).
27. Simons, E. L. 1989. Human origins. *Science* **245**:1343–1350.
28. Young, J. Z. 1971. An Introduction to the Study of Man. Oxford: Clarendon Press, p. 93.
29. The Gaia hypothesis, in Lovelock's words, " . . . supposed that the atmosphere, the oceans, the climate, and the crust of the earth are regulated at a state comfortable for life because of the behavior of living organisms. Specifically, the Gaia hypothesis said that the temperature, oxidation state, acidity, and certain aspects of the rocks and waters are at any time kept constant, and that this homeostasis is maintained by active feed back processes operted automatically and unconsciously by the biota" (1988, p. 19). See:

Lovelock, J. E. 1979. Gaia: A New Look at Life on Earth. Oxford: Oxford University Press.

Lovelock, J. E. 1988. The Ages of Gaia: A Biography of Our Living Earth. First edition. New York: W. W. Norton.

Margulis, L. 1981. Symbiosis and Cell Evolution: Life and Its Environment on the Early Earth. San Francisco: W. H. Freeman.

Resnik, D. B. 1992. Gaia: From fanciful notion to research program. *Persp. Biol. Med.* **35**:572–582.

30. After that sentence was written, I became aware that in 1955 Alan Gregg had developed that analogy in very explicit terms (A medical aspect of the population problem. Science 121:681–682). A more expanded version of this thought will be found in: Hern, W. M. 1990. Why are there so many of us? Description and diagnosis of a planetary ecopathological process. *Popul. and Environ.* **12**:9–39.

31. In 1970, Jacques Monod, a principal player and Nobel Laureate in the early days of molecular genetics, published a brilliant little book under the title, *Le Hasard et la Nécessité.* One of its principal themes was to enunciate the spiritual void created by the demonstration that evolution was "directed" by chance (*hasard*), throwing up all manner of DNA variation as a result of random mutation, and necessity, in the form of opportunistic selection, gathering up whatever of this output created a momentary advantage for the species. I am suggesting that now, 20 plus years later, for many that void can be replaced by the effort to find our proper place in the complex ecosystem of which we are a part.

Chapter 17

1. There is, of course, no lack of scientific and lay literature on the relation between diet and disease. A useful, well referenced compendium is "Diet and Health: Implications for Reducing Chronic Disease Risks," sponsored by the Committee on Diet and Health of the National Research Council (Washington, National Academy Press, 1989).

2. Chagnon, N. A. 1974. Studying the Yanomamö. New York: Holt, Rinehart, & Winston.

3. Crow, J. F. 1958. Some possibilities for measuring selection intensities in man. *Hum. Biol.* **30**:1–13.

4. See note 6, Chapter 10.

5. National Council on Radiation Protection and Measurements. 1987. Ionizing radiation exposures of the population of the United States. Bethesda (MD): National Council on Radiation Protection and Measurements.

6. Ames, B. N., Profet, M., and Gold, L. S. 1990. Nature's chemicals and synthetic chemicals: Comparative toxicology. *Proc. Natl. Acad. Sci. USA* **87**:7782–7786. See also: Swirsky, L., Slone, T. H., Stern, B. R., Manley, N. B., and Ames, B. 1992. Rodent carcinogens: Setting priorities. *Science* **258**:261–265.

7. Joint Committee on Atomic Energy. 1959. Biological and Environmental Effects of Nuclear War. Hearings before the Special Subcommittee on Radiation of the Joint Committee on Atomic Energy, Congress of the United States. Part 1. Washington: GPO.
8. Office of Technology Assessment, Congress of the United States. 1979. The Effects of Nuclear War. Washington: GPO.
9. Campbell, K., Carter, A. B., Miller, S. E., and Zraket, C. A. 1991. Soviet Nuclear Fission: Control of the Nuclear Arsenal in a Disintegrating Soviet Union. Cambridge (MA): Center for Science and International Affairs, John F. Kennedy School of Government.
10. Of the voluminous literature on this subject, I found "The Aging Society" (*Daedulus*, Winter, 1986) and "The Oldest Old" (Milbank Quarterly, vol. 63, no. 2, 1985, pp. 177–451) especially concise.
11. Preston, S. H. 1984. Children and the elderly: Divergent paths for America's dependent. *Demography* 21:435–457.
12. Richman, H. A., and Stagner, M. W. 1986. Children in an aging society: Treasured resource or forgotten minority. *Daedulus* **115**:171–189.
13. Miringhoff, M. L. 1991. Index of Social Health (1991). Tarrytown, NY: Fordham Institute for Innovation in Social Policy.
14. Spicker, S. F., Ingman, S. R., and Lawson, I. R. (eds.) 1987. Ethical Dimensions of Geriatric Care. Dordrecht (Holland): Reidel.
15. Wallace, A. R. 1891. Natural Selection and Tropical Nature. London: MacMillan and Co.
16. National Research Council. 1989. Lost Crops of the Incas: Little Known Plants of the Andes with Promise for Worldwide Cultivation. Washington, D.C.: National Academy Press.

Chapter 18

1. Advisory Committee on the Biological Effects of Ionizing Radiation. 1956. The Biological Effects of Atomic Radiation. Washington: National Academy of Sciences–National Research Council.
2. Russell, W. L., Russell, L. B., and Kelly, E. M. 1958. Radiation dose rate and mutation frequency. *Science* **128**:1546–1550.
3. Russell, W. L. 1965. Effect of the interval between irradiation and conception on mutation frequency in female mice. *Proc. Natl. Acad. Sci. USA* **54**:1552–1557.
4. Morton, N. E., Crow, J. F., and Muller, H. J. 1956. An estimate of the mutational damage in man from data on consanguineous marriages. *Proc. Natl. Acad. Sci. USA* **42**:855–863.
5. Advisory Committee on the Biological Effects of Ionizing Radiation. 1960. The Biological Effects of Atomic Radiation. Washington: National Academy of Science–National Research Council.
6. Priest, G. L. 1990. The new legal structure of risk control. *Daedulus* **119**:207–227.

7. Environmental Protection Agency. 40 CFR Part 191 (Proposed). Environmental Radiation Protection Standards for Management and Disposal of Spent Nuclear Fuel, High-level and Transuranic Radioactive Waste. Federal Register, Dec. 29, 1982, p. 58196. The term "transuranic waste" is applied to materials contaminated with transuranium nuclides emitting alpha particles with concentrations higher than 100 nanocuries per gram and half-lives longer than 20 years.

8. Eisenbud, M. 1988. Disparate costs of risk avoidance. *Science* **241**:1277–1278.

9. A precise timetable for these developments will be found in two articles by Don Scroggin: New guidelines face complicated problem of risk evaluation. *Legal Times* **13**:20–21 (1987), and Cancer-risk Assessments. *Trial Magazine* **23**:49,50,52,54 (1987).

10. Gardner, M. J., Snee, M. P., Hall, A. J., Powell, C. A., Downes, S., and Terrell, J. D. 1990. Results of case-control study of leukaemia and lymphoma among young people near Sellafield nuclear plant in West Cumbria. *Brit. Med. J.* **800**:423–429.

11. The reader may wish to refer to note 6, Chapter 4, and pp. 236–237 regarding terminology for radiation doses.

12. Nomura, T. 1982. Parental exposure to X-rays and chemicals induces heritable tumours and anomalies in mice. *Nature* **296**:575–577; Nomura, T. 1983. X-ray-induced germ-line mutation leading to tumors. Its manifestation in mice given urethan post-natally. *Mut. Res.* **121**:59–65; Nomura, T. 1986. Further studies on X-ray and chemically induced germ-line alterations causing tumours and malformations in mice. In: Genetic Toxicology of Environmental Chemicals, Part B, C. Ramel, B. Lambert, and J. Magnusson, eds. New York: Liss, pp. 13–20.

13. Jones, K. P., and Wheater, A. W. 1989. Obstetric outcomes in West Cumberland Hospital: Is there a risk from Sellafield? *J. Roy. Soc. Med.* **82**:524–527.

14. McLaughlin, J. R., Clarke, E. A., King, W., and Anderson, T. W. 1992. Occupational exposure of fathers to ionizing radiation and the risk of leukaemia in offspring—a case-control study. Atomic Energy Control Board Report INFO-0424; Urquhart, J. D., Black, R. J., Muirhead, M. J., Sharp, L., Maxwell, M., Eden, O. B., and Jones, D. A. 1991. Case-control study of leukemia and non-Hodgkin's lymphoma in children in Caithness near the Dounreay nuclear installation. *Brit. Med. J.* **302**:687–691 and 818; Kinlen, L. J., O'Brien, F., Clarke, K., Balkwill, A., and Matthews, F. 1993. Renal population mixing and childhood leukaemia: Effects of the North Sea oil industry in Scotland, including the area near Dounreay nuclear site. *Brit. Med. J.* **306**:743–748.

15. Parker, L., Craft, A. W., Smith, J., Dickinson, H., Wakeford, R., Binks, K., McElvenny, D., Scott, L., and Slovak, A. J. M. Childhood leukaemia and occupational radiation exposure: The geographical distribution of preconceptional radiation exposures associated with fathers employed at the Sellafield nuclear installation, West Cumbria, for births during 1950–1989. *Brit. Med. J.*, submitted.

Notes

16. Craft, A. W., Parker, L., Openshaw, S., Charlton, M., Newell, J., Birch, J., and Blair, V. 1993. Cancer in young people in the north of England 1968–85: Analysis by census wards. *J. Epid. Comm. Hlth.* **47**:109–115.

17. National Research Council, Committee on the Institutional Means for Assessment of Risks to Public Health. 1983. Risk Assessment in the Federal Government: Managing the Process. Washington, DC: National Academy Press.

18. A history of the efforts to obtain more uniform risk regulations will be found in: Institute for Regulatory Policy. 1991. Toward Common Measures. Washington, DC: Federal Focus.

19. Neel, J. V. 1970. Evaluation of the effects of chemical mutagens on man: The long road ahead. *Proc. Natl. Acad. Sci. USA* **67**:908–915.

20. Ames, B. N. 1983. Dietary carcinogens and anticarcinogens: Oxygen radicals and degenerative diseases. *Science* **221**:1256–1264; Ames, B. N. 1989. Mutagenesis and carcinogenesis: Endogenous and exogenous factors. *Environmental and Molecular Mutagenesis* **14**, Supplement 16:66–77.

21. Kuick, R. D., Neel, J. V., Strahler, J.R., Chu, E. H. Y., Bargal, R., Fox, D. A., and Hanash, S. M. 1993. Similarity of spontaneous germinal and *in vitro* somatic cell mutation rates in humans: Implications for carcinogenesis and for the role of exogenous factors in "spontaneous" germinal mutagenesis. *Proc. Natl. Acad. Sci. USA* **90**:7036–7040.

22. Documentation of this important point will be found in:

   Strong, L. C., Stine, M., and Norsted, T. L. 1987. Cancer in survivors of childhood soft tissue sarcoma and their relatives. *J. Natl. Cancer Inst.* **79**:1213–1220.

   de Vathaire, F., Schweisguth, O., Rodary, C., François, P., Sarrazin, D., Oberlin, O., Hill, C., Raquin, M. A., Dutreix, A., and Flamant, R. 1989. Long-term risk of second malignant neoplasm after a cancer in childhood. *Br. J. Cancer* **59**:448–452.

   Meadows, A. T., Baum, E., Fossati-Bellani, F., Green, D., Jenkin, R. D. T., Marsden, B., Nesbit, M., Newton, W., Oberlin, O., Sallan, S.G., Siegel, S., Strong, L. C., and Voûte, P. A. 1985. Second malignant neoplasms in children: An update from the late effects study group. *J. Clin. Oncology* **3**:532–538.

   de Vathaire, F., François, P., Hill, C., Schweisguth, O., Rodary, C., Sarrazin, D., Oberlin, O., Buertheret, C., Dutreix, A., and Flamant, R. 1989. Role of radiotherapy and chemotherapy in the risk of second malignant neoplasms after cancer in childhood. *Br. J. Cancer* **59**:792–796.

   Tucker, M. A., Coleman, C. N., Cox, R. S., Varghese, A., and Rosenberg, A. S. 1988. Risk of second cancers after treatment for Hodgkin's disease. *New Engl. J. Med.* **318**:76–81.

23. These calculations are based upon the data presented in: Shimizu, Y., Kato, H., and Schull, W. J. 1990. Studies of the mortality of a-bomb survivors. 9. Mortality, 1950–1985: Part 2. Cancer mortality based on the recently revised doses (DS86). *Rad. Res.* **121**:120–141.

A technical point: In principle, this calculation should, in parallel with the genetic calculation, be based on that fraction of endpoint (cancer) due to somatic cell mutation, i.e., should exclude cancer due to specific inherited predispositions. At present, there is no estimate of the magnitude of that component. The appropriate correction can only lower that figure of 2.5 Sv equivalents, but by an amount difficult to estimate.

Chapter 19

1. This highly organized approach to the societal problems listed in Chapter 17, with the kind of mobilization usually reserved for a military crisis, was to my knowledge first urged by John Platt (Perception and Change: Projections for Survival. Ann Arbor: University of Michigan Press. 1970).
2. A concise account of how this came about will be found in: Meadows, D. H., Meadows, D. L., and Randers, J. 1992. Beyond the limits. Post Mills, VT: Chelsea Green Publ. Co.
3. Gingerich, P. D. 1991. Fossils and Evolution. In: Evolution of Life, S. Osawa and T. Honjo, eds. Heidelberg: Springer-Verlag, pp. 3–20.
4. Born, M. 1968. My Life and My Views. New York: Scribner; Heilbroner, R. 1974. An Inquiry into the Human Prospect. New York: W. W. Norton; Thomas, L. 1983. The Youngest Science: Notes of a Medicine-Watcher. New York: Viking Press; see also Platt, op. cic.
5. Operations Evaluation Department, World Bank. 1991. The World Bank and Indonesia's Population Program. Washington, D.C.: The World Bank.
6. Freedman, R., and Blanc, A. K. 1991. Fertility transition: An update. In: Proc. Demographic and Health Surveys World Conference, Vol. 1. Columbia, MD: IRD/Macro International, Inc., pp. 5–24.
7. Horiuchi, S. 1992. Stagnation in the decline of the world population growth rate during the 1980s. *Science* **257**:761–765.
8. Anderson, R. M., May, R. M., Boily, M. C., Garnett, G. P., and Rowley, J. T. 1991. The spread of HIV-1 in Africa: Sexual contact patterns and the predicted demographic impact of AIDS. *Nature* **352**:581–589.
9. King, M. 1990. Health is a sustainable state. *Lancet* **336**:664–667.
10. Martin, J. 1993. Round Table: Would Machiavelli now be a better guide for doctors than Hippocrates? *World Health Forum* **14**:105–131.
11. Neel, J. V. 1970. Lessons from a primitive people. *Science* **170**:815–822; Neel, J. V. 1973. Social and scientific priorities in the use of genetic knowledge. In: Ethical Issues in Human Genetics, B. Hilton, D. Callahan, M. Harris, P. Condliffe, and B. Berkley, eds. New York: Plenum Press, pp. 353–368.
12. I am indebted to Professor Ron Freedman and Ms. Blair Cohen for the demographic modeling that permits that statement, and to Professor Freedman for drawing my attention to a number of the pertinent references to this chapter.
13. Two very readable treatments of the need for, and practice of, a more sus-

tainable agriculture are: Jackson, W. 1987. Altars of Unhewn Stone. San Francisco: North Point Press; Committee on the Role of Alternative Farming Methods in Modern Production Agriculture. 1989. Alternative Agriculture. Washington: National Academy Press.

14. Meadows, D. H., Meadows, D. L., Randers, J., and Behrens, W. W., III. 1972. The Limits to Growth. New York: Universe Books.
15. Groen, J., Smit, E., and Eijsvoogel, J. (eds.) 1990. The Discipline of Curiosity. Amsterdam: Elsevier.
16. A good discussion of current national policies with regard to population control will be found in: Keyfitz, N. 1989. The growing human population. *Scientific American* **261**:119–126.
17. Baulieu, E.-E. 1989. Contragestion and other clinical applications of RU-486, an antiprogesterone at the receptor. *Science* **245**:1351–1357.
18. Regelson, W. 1992. RU486: How abortion politics have impacted on a potentially useful drug of broad medical application. *Persp. Biol. Med.* **35**:330–338.
19. Jacobson, J. 1991. Coming to grips with abortion. In: State of the World, 1991, L. R. Brown, ed. New York: W. W. Norton, pp. 113–131.
20. Hardin, G. 1977. The Limits of Altruism: An Ecologist's View of Survival. Bloomington: Indiana University Press.
21. Gore, A. 1992. Earth in the Balance. Boston: Houghton Mifflin Co., pp. 407.
22. Ann Arbor News, July 27, 1992, p. A3.
23. Lederberg, J. 1963. Molecular biology, eugenics, and euphenics. *Nature* **198**:428–429.
24. The term "euthenics," with the same connotations, was in fact in use among the early eugenicists, the first usage known to me being by C. B. Davenport in 1911. (Heredity in Relation to Eugenics. New York: Henry Holt.)
25. Neel, J. V., Fajans, S. S., Conn, J. W., and Davidson, R. T. 1965. Diabetes mellitus. In: Genetics and the Epidemiology of Chronic Disease, J. V. Neel, M. W. Shaw, and W. J. Schull, eds. Washington, DC: Gov't. Printing Office, pp. 105–132.
26. Neel, J. V. 1962. Diabetes mellitus: A "thrifty" genotype rendered detrimental by "progress." *Am. J. Hum. Genet.* **14**:353–362; Neel, J. V. 1982. The thrifty genotype revisited. In: The Genetics of Diabetes Mellitus, J. Köbberling and R. Tattersall, eds. New York: Academic Press, pp. 137–147.
27. Five useful, well-referenced compendia on this subject are: Committee on Diet and Health of the National Research Council. 1989. Diet and Health: Implications for Reducing Chronic Disease Risk, Washington: National Academy Press; Eaton, S. B., Shostak, M., and Konner, J. 1988. The Paleolithic Prescription. New York: Harper and Row; McKeown, T. 1988. The Origins of Human Disease. New York: Oxford; Simopoulos, A. P., and Child, B. (eds.) 1990. Genetic Variation and Nutrition. Basel; Karger; Trowell, H. C., and Burkitt, D. P. (eds.) 1981. Western Diseases: Their Emergence and Prevention. Cambridge, MA: Harvard University Press.
28. For a cogent exposition of this concept, see: Fries, J. F. 1988. Aging, illness,

and health policy: Implications of the compression of mortality. *Persp. Biol. Med.* **31**:407–428.

29. Galton, Francis. 1892. Hereditary Genius: An Inquiry into Its Laws and Consequences. London: MacMillan and Co.
30. Durning, A. T., and Brough, H. B. 1992. Reforming the livestock economy. In: State of the World, 1992, L. R. Brown, ed. New York: W. W. Norton, pp. 66–82.
31. Neel, J. V. 1960. The genetic potential. In: The Nation's Children, E. Brinzberg, ed., Vol. 2. New York: Columbia University Press, pp. 1–23.
32. National Council on Radiation Protection and Measurements. 1987. Ionizing Radiation Exposure of the Population of the United States. Bethesda (MD): National Council on Radiation Protection and Measurements.
33. See Chapter 13, note 15.

Chapter 20

1. Five compendia that will provide explicit references for most of the statements of this and the following section are: Desnick, R. J. (ed.) 1991. Treatment of Genetic Disease. New York: Churchill Livingstone; Emery, A. E. H., and Rimoin, D. L. (eds.) 1983. Principles and Practice of Medical Genetics. Edinburgh: Churchill Livingstone; Friedmann, T. (ed.) 1991. Therapy for Genetic Disease. Oxford: Oxford University Press; Scriver, C. R., Beaudet, A., Sly, W. S., and Valle, D. (eds.) 1989. The Metabolic Basis of Inherited Disease. Two vols. New York: McGraw-Hill; King, R. A., Rotter, J. I., and Motulsky, A. G. 1992. The Genetic Basis of Common Diseases. Oxford: Oxford University Press.
2. Angastiniotis, M., Kyriakidon, S., and Hadjiminas, M. 1986. How thalassemia was controlled in Cyprus. *World Health Forum* **7**:291–297; Cao, A., Pirastu, M., and Rosatelli, C. 1986. The prenatal diagnosis of thalassaemia. *Br. J. Haematol.* **63**:215–220; Ristaldi, M. S., Pirastu, M., Rosatelli, C., Monni, G., Erlich, H., Saiki, R., and Cao, A. 1989. Prenatal diagnosis of beta-thalassaemia in Mediterranean populations by dot blot analysis with DNA amplification and allele specific oligonucleotide probes. *Prenat. Diagn.* **9**:629–638.
3. cf. especially: Beaudet, A. L. 1990. Carrier screening for cystic fibrosis. *Am. J. Hum. Genet.* **47**:603–605; Beaudet, A. L., and O'Brien, W. E. 1992. Advantages ofa two-step laboratory approach for cystic fibrosis carrier screening. *Am. J. Hum. Genet.* **50**:439–440; Biesecker, L., Bowles-Biesecker, B., Collins, F., Kaback, M., and Wilford, B. 1992. General population screening for cystic fibrosis is premature. *Am. J. Hum. Genet.* **50**:438–439; Congress of the United States, Office of Technology Assessment. 1992. Cystic fibrosis and DNA tests: Implications of carrier screening. Washington: U.S. Government Printing Office; Wilford, B. S., and Frost, N. 1990. The cystic fibrosis gene: Medical and social implications for heterozygote detection. *JAMA* **263**:2777–2783.

Notes

4. Lindeman, R., Hu, S. P., Volpato, F., and Trent, R. J. 1991. Polymerase chain reaction (PCR) mutagenesis enabling rapid non-radioactive detection of common beta-thalassaemia mutations in Mediterraneans. *Br. J. Haematology* **78**:100–104; Friedman, K. J., Highsmith, W. E., and Silverman, L. M. 1991. Detecting multiple cystic fibrosis mutations by polymerase chain reaction-mediated site-directed mutagenesis. *Cl. Chem.* **37**:753–755.
5. For a very thoughtful analysis of the problems which arose in the push for sickle cell trait and anemia screening, see: Bowman, J. E. 1991. Invited editorial: Prenatal screening for hemoglobinopathies. *Am. J. Hum. Genet.* **48**:433–438.
6. Committee for the Study of Inborn Errors of Metabolism. 1975. Genetic screening: Programs, principles, and research. Washington, DC: National Academy of Sciences; Holtzman, A. 1989. Proceed with Caution. Baltimore: Johns Hopkins University Press; Annas, G. J., and Elias, S. 1992. Gene Mapping. Oxford: Oxford University Press.
7. Holtzman, N. A., and Rothstein, M. A. 1992. Invited editorial: Eugenics and genetic discrimination. *Am. J. Hum. Genet.* **50**:457–459; Natowicz, M. R., Alper, J. K., and Alper, J. S. 1992. Genetic discrimination and the law. *Am. J. Hum. Genet.* **50**:465–475.
8. The data for this estimate are to be found in reference 1 of this chapter, as well as in: Brock, D. J. H., Rodick, C. H., Ferguson-Smith, M. A. (eds.) 1992. Prenatal Diagnosis and Screening. Edinburgh: Churchill Livingstone. (See also Carter, C. O. 1977. Monogenic disorders. *J. Med. Genet.* **14**:316–320.)
9. Editorial. 1992. From genetics to revelation? *Nature Genetics* **1**:77–78; Handyside, A. H., Leso, J. G., Tarin, J. J., Winston, R. M. L., and Hughes, M. R. 1992. Birth of a normal girl after *in vitro* fertilization and preimplantation testing for cystic fibrosis. *N.E. J. Med.* **327**:905–909; Simpson, J. L., and Carson, S. A. 1992. Preimplantation genetic diagnosis. *N.E. J. Med.* **327**:951–953.
10. Collins, F. S. 1992. Cystic fibrosis: Molecular biology and therapeutic implications. *Science* **256**:774–779.
11. Four illustrative and instructive publications are: Ciba Foundation Symposium 149. 1990. Human Genetic Information: Science, Law and Ethics. New York: Wiley; Friedman, T. 1989. Progress toward human gene therapy. *Science* **244**:1275–1281; Olson, S. 1986. Biotechnology: An Industry Comes of Age. Washington, D.C.: National Academy Press; President's Commission for the Study of Ethical Problems. 1982. Splicing Life. Washington, D.C.: U.S. Government Printing Office.
12. The precedent for this position traces back to Hippocrates (Aphorisms 6): "For extreme illnesses extreme treatments are most fitting," but see also William Shakespeare (Hamlet *IV iii* 9): "Diseases desperate grown by desperate appliance are relieved or not at all."
13. U.S. Congress, Office of Technology Assessment. 1984. Human Gene Therapy. Washington, D.C.: Government Printing Office (OTA-BP-BA-32).
14. For a description of the many issues confronting this Committee as it moved into these uncharted waters, see the Recombinant DNA Technical Bulletin

published by the National Institute of Health, USA, Department of Health and Welfare. For individual assessments of some of these issues, see: Carmen, I. H. 1992. Debates, division, and decisions: Recombinant DNA Advisory Committee (RAC) authorization of the first human gene transfer experiments. *Am. J. Hum. Genet.* **52**:245–260; Epstein, C. J. 1990. Editorial: Making history. *Am. J. Hum. Genet.* **47**:601–602; Walters, L. 1991. Human gene therapy: Ethics and public policy. *Human Gene Therapy* **2**:115–122.

15. As this book goes to press, the most concise description of the present status of this field is by W. F. Anderson (Human Gene Therapy. *Science* **256**:808–813, 1992).
16. Palmiter, R. D., and Brinster, R. L. 1986. Germ-line transformation of mice. *Annu. Rev. Genetics* **20**:465–499.
17. Neel, J. V. 1993. Germ line gene therapy: Another view. *Hum. Gene Ther.* **4**:127–128.
18. Miller, R. A. 1991. Gerontology as oncology. *Cancer* **68,** December 1 Supplement: 2496–2501.
19. The literature on aging is of course voluminous. I have found the following books and articles to contain especially good summaries of current thinking: Finch, C. E. 1990. Longevity, Senescence, and the Genome. Chicago: Universityof Chicago Press; Finch, C. E., and Schneider, E. L. (eds.) 1985. Handbook of the Biology of Aging, Second Edition. New York: Van Nostrand Reinhold; Franceschi, C., Crepaldi, G., Cristotalo, V. J., and Vijg, J. (eds.) 1992. Aging and Cellular Defense Mechanisms. Ann. N.Y. Acad. Sci. No. 663; Goldstein, S. 1990. Replicative senescence: The human fibroblast comes of age. *Science* **249**:1129–1133; Holliday, R. 1988. Toward a biological understanding of the aging process. *Persp. Biol. Med.* **32**:109–123; Johnson, J. E., Walford, R., Harman, D., and Miquel, J. (eds.) 1986. Free Radicals, Aging, and Degenerative Disease. New York: Liss; Regelson, W., and Sinex, F. M. (eds.) 1983. Intervention in the Aging Process, Parts A & B. New York: Liss; Warner, H. L., Butler, R. N., Sprott, R. L., and Schneider, E. L. 1987. Modern Biological Theories of Aging. New York: Raven.
20. Ning, Y., Weber, J. L., Killary, A. M., Ledbetter, D. H., Smith, J. R., and Pereira-Smith, O. M. 1991. Genetic analysis of indefinite division in human cells: Evidence for a cell senescence-related gene(s) on human chromosome 4. *Proc. Natl. Acad. Sci. USA* **88**:5635–5639.
21. Excellent reviews of the experimental modification of life span are to be found in: Sacher, G. 1977. Life table modification and life table prolongation. In: Handbook of the Biology of Aging, C. Finch and L. Hayflick, eds. New York: Van Nostrand, pp. 582–638; Yu, B. P. 1987. Update on food restriction and aging. In: Review of Biological Research in Aging, M. Rothstein, ed. New York: Liss. Vol. 3, pp. 495–505.
22. Specific examples of such altered gene action are reviewed in: Heyduri, A. R., and Richardson, A. 1992. Does gene expression play any role in the mechanism of the antiaging effect of dietary restriction. *Ann. N.Y. Acad. Med.* **663**:384–392.
23. See reviews in: Ames, B. N., Shigenaga, M. K., and Hagen, T. M. 1993.

Oxidants, antioxidants, and the degenerative diseases of aging. *Proc. Natl. Acad. Sci. USA* **90**:7915–7922; Harman, D. 1986. Free radical theory of aging: Role of free radicals in the origination and evolution of life, aging, and disease processes. In: Free Radicals, Aging, and Degenerative Diseases, J. E. Johnson, Jr., R. Walford, D. Harman, and J. Miquel, eds. New York: Liss, pp. 3–49.

24. Gelman, R., Watson, A., Bronson, R., and Yunis, E. 1988. Murine chromosomal regions correlated with longevity. *Genetics* **118**:693–704.

Chapter 21

1. Myers, N. 1988. Tropical forests and their species: Going, going . . . In: Biodiversity, E. O. Wilson, (ed.) Washington: National Academy Press, pp. 28–35.

2. Examples abound. I think of *Consumer's Digest* as a responsible publication, but then I read an article about gene therapy under the title: "A medical breakthrough that will change our lives" (*Consumer's Digest,* May/June 1992). Never have there been such high expectations on the basis of such limited progress.

3. Muller, R. T. 1991. In defense of abortion: Issues of pragmatism regarding the institutionalization of killing. *Persp. Biol. Med.* **34**:315–326.

4. Potter, V. R. 1988. Global Bioethics: Building on the Leopold Legacy. East Lansing: Michigan State Univ. Press; Potter, V. R. 1989. Getting to the year 3000: Can global bioethics overcome evolution's fatal flaw? *Persp. Biol. Med.* **34**:89–98.

Index

ABCC. *See* Atomic Bomb Casualty Commission
ABO blood group, 44, 192–93
"A-Bomb Mayor, The." *See* Hamai, Shinzo
Abortion
 antiabortionists, 30, 188, 350
 chromosomal abnormalities and, 169–70, 220
 frivolous genetic reasons for, 369
 gene therapy as alternative to, 374
 Japanese government policy, 86
 as prenatal counseling result, 30, 362, 363, 365, 369, 370–71, 393, 395
 spontaneous. *See* Miscarriages
 U.S. policy, 289, 350–51
 by Yanomama, 177, 178, 180
α-globin genes, 254, 255, 256, 261
Acid rain, *280*, 285
Acquired Immunodeficiency Syndrome.
 see AIDS
ADA. *See* Adenosine deaminase
Adaptive mechanisms, 208, 265, 279, 301–6
Adenosine deaminase (ADA), 375, 378
Adoption, 29, 112
Adrenocorticotrophic hormones, 260
AEC. *See* Atomic Energy Commission
Aeschylus, 356

Africa
 AIDS, impact on, 293–94, 345
 sickle-cell anemia studies, 47–54, 364, 401*n.20*
African-Americans
 gene pool, 349
 sickle-cell anemia, 41, 42, 47, 50, 364
Age structure of populations
 emerging gerontocracy, 310–12, 352, *360*, 382, 386, 392
 Yanomama pyramid, 174–78, *175*, 311
 U.S. pyramid, *175*
Aging, 310–12, 382–87
 biological, 382–83
 gene therapy for, 385–87, 393
 theories of, 383–87
AIDS, 267, 350
 as epidemic disease, 182, 292
 impact on Africa, 293–94, 345
Aid to Families with Dependent Children, 312
ALARA principle (radiation control), 31, 324, 329
Albumin, 153, 195
Alcoholism, 15
Aldosterone, 154, 155
Allan, William, 17, 24
Allen, G. E., 16

Italicized page numbers refer to figures and tables. Italicized numbers preceded by *n*. refer to notes.

Allison, A. C., 48
Alpha-Helix (ship), 142–43, 154, 159
Altland, K., 217
Amazon River, 137, 142
American College of Physicians, 31
American Film Festival, 142
American Foundation for Tropical Med-
    icine, 49
American Journal of Human Genetics, 129
American Society of Human Genetics, 24
Amerindians
    blood types, 192–94
    entry into New World, 406n.3, 414n.16
    gene pool, 349
    impact of European disease on, 139,
        163–65, 182
    missionaries to, 144
    physical appearance, 150
    post-Stone Age Culture, 139
    see also Tribal populations; specific
        tribal designations
Amerindian studies, 117–207
    background, 117–22
    demographic pattern, changes in,
        186–89
    ethical issues of, 171
    funding of studies, 123, 130
    genetic differences between villages,
        191–209, 231
    causes, 197–98
    tribal heterozygosity, 196–97
    genetic implications of headmanship,
        302, 303
    health/disease findings, 149–71, 182
    human evolution, implications for,
        198–207
    inbreeding, 135, 184–85, 187–88
    map, *141*
    and mutation-rate study, 218–20
    "private" tribal genes, 194–96
    researchers
        dangers faced by, 144–47
        pilot study organization, 122–25
        selection of field colleagues, 143–
            44
        transportation to field, 141–42, 147
    traumatic deaths, 305–6
    tribal demography, 173–89
    village microdifferentiation, 191–200,
        204–6, 231

Wright model, applicability of, 201
    see also Tribal populations; specific
        peoples
Ames, Bruce, 308
Amino acids, 44, 224, 231
    disorders of metabolism, 372
    nucleotide differences, 296
Amniocentesis, 30, 362, 364
Amoebic dysentery, 158
Anderson, Ray, 72
Anemia
    atomic radiation induced, 65
    iron deficiency, 41
    see also Cooley's anemia; Sickle-cell
        anemia; Thalassemia
Aniridia, 214
Ann Arbor (Mich.). *See* University of
    Michigan
Antibodies, 157, 161, 165, 170, 182
Antipollution laws, 335
Arboviruses, 157
Arends, Tulio, 133
Aristeides, 356
Aristophanes, 356
Asakawa, J., 235
Asano Library, 80, 81
Asch, Tim, 142
Ashkenazi Jews, 363
Association for the Aid of Crippled Chil-
    dren, 97
Athenia (ship), 8
Atherosclerosis, 304, 354, 358
Atomic Bomb Casualty Commission
    (ABCC), 59, 67, 79, 82, 83–84, 85,
        88, 90, 91, 109, 404n.8
    see also Child Health Study; Radia-
        tion Effects Research Founda-
        tion
Atomic bomb studies, 26–27, 35, 57–73,
    75–93
    administrative issues, 88–89
    Biochemical Genetics Study (1975–
        84), 234–47, 384, 385
    pilot study (1972–75), 229–34
    consanguineous/nonconsanguineous
        factors, 95–102
    control cities, 83
    cytogenetic program, 229–30
    ethical issues, 84–85
    first appraisal of genetic effects, 75–93

genetics program activation, 82–84, 87
hematological effects, 65
indicators of genetic effects, 83, 89–93, 98, 232, 234–247
Japanese complaints/cooperation, 85
military overtones, charges of, 84
mutation/electrophoretic variants data, 217
newborn questionnaire, 82–83, 95
new radiation effects program (1975–84), 229–47
prospective cohort studies, 338–39
psychology of victims, 78, 403–4n.8
radiation exposure estimates, 91–92, 416n.12
    new (1980), 236–37
    newest (1992), 416n.12
radiation exposure terminology, 402–3n.6
radiation-linked mutations, 70, 75, 92, 118, 210, 227, 230; see also Birth defects
radiation-related disease studies, 87–88
residual radiation, 404n.9
rogue cells, 167, 168
spectrum of genetic damage, 240
stillborn/neonatal autopsy program, 86, 92
survivors' radiation sickness, 78–79
see also Hiroshima and Nagasaki
Atomic Energy Commission (AEC), 68, 70, 84, 87, 88
    "Effects of Atomic Weapons, The," 91
    frankness on health issues, 89
    funding of Amerindian studies, 123, 130
    funding of Child Health Study, 97, 108
    successor agencies, 402n.4
Australian aborigines, 121, 200, 349
Autopsies, 86, 92
Awa, Akio, 138, 168, 170, 230
Aztec civilization, 315

Bacteria
    Yanomama exposure, 157–58, 159–60
    see also Antibodies; specific kinds
Baird, P., 99

Balanced polymorphism, 47, 48, 52, 364
Baltimore, David, 267
Baniwa Indians, 140, 143
Barbosa, Pimental, 123, 126
Basho (Japanese poet), 352
Beadle, G. W., 7, 44, 71, 90, 318
BEAR Report (1956). See "Biological Effects of Atomic Radiation, The"
Beaver, P. C., 159
Belgian Congo, 52
"Bells of Urakami, The" (Nagai), 78
β-endorphin. See Endorphin
Bering Land Bridge, 121
Berry, George, 22
Beser, J., 77
BGS. See Biochemical Genetics Study
Bienzle, U., 52
Biochemical Genetics Study (BGS; 1975–84), 234–47
"Biological Effects of Atomic Radiation, The" (BEAR report; 1956), 318–21, 358
"Biomedical Challenges presented by the American Indian" (symposium), 171
Biopesticides, 287
Birth control
    breast feeding as, 177
    egalitarian policy, 346–352, 391
    eugenics and, 15
    Japanese, 86, 90, 345
    methods, 350
    population growth and, 289, 345–42
    see also Abortion
Birth defects
    American vs. Japanese, 92–93
    analysis (1953), 89–93
    in children conceived after atomic bombing, 61–64, 66, 67, 70, 73, 75, 83, 89–93, 98, 237
    consanguinity factor, 70–71, 92, 95–96, 99–114
    infanticide for, 305
    mutation-linked, 212–16, 338–39
    prenatal diagnosis, 29–30, 361–71, 379, 392
    stillbirth/neonatal death autopsy program, 86
    survival of children with, 304
    Yanomama infanticide for, 176

Birth defects (*cont.*)
  see also Child Health Study; specific
    defects and diseases
Birth rates, 86–87, 90, 98, 176–77, 178,
  180, 182–83, 282
Birth weight, 91
β-lipotrophin, 260
Block, Melvin, 57, 64–65
Blood
  clotting, 258, 262
  diseases. *See* Cooley's anemia; Hem-
    ophilia; Sickle-cell anemia;
    Thalassemia
  serum, 153
  types, 44, 192–96
  see also Hemoglobin headings
Blood pressure
  Amerindian studies, 150–51
  see also Hypertension
Bloom, Arthur, 166, 230
Blowfly larvae, 6–7
BNFL. *See* British Nuclear Fuel
Borges, Wayne and Jane, 404*n.12*
Born, Max, 344
Boulding, Kenneth, 290
β-polypetide of hemoglobin, 250–53, *251,*
  255, 256
Bradford, W. L., 40
Brain
  hormones. *See* Endorphin
  primitive human vs. modern, 313–15
Brazil
  as genetic studies site, 120–21
  government Indian policy, 207
  see also Amerindian studies
Brazilian Air Force, 124–25, 142
Brazilian Indian Protective Service, 122
Brazilian Revolution of 1964, 133
Breast feeding, 153, 155, 161, 164, 177
Breeding structure. *See* Demography, tri-
  bal
Brewer, Charlie, 146, 152, 408*n.8*
Brewer, Ken, 81
Brewer, Richard, 72
Bridges, Calvin, 5
Briggs, Lord Asa, 290
British Nuclear Fuel (BNFL), 331–32
Bronk, Detlev W., 318
Brues, Austin, 57, *60*
Bubonic plague, 291

Buddhism, 76, 107, 108
Building-block reuse in evolution, 262–63
Buraku system, 110
Buruti race (relay race), 150
Bush, George, 289, 352
Butterflies, 208–9

California Institute of Technology, 7, 43–
  44, 45
Cambridge University, 44
Caminopetros, 40
Canadian Academy (Kobe), 109
Cancer
  aging and, 284, 383
  childhood chemotherapy genetic ef-
    fects, 336–37
  childhood leukemia, 331–38, 377
  chromosomal rearrangements, 169,
    238
  cohort studies, 338–39
  diet link, 354
  gene therapy for, 377, 379
  proto-oncogenes and, 259–60
  radiation-linked, 75, 241, 245, 327–28,
    331, 359
  see also Retinoblastoma
Carcinogenicity Guidelines Group, 330
Caries epidemiology, 152
Carnegie Institution (Washington, D.C.),
  14, 16
Carnegie Institution Laboratory for the
  Study of Experimental Evolution
  (Cold Spring Harbor, N.Y.), 13
  Eugenics Record Office, 14–18, 24, 25,
    319, 381
Carrier states, 41–42, 188, 363, 365, 366,
  368, 369, 371
Casiquiare River, 162
Catholicism, 106, 107–8
Cavalli-Sforza, L. L., 290
Cayapo Indians, 141
Cell division. *See* Mitosis
Center for Disease Control, U.S., 158
Central America. *See* South and Central
  America
Centromeres, 269
Chagnon, Napoleon, 134, 140, 146, 174,
  176, 179, 182, 185, 306
Chakraborty, Ranajit, 219

Chandler, J. H., 213
Chaos theory, 277, 419*n.12*
Charles, Don (D. R.), 6, 71, 90, 92
Chelyabinsk disaster, 247, 406*n.7*
Chemicals
    genetic risk assessment, 317–18, 335–38, 339
    as natural mutagens, 275, 307–8, 317–18, 359
    see also Pesticides and herbicides
Chemotherapy, 336–37, 339, 377
Chernobyl disaster, 168, 247, 406*n.7*
Chibcha-speaking tribes, 141
Child abuse, 313
Child Health Study (1958–60), 97–114, 229, 332, 333
    Japanese and American personnel, 98, 108
    participation rate, 98
Childhood diabetes, 304, 375
Childhood leukemia, 331–38, 377
Children. *See* Biochemical Genetics Study; Birth defects; Child Health Study; Newborns
Children's Hospital (Detroit, Mich.), 43
Chimpanzee, 296, 382, 384, 421*n.26*
China, 345
Chinnock, F. W., 77
Chlorofluorocarbons, 287, 342
Cholesterol values, 153, 304, 354
Chorionic villi biopsy, 362
"Christian Century," 107–8
Chromosomes
    abnormal cell division, 165–70, *167*, 220–21, 230, 238, 268, 272
    abnormality-disease link, 99, 169, 257, 355, 362
    balanced reciprocal exchanges, 241
    centromeres, 269
    changed concept of, 221
    cross-over events, 272–73
    cytogenetic studies, 165–70, 229–30, 241
    DNA coiling, *248*
    ease of *Drosophila* study, 5, 9, 17
    gene therapy and, 377
    gross mutations, 220–21, 224
    nucleotide sequence in human, 389
    telomeres, 269
    viral effects, 170, 219, 267

    see also DNA; Gene; Mutations; Nucleotides
Cimon, 356
City of Flint (ship), 8
Cleft lip/palate, 93
Cleveland Museum of Natural History, 4, 50
Club of Rome, 347
Cohen, Blair, 426*n.12*
Cohen, M. N., 182
Colaert, J., 49
Cold Spring Harbor (New York). *See* Carnegie Institution Laboratory for the Study of Experimental Evolution
Collective genetic material. *See* Gene pool
College of Wooster, 1, 3–4
Colon, multiple polyposis of, 214
Color blindness, 15, 151
Columbia University, 5, 12–14
Commission for Technical Cooperation in Africa South of the Sahara, 49–50
Committee on Atomic Casualties, 68, 69
    Subcommittee on Genetics, 68, 71–72
Committee on Genetic Effects (National Academy of Sciences), 318–25
    Wright's Appendix, 323–24
Common cold, 164
Complexity theory, 419*n.12*
Comprehensive Environmental Response Compensation and Liability Act of 1980, 326
Computer simulation, 179–80, 185, 186, 219
Congenital defects. *See* Birth defects
Conn, Jerome, 353
Consanguinity effects
    Amerindian, 135, 184–85
        consequences of, 187–88
    Japanese study, 70–71, 92, 95–114, 229, 247, 278
        controls, 97
        fertility, 102, 113–14
        findings, 99–102, 113–14, 226
        implications, 95–97, 278, 307
        participation, 99
        socioeconomic status, 99, 101
Contraception. *See* Birth control
Cooley, T. B., 37
Cooley's anemia 21, 37–40; *see also* Thalassemia

Coordination of genetic research activity, 261
Coronary heart disease, 354, 383
Costa Rica, 141
Cost-effectiveness of genetic procedure, 368–69
Cotterman, Charles, 24, 25
Council of Trent, 108
Counseling. *See* Genetic counseling
Cousin marriages. *See* Consanguinity effects
Craig, C. C., 90
Cramer Fellowship, 14
Creationism, 209
Cretinism, 372
Cro-Magnon Man, 313
Croonian lectures (1908), 44
Crow, J. F., 306, 318, 322
Crowe, Frank, 213, 214–15
Cultural anthropology, 134, 173, 202
Cyprus, 365
Cystic fibrosis, 365–67, 371, 373, 379
Cytogenetics, 165–70, 204, 234
    chromosomal mutation detection, 220, 229–30

Dameshek, William, 21, 39–40
Danforth, C. H., 71
Darling, George, 233
Dartmouth University, 10–12, 14, 24, 211
Darwin, Charles, 208–9, 302, 314
    see also Natural selection
Davenport, C. B., 14, 15, 16, 427n.24
"Dead end" cells, 268
Death causes
    "natural," species variation in, 382
    ten U.S. leading (1900; 1988), 283–84, *283*, 304
    trauma, 182, 305–6, 313–14
    Yanomama Indians, 182
Death rates
    of Amerindian infants, 153, 180, 303–4
    of children of inbred parents, 100, 101, 113–15
    Hiroshima and Nagasaki casualties, 78
    of Indian (Asian) infants, 182, 183
    of Japanese atomic bomb survivors' children, 90, 91

and population explosion, 282
    in United States (1900; 1988), *283*, 284
    see also Death causes
Defense, U.S. Department of, 91
Deforestation, 285, 287
Demerec, Miloslav, 13, 18, 24, 318
Demography
    changing patterns, genetic implications of, 186–89
    mutation-disease studies, 212–13, 215–16
    population explosion effects, 282–99
    tribal, 173–86
    see also Birth rates; Death rates; Life expectancy and survival; Population
Denevan, W. M., 163
Dental caries and malocclusions, 151–52
Deoxyribonucleic acid. *See* DNA
Detroit (Mich.), sickle-cell anemia studies, 42–43, 50
Developing nations, 284, 287, 289, 304, *305*, 342
    population control problem, 351
    total fertility rate decline, 345
Diabetes mellitus, 153–60, 284, 304, 353, 354, 375, 392
    "thrifty genotype" hypothesis, 353–54
Dice, Lee R., 24
Diet and nutrition
    Amerindian, 152–54
    chemical mutagens in, 317–18, 336
    disease relationship, 394–95, 304–5, 353, 354, 355
    euphenic practices, 357, 396
    famine/malnutrition, 188, 286, 291, 294–95, 345–46
    free-radical production, 384, 385
    as genetic metabolic disorder therapy, 372–73
    growing population needs, 286, 299
    life-span prolongation, 392
    natural mutagens in, 219–20, 318
    natural selection relationship, 118, 304–5
    Paleolithic/Third-World/modern Western comparison, *305*
    see also Breast feeding
Dinman, Bertram, 155

Dipetalonema perstans (microfilaria), 159
Diphtheria, 157, 284
Disease–mitochondrial proteins defects
    link, 270
Disease predisposition, genetic, 99, 354,
    365, 392
DNA (deoxyribonucleic acid)
    and aging, 383, 384–87
    coiling, single chromosome, *248*
    complexity/potentiality for error, 271–
        75
    described, 221, *222*
    diagnostics, 363, 364, 365, 367–68,
        373
    double stranded molecule, 221, *222*,
        223
    factor VIII sequence, 258–59, 262
    gene pool varieties, 266–75
    genetic code, 265, 276
    genetic polymorphisms, 192, 195
    genetic variations, 276–77
    government oversight of research ap-
        plications, 378
    hemoglobin disease studies, 45, 54,
        253, 258
    of hemoglobin gene, 251–56
    human similarity with other life forms,
        296–98
    instability of, 272
    in introns, 254, 273–75
    "junk"/"selfish," changing view of,
        265, 270
    meiotic risks, 272–73
    mitochondrial, 270
    mitotic risk, 272
    mutations, 223, 224–26, 246, 254, 259,
        268, 274–75
    packaging in nucleus, *248*, 271–72
    and pseudogenes discovery, 256
    repair mechanisms, 245, 275–76, 384
    repetitive, 268–69, 273
    retroviral sequences, 225, 226, 267,
        377–78
    structure, 271–75
    see also Gene therapy
Dobzhansky, Theodosius, 13, 17, 18, 322,
    323
Domestic violence, 136, 313
Donnelly, Charles, 152
Dosage System, 1986 (DS86), 236–37, 245

Double helix, 221, *222*, *223*
Dounreay Nuclear Reprocessing Plant
    (Scotland), 332
Down syndrome, 31, 362, 368
Drew, A. L., 98
Drosophila
    hair pattern, 7, 71
    insect hormone action study, 6
    mutation findings, 11–12, 13, 71, 79,
        170, 211, 225, 235, 242, 268,
        322
    population bottlenecks, 2
    usefulness in genetic studies, 5, 17–18
Drosophila funebris, 3
DS86. *See* Dosage System, 1986
Duchenne's muscular dystrophy, 257–58
Duffy blood group, 53
Dulbecco, Renato, 267
Dunn, L. C., 13, 18
Dutch East India Company, 107
Dwarfism, 375
Dysentery, 157, 158

Earth
    gene pool from afar, *264*
    see also Environment
Earth in the Balance (Gore), 352
Ecdysone (hormone), 6
Ecosystem. *See* Environment
Edington, George, 49
"Effects of Atomic Weapons, The" (Los
    Alamos Laboratory), 91
"Effects of Nuclear War, The" (OTA), 308
Egremont North (England), 333
Ehrlich, P. R., and A., 288, 289
Eisely, Loren, 189
Eisenbud, Merle, 87, 88, 330
Eisenhower, Dwight D., 289
Elderly. *See* Age structure; Aging
Eldredge, N., 201
Electrophoresis technique
    Amerindian studies, 43, 45, 194, 196,
        203, 216–20, 231
    Japanese radiation-effects study, 231–
        32, 233, 236, 238, 241
    sickle-cell hemoglobin detection, 43,
        364
    variants, 218–19
Embryology, 5–6

Encephalitis, 157
Endangered species, 295–98, 389
Endemic disease, epidemic vs., 160–61
Endorphins, 260
Energy, U.S. Department of, 389, 402n.4
Energy and power needs, 287, 289, 299, 351–52
Energy Research and Development Administration, 402n.4
Entamoeba histolytica, 145, 158
Enteroviruses, 157
Environment
    chemical risk assessment, 317–18, 335–38, 339, 359
    ecosystem impoverishment, *280*, 284–87, 295–97
    effects on gene pool, 118, 201, 270, 291–98
        increasing exposure to mutagens, 306–10
    genetic risks, 275, 306–10, 317–39, 359
    human crisis and, 389
    international agreements, 342
    people-resources conflict, 281–99, 345
    radioactive risk assessment, 317–25, 359
    regulatory agencies, 325–31
    regulatory uniformity, need for, 335
    see also Atomic bomb studies; Population; Radiation
Environmental Protection Agency (EPA), radioactive waste policy, 325–31
Enzymes, 235, 271, 275–76, 355, 363, 375
    repair, 275–76
EPA. *See* Environmental Protection Agency
Ephrussi, B., 7
Epidemic diseases
    endemic vs., 160–61
    and longevity curve, 182
    new, 293
    population vulnerability to, 291–93
    see also AIDS; Measles epidemic
Erosion. *See* Soil erosion
Escherichia coli, 159–60
Esmeralda (Orinoco outpost), 134, 138, 140, 147, 407n.2
Espirito Santo (Brazil), 120

"Essay on the Principle of Population" (Malthus), 284
Ethical issues
    abortion, 30, 188, 350
    aging-related, 386
    Amerindian studies, 171
    atomic bomb use, 84–85
    eugenics-related, 10, 15–16, 17
    gene therapy-related, 376–80, 393
    genetic counseling-related, 361–62, 368, 369, 370–71
    human relationship with other life forms, 296–97
Ethiopian famine, 291, 345
Ethnic groups
    gene pools, 294, 349
    genetic defect/disease link, *36*, 37–55, 355, 363, 364
    world population growth, 347, *348*, 349
    see also African-Americans; Amerindian studies
Eugenics, 10, 15–16, 17, 312, 319, 394, 427n.24
    germ-line therapy as new, 381
Eugenics Record Office (Carnegie Laboratory), 14–18, 24, 25, 319, 381
Euphenics, 352–58, 385, 387, 392–93
Euripides, 356
Euthanasia, 17
Euthenics, 427n.24
Eveland, W. C., 159
Evolution
    adaptive mechanisms, 208, 265, 279, 301–6
    Amerindian studies, implications for human, 119, 121, 160, 185–90, 198–207
    changing determinants, human, 198–207, 301–6
    changing pace of human genetic, 301–6
    and genetic coadaptation, 209
    and genetic equilibrium, 199, 201–7
    and genetic rearrangements, 170, 225–26, 275
    "kinship effect" in human, 203
    mammalian life-span variations, 382
    and primitive mind of modern humans, 313–15

pseudogene participation in, 266
punctuated equilibrium concept, 201–7
  rogue cells and, 204
  species duration, 344
  species losses, 295–98
  step-wise, 203
  Wright model of rapid, 200–201, 206
  Yanomama imperfectly mirroring earlier human stages of, 139
  see also Natural selection
"Evolution in Mendelian Populations" (Wright), 413*n.11*
Exons, 252, 253, 258, 274
Eye
  Amerindian vision, 151
  aniridia, 214
  color, 15
  visual defects, 86
  see also Retinoblastoma

Factor VIII, 258–59, 262
Fajans, Stefan, 353
Falls, Harold (H. F.), 25, 212, 213
Family Registration Law of 1871 (Japan), 62
Family size goal. *See* Two-child families
Famine, 286, 291, 345–46
Father role
  and human socialization, 202–3
  Yanomama, *172*
Federal Radiation Council, U.S., 324, 358
Feeblemindedness. *See* Mental deficiency
Fels Research Institute (Yellow Springs, Ohio), 24
Female infanticide, 176
Ferrell, Robert, 233
Fertility
  consanguineous marriage, 102, 113–14
  decline in total rate, 345
  genetic significance of differentials, 118, 186–87, 197–98, 302–3
  Indian headmen, 179, 197, 302
  Yanomama men, 179–80
  Yanomama women, 176–78
  see also Birth control; Birth rates

Fetus. *See* Abortion; Birth defects; Prenatal diagnosis
Feudalism, 104, 107, 110, 111–12
Fiber, dietary, 158, 354
"Fire ants," 144
Firestone Tire and Rubber Co., Plantation Medical Service, 48–49, 50
First-cousin marriages
  in Yanomama patrilineal kinship system, 135, 184–85
  see also Consanguinity effects
Firth, Raymond, 178
Fisher, R. A., 9, 323
Flea larva, 145
Forsyth, A., 144
Fossil fuels, 287
406th Medical Laboratory, 59, *60*
Fox, D. P., 166
Francis, Thomas, 88
Freedman, Ron, 426*n.12*
Free radicals, aging and, 383–84, 385
Fruit flies. *See* Drosophila
Fujita, M., 235
Fukuoka (Japan), 60
Funaoka-yama, Battle of (1511), 103
Functional genes, 249–61, 266
Functional number of genes, 221
Furusho, T., 230

Gaia hypothesis, 297, 421–22*n.29*
Gai-sen-kan (Triumphal Return Building), 80, 81
Galton, Francis, 356
Galton Laboratory (London), 217
Gamma globulin, 153
Gamma radiation, organ exposure, 237
Gardner, Martin, 332, 333, 334
Garrod, Sir Archibald, 44
Gene
  absence of standard, 263
  activity coordination, 261
  building-block reuse, 262–63
  chances of specific gene's survival, 185–86
  complexity of, 249–61, 266
  components of, 221
  creation of "new," 274
  early view of, 271
  factor VIII, 258–59

Gene (*cont.*)
  jumping, 170, 225, 259, 266, 267
  lethal, 101–2, 114
  "master," 261
  multiple repeats, 226, 255
  problem of defining, 255
  proto-oncogene classes, 259
  size, 265
  survival chances, 185–86
  two functions, 271
  types of, 266–75
  see also DNA; Pseudogenes
Genealogy, 111–12
Gene pool, *264,* 265–79
  complexity of, 378
  consanguinity study implications, 96,
    101
  definition of term, 117, 249, 265
  DNA types, 266–70
  early tribal diversity, 199
  error potential, 271–75
  ethnic pools, size of, 294, 349
  gene therapy, implications for, 378
  genetic medicine for, 341–59
  genetic variation, amounts, 276–77
  gerontocracy emergence, 310–12, 352,
    360, 386
  homogenization trend, 302
  human problems, 343–44, 389–97
  long-range problems, 301–15
  nature of, 117–19
  stabilization, 391; *see also* Birth con-
    trol
  threats to integrity of, 118, 201, 281–82
    adverse environmental changes,
      291–99
    mutagen exposure, 306–10
    nuclear war, 309–10
    successful genetic disease therapy,
      379–80
  time scale for change, 312
  toughness of, 278
  turmoil in, 277–79
  see also Genome
Genesee Street (Rochester, N.Y.), 26
Gene therapy, 267, 370, 371, 373–82
  and aging, 385–87, 393
  discussion of, 373–76
  first protocol, 378
  as genetic priority, 393

  germ-line, 380–82, 394–95
  image problem, 393–94
  issues created by, 376–80, 393
  techniques, 374
  three approaches to, 375–76
Genetic coadaptation, 209
Genetic code, 265, 276
Genetic counseling, 29–31, 35, 41
  consanguinity study and, 96, 99, 101
  examples of, 362–71
  function of, 361–62
  importance of, 393
  see also Birth defects; Prenatal diag-
    nosis
Genetic disease, 9, *36,* 37–55, *53, 210,*
    211–16, 226, 257–59, 259–60, 355,
    363, 364, 365, 401*n.20*
  recessively inherited, 9, 30, 47, 95, 96,
    99, 187, 216, 355, 362, 365–66,
    371, 380
  treatment of, 371–73; *see also* Gene
    therapy
  see also specific diseases
Genetic equilibrium, 199
"Geneticist Looks at Modern Medicine,
    A" (lecture; 1960–61), 32–33
Genetic polymorphism. *See* Polymorph-
    ism, genetic
Genetic risk assessment, 317–39
Gene transfer, 375, 376
Genius, 356
Genocide, 347
Genome
  early view of, 271
  imperfect understanding of, 278, 279,
    380
  potential chromosomal mispairing,
    273
  retroviral dangers, 275
  tapestry of improvisations in, 378
  U.S. federally funded project, 18, 389,
    390
  see also Gene pool
Genotype, 4, 9
  and chronic malnutrition, 304
  number of potential, 277
  optimization of, 352–53, 391–93
Germany
  eugenics, 10, 17
  spontaneous mutation data, 217

Germ-line gene therapy, 380–82, 394–95
Gerontocracy, 310–12, 352, *360*, 382, 386, 392
Glass, Bentley, 13, 18, 318
Global 2000 Report to the President (1980), 288
Global warming, 287, 357
Glucose-6–phosphate dehydrogenase, 355
Glucose tolerance test, 143
Glycoprotein, 258
Goiter, 155, 156
Gold mining among Yanomama, 207
Gonadal radiation doses, 307, 308, 320, 335
   mouse experiments vs. Japanese doses, 242
   new estimates, 236–38
Goodpasture, Ernest, 87, 88
Gore, Al, 352
Goriki, K., 235
Gould, S. J., 201
Gout, 392
Government
   DNA research oversight, 378–79
   regulatory agencies, 325–31, 335
   see also specific agencies and laws
Great Britain
   radiation genetics case, 331–35
   spontaneous mutation data, 217
Great Depression, 2
Greece, ancient, 356
Greek ancestry, 37, 40, 365
Green, M. M., 225
Greenhouse effect, 287, 357
Green Revolution, 285–86, 299
Gregg, Alan, *422n.30*
Groves, Leslie, 77
Guam, *59*
Guaymi Indians, 141
Guiana, 140
Gulf War, 395, 396
Gustafson, James, 290

Hadorn, Ernst, 6, 18
Hair color, inheritance of, 16
Haldane, J. B. S., 9, 47, 92, 224, 323
Haller, M. H., 16
Hallucinogens, 135, 220

Hamai, Shinzo, 403–4*n.8*
Hamamoto clan, 104
Handler, Philip, 290, 298
   proposed "white paper," 298–99
Hardin, Garret, 351
Hardy-Weinberg Law, 40, 47
Harriman, Mrs. E. H., 15
Harris, H., 217
Harvey Society of New York City, lecture (1960–61), 32–33, 352
Hazama, R., 235
Headmanship, 178–80, 186–87, 192, 197–98
   genetic implications of, 302, 303
Health and disease
   aging and, 382–83
   Amerindian deterioration, 182
   Amerindian study, 126, 127, 128, 149–71
   Amerindian susceptibility to European diseases, 139, 163–65, 182
   childhood leukemia incidence, 331–38, 377
   "civilization"–linked, 304, 353, 354–55, 392
   emergent diseases, 293
   environmental impact on, 291–94
   genetic, treatment of, 371–73; *see also* Gene therapy
   genetic carriers. *See* Carrier states
   genetic improvement, 353
   hemoglobin disorders studies, 45, 54, 253, 258
   high-level radioactive waste effects, 326, 327–29
   low-level radioactive waste effects, 329–31
   mutation and human, 211–16, 226
   nutritional effects, 294–95, 304–5, 353, 354, 355, 392
   see also Death causes; Genetic disease; specific conditions
Hearing defects, 86
Heart disease, 354, 383
Heilbroner, Robert, 344
Hematology, 37–44, 65
Hemoglobin
   disease-linked abnormalities, 37, 44, 45, 46, 47, 54, 247, 364
   Lepore-type, 253

Hemoglobin C, 51, 52, 53, 224
Hemoglobin E, 52
Hemoglobin genes
  coordination of loci, 261
  and natural selection, 118
  structure of, 250–59, *251*
Hemoglobin S, 52, 53
Hemophilia, 258–59, 262, 304, 371
Henshaw, Paul S., 57, 59, *60*, 68
Hepatitis B virus, 157
Herbicides. *See* Pesticides and herbicides
Hereditary Genius (Galton), 356
Heredity Clinic (University of Michigan),
  24, 29–31, 212
Herndon, C. Nash, 24, 25
Heron, D., 15
Herpes simplex virus, 376
Heterozygotes
  carriers, 188
  genetic testing, 363
  index, 196–97
Hexosaminidase A deficiency, 362, 363
Hibakusha (atomic bomb survivors),
  403*n.8*
Hiernaux, Jean, 49
High-level radioactive waste, EPA policy
  on, 325–29
High Plains (U.S.), 285
Hijiyama Hill, 81–82, 404*n.8*
Hippocrates, 344, 345, 429*n.12*
Hirado island, *94*
  consanguinity study (1964–65), 108–
    14, 229
  "hidden Christians," 107–8
  living arrangements, 109, 112, 115
Hiroshima and Nagasaki
  casualties, 78
  Child Health Study (1958–60), 99–
    114
  choice as atomic bomb targets, 77–78,
    103
  clinic-laboratory building, 80–82
  consanguineous marriage. *See sub-*
    head Child Health Study *above*
  differences between, 87–88
  extent of devastation, *56*, 61, *62*, *63*
  follow-up studies, 26, 27, 59–73, 75–
    93, 229–47
    analysis, 89–93
  gonadal-dose distribution, 242

history/population components, 76–80,
    107
  Nagasaki atomic bomb marker, *63*
  new estimates of survivors' radiation
    exposures, 236–37, 307, 416*n.12*
  pregnancy termination studies, 86–87,
    91–92
  radiation sickness in survivors, 80
  registered births, 65–66, 98
  rogue cell incidence, *167*, 168
Hitchcock Clinic (Dartmouth), 14
Hitler, Adolf, 10
HIV viruses. *See* AIDS
Hogben, L. T., 9
Hollifield, Chester, 308
Holocaust, 17
Homeostasis, 209, 297, 421*n.29*
Homozygosity, 95–96, 101–2, 187–88,
    364
Honshu island (Japan), 76, 103; *see also*
    Hiroshima and Nagasaki
Hookworm, 158
Hopkinson, D. W., 217
Hormones
  genetic basis, 260–61
  insect action study, 6
  see also specific hormones
Hosojima island (Japan), consanguinity
    study, 103–6
Hosokawa clan, 103
House of Representatives Committee on
    Immigration and Naturalization,
    U.S., 15, 319
"How Humans Adapt: A Biocultural
    Odyssey" (Smithsonian sympo-
    sium), 290–91, 295
Howland, Joe, 26–27
HPRT. *See* Hypoxanthine-guanine phos-
    phoribosyl transferase
Human demographic patterns, 186–89
Human evolution. *See* Evolution
Human Gene Therapy subcommittee pro-
    tocols, 378–79
Human Genome Project, 389, 390
Humboldt, Alexander, 138, 407–8*n.2*
Huntington chorea, 15, 214, 226, 367
Hybrid dysgenesis, 12
Hybridization of orchids, 209
Hypertension, 151, 154, 304, 353, 354,
    392

Hypoxanthine-guanine phosphoribosyl transferase (HPRT), 375

IBP. *See* International Biological Program
Identical twin studiess, 9, 156
Ie concept (stem family), 112
"Immense Journey" (Eisely), 189
Immigration controls, 15, 319
Immunization, 162
Inbreeding. *See* Consanguinity effects
Inca civilization, 315
India
    life expectancy/survival curve, 182, *183*
    Travencore radiation exposure, 102–3
Indian Protective Service (Brazil), 123, 124
Indonesia, 345
Infanticide, Yanomama, 176, 177, 178, 180, 305
Infant mortality
    Amerindian, 153, 180, 303–4
    India, 182, 183
Infants. *See* Newborns
Influenza epidemic, 292
Ingram, Vernon, 44, 253
Innoshima island (Japan), 103, 104
Insanity, 15
Insect hormone action study, 6
Insecticides. *See* Pesticides and herbicides
Insulin, 375, 376
Intellectual functioning
    euphenic principles and, 355–56, 357–58, 392–93
    Japanese inbreeding study findings, 100, 101, 113
Intelligence tests, 100, 101, 103
International Atomic Energy Agency, 168
International Biological Program, 130
International Christian University (Tokyo), 109
International Commission on Radiation Protection, 324, 358
International Council of Scientific Unions, 130
International Union of Biological Sciences, 130

Introns, 251–52, 254, 256, 257, 258, 265
    and potential for genetic error, 273–75
In utero diagnosis. *See* Prenatal diagnosis
In vitro fertilization, 371
Iodine metabolism, 155–56
Iron deficiency anemia, 41
Irrigation, 285–86
Ishikuni, Norio, 103
Itano, H. A., 43–44, 45, 46
IUBS. *See* International Union of Biological Sciences
IVIC. *See* Venezuelan Institute for Scientific Investigations

Jaguar, 145
Japan
    atomic radiation effects studies. *See* Atomic bomb studies; Hiroshima and Nagasaki; Radiation Effects Research Foundation
    "Christian Century," 107–8
    consanguineous marriage decline, 247
    consanguineous marriage effects, 70–71, 92, 95–114, 226, 229, 247, 278
    feudal institutions, 104, 107, 110, 111–12
    life expectancy/survival curve, 182, *183*
    ministries, 67
    morality of atomic bombing of, 84–85
    National Institute of Health, 66–67, 68
    population control program, 86, 90, 345
    social structure, 111
    tea ceremony, 112
    U.S. Occupation, 59–60, 61, 62, 66, 84
    vital statistics registration, 61–62, 65–66
JCV virus, 170
Jesuit missionaries, 76, 107, 108
Jews, 363
Joffre, Maréchal, 88
Johnson, Lyndon B., 289
Joint Commission (atomic bomb effects), 58, 65, 79, 80
Jones Laboratory (Cold Spring Harbor, N.Y.), 13
Jumping genes, 170, 225, 259, 266, 267

Junqueira, Pedro, 122, 123, 124, 126, 129
Juntendo University, 108

Kaback, M., 363
Kageoka, T., 235
Kakizoe, Shinobu, 109
Kakure kirishitan (hidden Christians), Hirado hamlets, 94, 107–8, 110
Kamaboko-jo ("fish cake palace"), 81
Kanamari Indians, 141, 143, 205
Kato, H., 238
Keiter, Friedreich, 122
Keloids study, 64–65
Kennedy, John F., 289
Kenya, ecosystem losses, 295
Kerala State (India), 102–3
Kevles, D. J., 16
Kida, F., 67, 68
Kimura, M., 218, 219
Kinetochore, 169
King, Alexander, 347
King, M., 346
King, Martin Luther, 102
Kinship system, 135, 136, 184–85, 199, 202–3
Knudson, A. G., 213
Kodani, Masuo, 72
Kokura (Japan), 77
Komai, Taku, 115
Koseki system, 111
Kraho Indians, 141
Krzywicki, L., 173
Kucho (administrator), 110–11
Kurata, Robert, 404n.12
Kure (Japan), 60, 83, 84
Kuroshima island (Japan), 106
Kyoo Wan Choi, 159
Kyoto (Japan), 60, 77
Kyoto Prefectural University School of Medicine, 108
Kyushu (Japan), 76, 107
Kyushu University, 108, 109, 111

Lactation. see Breast feeding
Lambotte-Legrand, J. and C., 47
Landsteiner, Karl, 44
Laughlin, Harry (H. H.), 15, 319
Lawrence, Dale, 157

Lawrence, John S., 38–39, 45
Lawrence D. Buhl Building (Ann Arbor, Mich.), 33
Layrisse, Miguel, 133
LDL. See Low density lipoprotein receptor
Lederberg, J., 352
Legal action. See Suits and claims
Le Hasard et la Nécessité (Monod), 422n.31
Lehmann, H., 47, 138
Leigh, Day, and Co., 332
Leitão, Ismael, 125–26
Lepore-type hemoglobin, 253
Lesch-Nyhan disease, 375, 376
Lethal gene, probability of, 101–2, 114
Leukemia, 258, 331–38, 377
Lewis, Ed, 13, 18
Li, Francis, 179
Liberia, 48–49, 50–53, 247
Liberian Institute for Tropical Medicine, 49
Life expectancy and survival rates
  Amerindian, 180–82, 183
  children of Japanese atomic bomb survivors, 241
  dietary prolongation, 392
  genetic differentials, 302, 303–6
  newborn (1990), 385
  United States (1900;1988), 284
Life span, 382, 383, 392
Life table, Yanomama, 180–84, 181
"Limits to Growth, The" (report), 288, 347
Lindahl, T., 272
Lineage exogamy, 135
LINEs (long interspersed nuclear elements), 268, 269
Livestock, 357
Livingstone, Frank, 49, 50, 51
Lovelock, J. E., 297, 421n.29
Low density lipoprotein (LDL) receptor, 262, 273
Low-level radiation exposure. See under Radiation
Low-level radioactive waste, EPA policy on, 329–31
Ludmerer, K. M., 16
Luria, Salvador, 18–19
Lymphocytes
  genetically engineered. 379

presence of abnormal, 165–70, 267–68
retroviral infection, 267
Lysenkoism, 405n.15

MacArthur, Douglas, 88
MacCluer, Jean, 179
MacDonald, Duncan, 89
Machle, Willard, 87–88
Macklin, Madge, 17
MacMillan, Harold, 49
Macushi Indians, 141
Maize, 153
Major, Eugene, 170
Makiritare Indians, 140, 146, 155, 194
Malaria
    eradication campaign, 292
    impact on Yanamama, 157, 182
    sickling trait/resistance to, 47, 48, 51,
        52–54, 364
    as tropical rain forest hazard, 145
Male-descent lineage. *See* Patrilineal kin-
    ship system
Malformations, 92–93
Malignant tumors. *See* Cancer
Malnutrition, 188, 294–95, 345–46
Malthus, T. R., 284
Mandan Indians, 204
Manhattan Engineering District, 26, 27,
    35, 57, 58, 71, 77, 78, 79, 402n.4;
    see also Atomic bomb studies;
    Atomic Energy Commission
Manioc, 153
Mansonella ozzardi (microfilaria para-
    site), 159
Margulis, L, 297
Marine Biological Laboratory (Woods
    Hole, Mass.), 6–7
Marriages, consanguineous. *See* Con-
    sanguinity effects
Marriages, Yanomama
    age at, 177
    discouragement of intertribal, 195–96
    domestic violence, 136
    patrilineal kinship system, 135, 184–
        85
    polygamy, 179, 186–87
Martin, J., 346
Maruyama (Nagasaki), 86–87
Marx, J. L, 77

Mato Grosso (Brazil), 123, 124, 127, 130
Matsubayashi, Ikuso, 64, 66
Matsumoto clan, 103
Matsuura clan, 107, 109–10, 111–12
Maya civilization, 315
Maybury-Lewis, David, 123–24, 125–26,
    127, 128–29, 134
McClintock, Barbara, 12, 13, 266
McIntosh, R., 99
Measles epidemic, 161–65
Medical Research Council Radiobiology
    Unit (U.K.), 93
Mediterranean anemia. *See* Cooley's
    anemia
Mediterranean region, Thalassemia in-
    cidence, 37, 40, 45, 364–65
Meiji Restoration, 104, 105, 107
Meiotic divisions, 272–73
Mello, Nunes de, 159
Mendel's Laws, 283
Menopause, 178
Mental deficiency, 15, 86, 362, 372, 373,
    375
Mentors, importance to scientists, 18–19
Mercury levels in Amerindians, 155
Metabolic disorders, newborn screening
    for, 372
Mice
    germ-line transgenes, 380, 383
    leukemia in, 258
    life span, 382, 383
    radiation effects studies, 71, 238, 242–
        44, 245, 319, 321–22, 325, 332
Michigan, University of. *See* University of
    Michigan
Microfilaria, 159
Midgley, Mary, 290, 295
Miller, R. A., 383
Miller, R. W., 98
Miltiades, 356
Mineral metabolism, 154–56
Miscarriages, 86, 220–21, 226, 369
Missionaries, 144, 150, 165, 174, 176
    Jesuit, 76, 107, 108
Missionary Aviation Fellowship, 130
Mitochondrial DNA (mtDNA), 270
Mitosis, 272
Mitsubishi Arms Manufacturing Plant,
    62, 76, 77
Mitsubishi shipyards, 77

Miyata, K., 144
MN blood group, 44
Mobile genetic elements, 266–68, 270
Mohrenweiser, Harvey, 235
Molecules, three-dimensional structure
  of, 221, *223*
Monod, Jacques, 422*n.31*
Morality. *See* Ethical issues
Morgan, T. H., 5, 7
Morphogenesis, 209, 271
Mortality. *See* Death rates
Morton, Newton, 89–90, 322
Mount Fuji, 115
Mozambique, famine in, 345–46
mtDNA. *See* Mitochondrial DNA
Muller, H. J., 5, 13, 18, 71, 118, 224, 227,
  242, 278, 318, 322, 323, 324
Muller, R. T., 395
Murphy General Hospital, 27
Muscular dystrophy, 379
Muskie, Edmund, 288
Mutation rates
  average per generation, 221, 223, 272
  of neurofibromatosis gene, 257–58
  new perspective on human, 238–39,
    307, 309, 321–22
  protection against increases, 358
  spontaneous and induced human, 381
  studies, *210*, 211–20, 227, 231–32
  sustaining recessively inherited child-
    hood disease, 380
  viral role, 268
Mutations
  biochemical studies, 216–20
  chemotherapy-linked, 337
  chromosomal, 220–21, 224, 267–68
    cross-over and, 273
  and DNA, 223, 224–26, 246, 254, 259,
    268, 274–75
  Drosophila findings, 11–12, 13, 71, 79,
    170, 211, 225, 235, 242, 268,
    322
  dynamics of, 119
  environmentally linked, 275, 306–10,
    359
  and genetic diseases, 53, *210*, 211–16,
    226, 257–59, 259–60, 363, 365,
    401*n.20*
  genetic repair, 275–76
  induced, 381

and male-female birth rates, 90
and malignant tumors, 259–60
as mitotic risks, 272
natural mutagens, 219–20, 275
neutral, 219
and "new" population variation, 203–4
and population diversity, 324
and private polymorphisms, 195
radiation-linked, 75, 79, 92, 118, *210*,
  227, 231–32, 278–79, 309
  British childhood leukemia study,
    331–38
  doubling dose calculation, 230,
    239–42, 241–44, 245, 246,
    359
and RNA sequence change, 221–22
spontaneous, 211–27, 231–32, 241, 381
retroviral, 225
Myers, Norman, 389

Nagai, Takashi, 78
Nagasaki. *See* Hiroshima and Nagasaki
Nagasaki Medical School and Hospital,
  78
National Academy of Sciences, 18, 27,
  39, 57, 58, 68, 69, 89, 232
  Biological Effects of Atomic Radiation
    (BEAR) report, 318, 358
  Committee on Genetic Effects, 318–25
  Japanese atomic bomb effects re-
    search project, 72–73, 87, 88,
    187
National Cancer Institute, 337
National Council on Radiation Protection
  and Measurements, 324
National Institute of Dental Research, 152
National Institute of Health, Japanese,
  66–67, 68
National Institute on Aging, Advisory
  Council, 310–11, 382, 386
National Institutes of Health
  Child Health Study funding, 97
  Human Genome Project funding, 389
  sickle-cell research funding, 42
National Neurofibromatosis Foundation,
  215
National Research Council, 58, 68, 73,
  290, 335
  fellowship, 10, 12–14, 21

National Science Foundation, 130
  Alpha-Helix ship, 142–43
National Sickle-Cell Anemia Control Act
  of 1972, 55, 369
Natural mutagens, 219–20, 275
Natural selection
  disease as agent of, 149–70
  in genetic terms, 185–86, 199, 278
    coadaptation, 209
    differential fertility/survival factors,
      302–6
    introns and, 274
    mutations and, 117–19, 186, 198–
      99, 203–4, 206, 226, 279
    Wright model, 200–201
  raids and wars as agents of, 310
  Yanomama men vs. women as agents
    of, 187
Neanderthal man, 313
Neel, Alex, 109, 115, 235
Neel, Frances, 26, 27, 73, 109, 115, 235
Neel, James V., Jr., 109, 115, 142, 235
Neel, Priscilla, 22, 25, 27, 28, 73, 82, 102,
      109, 112, 114–15, 143, 208, 235
  calligraphy by, *388*
  identical twin, 247
  participation in Yanomama expedi-
    tions, 139–40
Nervous system birth defects, 93
Neurofibromatosis, 257–59
  mutation-rate study, *210*, 214–16
Neuromuscular development, con-
    sanguinity study, 100, 101
Neutral mutations, 219
Neutron bomb, 80
Neutron radiation, 237
Newborns
  chromosomal abnormalities, 170, 220
  life expectancy/survival rates (1990),
    385
  measles epidemic impact on, 164
  metabolic disorders, routine screening
    for, 372
  passive immunity, 161
  radiation effects study, 82–84, 92–93,
    404*n.*12
  see also Breast feeding; Infanticide;
    Infant mortality
New Guinea tribes, 200
New Tribes Mission, 130

Nigeria, 51
Niger River, 51
Niigata (Japan), 77
Nishihara surname, 104
Niswander, J. D., 98
Nixon, Richard M., 54–55, 289
Nomadism, 15
Norplant System (contraception), 350
NRC fellowship. *See* National Research
    Council, fellowship
Nuclear disarmament, 84, 309
Nuclear Waste Policy Act of 1982, 316
Nuclear weapons
  and potential for genetic problems,
    308–10
  radioactive waste policies, 325, 326–31
  testing, fall-out, 89, 307, 319, 320
  war use, biological/environmental ef-
    fects of, 308–10
  see also Atomic bomb studies; Nuclear
    disarmament
Nucleic acid. *See* DNA; RNA
Nucleotides, 221, 222–23, 224
  components, 265, 272
  and DNA coding, 273, 274, 276
  human-chimpanzee comparison, 296
  multiple repeats of specific sequences,
    226, 255
  sequence in human chromosomes, 389
  in sickle-cell genetic code, 253
  variation in, 276–77, 278
Nucleus, *248*, 271–72, 374; *see also* Gene
    therapy
Nursing. *See* Breast feeding
Nutels, Noel, 165
Nutrition. *See* Diet and nutrition

Oak Ridge National Laboratory, 57, 79,
    92
Obesity, 152, 154, 304, 353, 354, 392
Ocamo River, 140, 152, 162
Occupational hazards, 307–8, 336
Occupation of Japan, U.S., 59–60, 61, 62,
    66, 84
Office of Radiation Programs (ORP), 326,
    328, 330
Office of Technology Assessment (OTA),
    Advisory Panel on the Effects of
    Nuclear War, 308

O-furo (hot tub), 109
Ohio Academy of Science, 4
Ohta, T., 218, 219
Oil consumption rates, 287, 289
Older people. *See* Age structure; Aging
Olesen, E. B. and K., 51–52
Oliver, William, 154
Oncogene, 259–60
Ontario (Canada), 333
Oral contraceptives, 350
Orchids, 208–9
Orihel, T. C., 159
Orinoco River, 133, 134, 137, 138, 140,
    145, 152, 162, 188
ORP. *See* Office of Radiation Programs
Ortner, D. J., 290
Osaka (Japan), 60
Osler, William, 5
OTA. *See* Office of Technology Assess-
    ment
Otake, M., 230
Oughterson, A. W., 58
Overpopulation, 188, 282–99, 391
Ozone layer hole, 287, 342

Paleolithic diet, *305*
Panama, 141
Pan American Health Organization, 171
Panoa Indians, 141, 143
Parasites, 158–60
Parima Mountains, 137
Patrilineal kinship system, 135, 184–85
Pauling, Linus, 43–44
Pediatrics, 86, 404*n.12*
Penrose, L. S., 9
Pericles, 356
Perimetral Norte road, 207
Persian Gulf War, 395, 396
Pesticides and herbicides, 286–87, 292,
    317–18
Pfeiffer, Sig, 80
Phelps, L. V., 65
Phenylketonuria, 304, 372–73, 379
Phidias, 356
Phosphoglycerate kinase, 273
Physical measurements, consanguinity ef-
    fects, 100, 101
Piaroa Indians, 141
Pima Indians, 153, 154

Pituitary gland, 260, 375
Plantain, 135, 137, 139, 153, 184
Plant species losses, 295
Plasmodium falciparum, 48, 52–54, 145,
    157, 182
Plasmodium malariae, 157
Plasmodium vivax, 53, 157
Plato, 356
Platt, John, 344, 426*n.1*
Pleistocene Period, 183–84
Plummer, George, 404*n.12*
PNP. *See* Purine nucleoside phos-
    phorylase
Point mutations, 220
Poliomyelitis, 157
Pollution, 299
Polygamy, 179, 186
Polymorphism, genetic, 192, 193, 194–95,
    197–98, 231, 322
    balanced, 47, 48, 52, 364
Polynesians, 121, 349
Polypeptides, 221, 260
    factor VIII, 258
    in hemoglobin molecule, 250–55
    incomplete, 217–18
POMC locus, 260
Population
    collective genetic material of. *See*
        Gene pool
    egalitarian growth control, 346–52,
        391
    genetic bottlenecks, 2
    genetic medicine for, 341–59
    gerontocracy emergence, 310–12, 352,
        360, 382, 386, 392
    and mutation balance, 226
    mutation impact on, 239, 324
    overpopulation impact on, 188, 282–
        99, 391
    primitive, 120
    scheduled urban growth, 284
    "steady-state," 352
    ultimate numerical goal, 347
    world, at Pleistocene Period, 183–84
    world, estimate by regions, *348*
    see also Demography; Tribal popula-
        tions
Population control. *See* Birth control
Population Index of Potential Selection,
    306

Portuguese traders, 76, 106–7
Positive selection, 186
Potter, V. R., 395
Poverty, modern U.S., 311–12
Powell, W. N., 45
Pregnancy, 153, 155
    prenatal diagnosis, 30, 361–71, 392,
        393, 394, 395
    Yanomama rates and spacing, 174,
        176, 177–80, 180, 187
    see also Birth defects; Birth rates; Fer-
        tility; Genetic counseling
Pregnancy termination. See Abortion;
    Miscarriages; Stillbirths Prejudice,
    eugenically based, 15
Prenatal diagnosis, 29–30, 361–71, 379,
    392
    ethics of abortion recourse, 30, 370–
        71, 395
    fetal risk from, 369
    importance of, 393, 394, 395
    in vitro fertilization and, 371
    large-scale screening programs, 368–
        69
    see also Birth defects; Genetic coun-
        seling; specific diseases
Primitive populations
    mind compared with modern humans,
        313–15
    technical sense of term, 120
    see also Amerindian studies; Tribal
        populations
Primogeniture, 104
Priorities in human genetics, 390–97
Private polymorphisms, 195, 197
Problems in study of human genetics, 9–
    10, 17–18
Protein coding. See DNA
Protein sources, Amerindian dietary, 153,
    161
Proto-oncogenes, 259–60
Pseudogenes, 255–56, 266, 267, 273
Psychology, of atomic bomb victims, 78,
    403–4n.8
Public Health and Welfare Section (GHQ,
    SCAP), 59, 64
Punctuated equilibrium, concept of, 201–
    7
Purine nucleoside phosphorylase (PNP),
    375

Purines, 221, 222, 272, 276
Pyrimidines, 221, 222, 272, 276

Quarton, Gardner, 247

Rackham Foundation, 24
Radiation
    cancer risks, 75, 241, 245, 327–28,
        339, 359
    childhood leukemia case, 332–38
    doubling dose calculations, 230, 239–
        42, 241–44, 245, 246, 359
    exposure disasters, 168, 247, 406n.7
    exposure terminology, 402–3n.6
    genetic risk assessment, 317–39
        new perspective on data, 238–39
        reevaluation of experimental data,
            236–38, 242–47, 278–79, 307,
            321–22, 330
        vs. risk management, 329
    low-level exposure, 71, 241–42, 307,
        329–31; see also X-ray radiation
        doses
    maximum permissible exposures, 324,
        329, 331, 358
    new estimates, 236, 244–47, 359
    megaton nuclear weapon, 308–10
    natural, 120, 318
    public anxiety over, 325
    spectrum of damage, 240
    U.S. government policy development,
        325–31
    see also Atomic bomb studies; Gona-
        dal radiation doses; Mutations
Radiation Effects Research Foundation
    Biochemical Genetics Study, 234–47,
        336
    building, 228
    DNA-mutations study, 246
    establishment and funding, 234
Radioactive waste, EPA policy on, 325–
    31
Radon, 307, 320
Rain forest. See Tropical rain forest
Ramsey, Norman, 78–79
Raper, A. B., 47, 49
Reagan, Ronald, 80, 288, 289
Reagan, Tom, 295

Recessively inherited traits
  disease, 9, 30, 47, 95, 96, 99, 187, 216,
    355, 362, 365–66, 371, 380
  mutation rates, 216–17, 380
  probability of, 96
  red headedness, 16–17
  see also Carrier states
Reciprocal translocations, 230
Recombinant DNA Advisory Committee,
    378–79, 381
Red blood cells
  sickle-cell anemia, 36, 41
  see also Hemoglobin
Red headedness, 16–17
Reed, T. E., 17, 213
Renin, 153, 154
Repair enzymes, 275–76
Repetitive DNA, 268–69, 273
Reproduction. *See* Birth rates; Pregnancy;
    Fertility
Reproductive advantage, 187
Reproductive compensation, 114
RERF. *See* Radiation Effects Research
    Foundation
Retinoblastoma, 238, 259–60
  mutation-rate study, *210*, 212–13
Retrogenes, 266, 267
Retrotransposons, 170, 225, 266, 267,
    268
Retroviruses, 170, 225, 267, 275
  gene therapy use of tissue-specific,
    376, 377–78
Reverse transcriptase, 267
Rh blood group, 44, 192, 193
Ribonucleic acid. *See* RNA
Ribosomes, 270
Richards, P. W., 144
"Right-to-lifers," 30, 188, 350
Rio Negro, 162
RNA (ribonucleic acid), 221–22, 251, 252,
    253, 254, 256, 273
  aging and, 383
  molecules functioning without transla-
    tions, 270
  retroviral, 267
Robson, E., 217
Roche, Marcel, 155, 156
Rochester (N.Y.). *See* University of Ro-
    chester
Rockefeller Foundation, 97, 102, 103

Rodarte, J. G., 45
Roentgen abbreviation, 237, 402–3n.6
Roe vs. Wade, 350
Rogue cells, 166–70, *167*, 204, 225, 267,
    390
Roman Catholicism, 106, 107–8
Roraima Territory (Brazil), 207
Roundworm, 158
Royal High Courts of Justice (England),
    316, 332, 333
r (roentgen abbreviation) terminology,
    237, 402–3n.6
RU-486, 350
Ruckelshaus, William, 330
Rucknagel, Don, 54
Russell, W. L., 244, 318, 321, 322, 324

Sahel famine, 345
Sakuri surname, 104
Salesian Mission, 130, 140
Salmonella, 157
Salt
  hypertension link, 354
  Yanomama balance, 154–56
Salzano, Francisco, 120, 121, 122, 123,
    128, 129
Sams, C. F., 66
Samurai, 103, 104
São Marcos (Brazil), 130
Sardinia, 365
Sasebo (Japan), 83
Satoh, Chiyoko, 235
Schizophrenia, 382
School-performance and inbreeding re-
    cords, 100, 101, 113
Schull, Jack (W. J.), 86, 89–90, 91, 96,
    102, 106, 107, 108, 109, 110, 115,
    213, 214, *228*, 232, 238, 244–45
Schull, Vicki, 109, 115
Science (journal), 43, 45, 72–73
Science Advisory Board (EPA), 331
Scientific Group on Research in the Pop-
    ulation Genetics of Primitive
    Groups, 130
Screening programs for genetic disease,
    368–69
Scrimshaw, Nevin, 290
Seascale (West Cumbria, Eng.), 331

Sellafield nuclear reprocessing installation (England), 332
and childhood leukemia cases, 332–38
Semipalatinsk disaster, 247, 406n.7
Senior citizens. *See* Age structure; Aging
Serine proteases, 262
Serum albumin, 153, 195
Seventh International Congress of Genetics (Edinburgh), 7–8
Sex chromosome abnormality, 230, 241, 355; *see also* X-chromosome
Sex ratio
abortion based on sex of fetus, impact on, 369
female infanticide and, 176
of Japanese atomic bomb survivors' children, 90–91
of Yanomama Indians, 176, 180
Shakespeare, William, 429n.12
Shavante Indians, *116*, 122–29
appearance, 150
enlarged thyroid, 156
fieldwork with, 125–31
headman fertility, 179, 197, 302
health study of, 126, 127, 128, 149–71
low cholesterol, 153
numbers, 134
prehistory, 123
village pedigree, 126, 128–29
Shaw, M., 213
Shigei settlement (Japan), 104
Shintoism, 115
Shishi (Japan), *94*
Shizuoka (Japan), 101
Short interspersed nuclear elements. *See* SINEs
Sickle-cell anemia, 41–55, 224, 355
distribution/frequency studies, 47–48, 50, 51, 53
gene therapy for, 374, 379, 385
genetic code, 253
mutation types, 401n.20
prenatal testing for, 364
public recognition of, 54–55
red blood cells, *36*, 41
sickle-cell trait confused with, 369
U.S. government program, 368–69
Silvestroni, E., 41
Simões, Girlay, 124
Simons, E. L., 296

SINEs (short interspersed nuclear elements), 268–69
Singer, S. J., 43–44
Skin
infections, 157
neurofibromatosis, 214–16
Slash-and-burn agriculture, 139, 184, *202*
Smallpox, 163
Smithsonian Institution, 290–91, 295
Snell, Fred, 64, 65
Snyder, Lawrence (L. H.), 4, 17, 24, 71, 90
Socrates, 356
Soil erosion, 284–85, 347
Somalian famine, 291, 346
Sophocles, 356
South Africa, 49, 50, 200
South and Central America, Amerindian studies, 117–47, *141*
Soviet Union
Neel-Schull genetics book translation, 405n.15
nuclear energy disasters, 168, 247, 406n.7
nuclear strength, 309
nuclear testing, 319
Species
duration, 344
human crisis, 389
life-span variations, 382
losses problem, 295–98
Spencer, Warren P., 2–4, 5, 18, 26, 71, 79
Spontaneous abortion. *See* Miscarriages
Spontaneous mutation. *See* Mutations, spontaneous
Spuhler, J. N., 98
Squire, W., 165
Station for the Experimental Study of Evolution (Cold Spring Harbor, N.Y.), 14–18
Statistical techniques, 9
Steel tools, 139
Step-wise evolution, 203
Sterilization, 350
Stern, Curt, 5, 6, 9, 10, 11, 17, 18, 25, 26, 71, 79, 90
Stillbirths, 86, 91, 92, 180, 237
Stingray, 145
Stone Age Culture, 139, 313
Stools, 157–60

Strategic Defense Initiative, 80
Strauss, M. B., 21
Streptococci, 157
Streptococcus mutans, 152
Stroke, 383
Strong Memorial Hospital, 23, 25
"Study of natural selection in primitive
    and civilized populations, The",
    118
Sturtevant, A. H., 5, 7, 318
Suaw Shrine (Nagasaki), 56
Sudanese famine, 291, 346
Suicide, 306
Suits and claims, radioactivity exposure,
    325, 331–35
Sumié painting, 114–15
Superfund. *See* Comprehensive En-
    vironmental Response Compensa-
    tion and Liability Act of 1980
Superoxide, 384
Supreme Command Allied Powers, Gen-
    eral Headquarters, Public Health
    and Welfare Section Survival rates.
    See Life expectancy and survival
    rates
Survival patterns, basis for changes, 302,
    303–6
Suzue surname, 104
SV40 virus, 170
SV (sieverts) terminology, 237
Syphilis, 163

Taboos, 174, 177
Takahashi, N., 235
Takashima, Koji, 83
Tanis, Robert, 233
Tatami mats, 115
Tatum, E., 7, 44
Tawn, Janet, 166, 168
Taylor, Paul, 295
Tay-Sachs disease, 362–63, 368, 369
Tea ceremony, 122
Teeth, 151–52
Telomeres, 269
Temin, Howard, 267
Tentative 1965 Dosage, Revised (T65DR),
    236
Tessmer, Carl, 83–84
Thailand, 345, 364

Thalassemia
    gene therapy, 374, 379, 385
    genetics of, 40–41, 45, 250, 253–54,
        259, 364
    major/minor, 40–41
    as malaria protection, 46–47, 364–65
    naming of, 40
    observed in Africa, 52, 53
    prenatal testing for, 364–65
Themistocles, 356
Third World. *See* Developing nations
Thomas, Lewis, 344
Thompson, Elizabeth, 219
Thucydides, 356
Thyroid, 155–56, 372
Ticuna Indians, 141, 143
Tikopia island, 178
Time scale for genetic change, 312
Tokugawa, Ieyasu (Shogun), 107
Tokyo (Japan), fire-bombing casualties,
    84
Tokyo Medical and Dental University,
    108
Toulmin, Stephen, 290
Toxic Substances Control Act of 1971,
    326
Trace metals, 155
Traumatic death, 182, 305–6, 313–14
Travencore, radioactive sands of, 102–3
Tribal populations
    boundary setting, 192
    brain/mind compared with modern
        humans, 313–15
    civilization, effects on, 186–89
    common stereotypes, 192
    definition of tribe, 191
    demography, 173–86
    destruction/acculturation, 247
    genetic reshuffling, 199–200, 203
    heterozygosity, 196–97
    infanticide/congenital defects link, 305
    kinship system, 136, 199, 202–3
    mutation rates, 219–20
    "private" genes, 194–96
    technical sense of "primitive," 120
    traumatic deaths, 305–6
    village microdifferentiation, 191–94,
        197–200, 231
    Wright model, 201
    see also Amerindian studies

"Tropical Nature" (Forsyth and Miyata), 144
Tropical rain forest
    dangers to researchers, 144–47
    environmental endangerment, 285, 295
Trout fishing, 143
Truman, Harry S., 58, 289
Tsuzuki, Masao, 59–60, *60*, 62, 402*n*.2
Tuberculosis, 163, 284, 292–93
Twin studies. *See* Identical twin studies
Two-child families, 347, 371, 391, 395

Ueda, N., 235
Ukraine, cytogenetic study, *167*, 168
Ulrich, Frederick, 57, 64
Unevangelized Field Mission, 130
Union of Concerned Scientists, 288
United Nations, 341, 345, 349, 396
University of California (Berkeley), 79
University of Chicago, 79
University of Michigan
    attractions, 13, 25
    Child Health Study data, 97, 113–14
    Hospital—Medical Center, 24, 25, 28, 32
        Heredity Clinic, 24, 28–31
        Women's Hospital study, 217
    human genetics program, 24–25, 32, 120, 121, 232
    Mental Health Research Institute, 247
    physical facilities for human genetics, *20, 33*
    reaction to Hardin's population control speech, 351
    sickle-cell anemia studies, 42–43
    spontaneous mutation studies, 211, 217
University of Minnesota, 29
University of Rochester
    biomedical effects of radiation studies, 79
    genetically related blood disease studies, 39, 40–41, 42
    graduate studies at, 4–7
    Manhattan Engineering District, 26–27, 58
    medical studies at, 13–14, 21, 22–23

University of Texas (Houston), 232
Untoward Pregnancy Outcome (UPO), 237, 241, 243, 244
Urakami Cathedral, 78
Urakami Valley (Nagasaki), *62*, 78
Urban population
    epidemic disease susceptibility, 291
    growth, 284, 299
U.S.S.R. *See* Soviet Union

Valentine, Bill, 21, 39, 47–48
Valine, 44
Vandepitte, J., 48, 49
Variable-number of tandem-repeats. *See* VNTRs family
"Various Contrivances by Which Orchids are Fertilized by Insects, The" (Darwin), 208–9
Vasectomy, 350
Velasquez (Makiritare chief), 138
Venezuela 133–40; *see also* Yanomama Indians
Venezuelan Air Force, 134, 138, 140, 147
Venezuelan Institute for Scientific Investigations (IVIC), 133, 134, 147, 155, 166
Village microdifferentiation, 191–94, 195–96, 231
    causes, 197–98, 199
    examples, 204–6
Violence
    Yanomama, 135–36
    see also Traumatic death
Viruses
    AIDS epidemic, 283–84, 292
    Amerindian exposure to, 157
    Amerindian responses to, 164
    and chromosomal rearrangement, 170, 219, 267, 268, 275
    and gene therapy, 376, 377–78
    see also Antibodies; Retroviruses
Vision. *See* Eye
VNTRs family (variable-number of tandem-repeats), 269

Wallace, A. R., 314, 315
Waltham General Hospital, 27

Wapishana Indians, 141
Ward, R., 194
Warren, Shields, 58
Warren, Stafford, 26, 27
Wasps, 144–45
WASPS (White, Anglo-Saxon, Pro-
   testants), 10
Waste disposal. *See* Radioactive waste
Water supply problems, U.S., 285–86,
   299
Weapons testing. *See* Nuclear weapons
Weaver, Warren, 318–19, 320
Weiss, Ken, 180
Wells, I. C., 43–44
Whipple, George H., 14, 40
WHO. *See* World Health Organization
Whooping cough, 163
Wife abuse. *See* Domestic violence
Willcox, M. C., 52
Williamson, George, 4
Willier, B. H., 5, 10, 11
Wilms' tumor, 238
Wilson, P. J., 202
Wintrobe, Max (M. M.), 21, 37, 39
Wivel, Nelson, 381
Wood, John, 404*n.12*
Woods Hole (Mass.), 6–7
Wooster (Ohio), 1
Wooster College. *See* College of Wooster
Works Progress Administration (WPA),
   2
World Health Forum roundtable, 346
World Health Organization, 120–21, 123,
   124, 130–31, 292, 293, 346
World War II, 8, 10, 13, 17, 25–26, 61
   Hiroshima population, 76
   morality of atomic bomb use, 84–85
   see also Atomic bomb studies
Wright, Sewall
   Committee on Genetic Effects report,
      318, 323–24
   model of evolution, 200–201, 206,
      419*n.12*
Wright, Stanley and Phyllis, 404*n.12*
Wyden, P., 78, 79, 403–4*nn.7, 8*

Xanthippus, 356
X-chromosome, 90, 257, 258, 355

Xenophon, 356
Xingu National Park (Brazil), 165
X-ray radiation doses, 71, 79, 242, 307,
   320, 344–45, 402–3*n.6*

Yamamoto, Manabu, 111
Yamazaki, James, 404*n.12*
"Yanomama: A Multidisciplinary Study"
   (film), 142
Yanomama Indians, 133–40, 146
   age pyramid, 174–78, *175*, 311
   appearance, 150
      microdifferentiations in, 204–6
   as approximation of earliest societies,
      139, 200
   blood types, 192–94
   characteristics, 134–38, 142, 143
   children playing, *190*
   chromosomal damage, 166, *167*
   color blindness, 151
   current plight, 207
   diet, 153, 154
   E. coli, 160
   ethnographer, 142
   father and son, *172*
   genetic differentiations, 193–94, 197,
      199–200
   genetic flux, 199
   health, 149–71
   hepatitis B virus, 157
   inbreeding, 184–85
   kinship system, 136, 199
   life table, 180–84, *181*
   low cholesterol, 153
   male adults, *300*
   measles epidemic, 161–65
   mother and children, *202*
   nature of fieldwork with, 139–40
   numbers, 134, 138
   parasites, 159
   private polymorphisms, 195
   reproductive patterns, 176–80, 182–83,
      186–88, 305
   salt balance, 154–55
   stools, 158
   stresses, 310
   teenager, *205*
   territory, 136–38, *137*

young children, *132*

young male, *148*

Yellow fever, 146, 157

Yorkshire Television, 331

Yoshi (adopted son), 112

Yoshimoto, Y., 237–38

Young, J. Z., 297

"Young people's group" (seinendan), 105

Zero population growth, 352

Zuelzer, Wolf, 43, 48, 49